Lecture Notes in Physics

Founding Editors

Wolf Beiglböck

Jürgen Ehlers

Klaus Hepp

Hans-Arwed Weidenmüller

Volume 1006

The series Lecture Notes in Physics (LNP), founded in 1969, reports new developments in physics research and teaching - quickly and informally, but with a high
quality and the explicit aim to summarize and communicate current knowledge in
an accessible way. Books published in this series are conceived as bridging material
between advanced graduate textbooks and the forefront of research and to serve
three purposes:

- to be a compact and modern up-to-date source of reference on a well-defined
  topic;
- to serve as an accessible introduction to the field to postgraduate students and
  non-specialist researchers from related areas;
- to be a source of advanced teaching material for specialized seminars, courses
  and schools.

Both monographs and multi-author volumes will be considered for publication.
Edited volumes should however consist of a very limited number of contributions
only. Proceedings will not be considered for LNP.

Volumes published in LNP are disseminated both in print and in electronic for-
mats, the electronic archive being available at springerlink.com. The series content
is indexed, abstracted and referenced by many abstracting and information services,
bibliographic networks, subscription agencies, library networks, and consortia.

Proposals should be sent to a member of the Editorial Board, or directly to the
responsible editor at Springer:

Dr Lisa Scalone
Springer Nature
Physics
Tiergartenstrasse 17
69121 Heidelberg, Germany
lisa.scalone@springernature.com

Urko Reinosa

Perturbative Aspects of the Deconfinement Transition

Beyond the Faddeev-Popov Paradigm

 Springer

Urko Reinosa
Center for Theoretical Physics
École Polytechnique
Palaiseau, France

ISSN 0075-8450 ISSN 1616-6361 (electronic)
Lecture Notes in Physics
ISBN 978-3-031-11374-1 ISBN 978-3-031-11375-8 (eBook)
https://doi.org/10.1007/978-3-031-11375-8

This Springer imprint is published by the registered company Springer Nature Switzerland AG
The registered company address is: Gewerbestrasse 11, 6330 Cham, Switzerland

To my parents,

Preface

In the case of non-abelian gauge theories, the standard Faddeev-Popov gauge-fixing procedure in the Landau gauge is known to be incomplete due to the presence of gauge-equivalent gluon field configurations that fulfill the gauge condition, also known as Gribov copies. A widespread belief is that the proper analysis of the low energy properties of non-abelian theories in this gauge requires, therefore, the extension of the gauge-fixing procedure, beyond the Faddeev-Popov recipe.

This manuscript reviews various applications of the Curci-Ferrari model, a phenomenological proposal for such an extension, based on the decoupling properties of Landau gauge correlators as computed on the lattice. In particular, we investigate the predictions of the model concerning the confinement/deconfinement transition of strongly interacting matter at finite temperature, first in the case of pure Yang-Mills theory for various gauge groups, and then, in a formal regime of quantum chromodynamics where all quarks are considered heavy. We show that most qualitative aspects and also many quantitative features of the deconfinement transition in these theories can be accounted for within the Curci-Ferrari model, with only one additional parameter, adjusted from comparison to lattice simulations. Moreover, these features emerge in a systematic and controlled perturbative expansion, as opposed to the ill-defined perturbative expansion within the Faddeev-Popov model in the infrared.

The applications of the Curci-Ferrari model at finite temperature and/or density require one to consider a background extension of the Landau gauge, known as the Landau-DeWitt gauge. Therefore, besides the above-mentioned applications, the manuscript is intended as a thorough but pedagogical introduction to these techniques at finite temperature and/or density, including the rationale for considering self-consistent backgrounds and the associated background-field effective action, the role of the Weyl chambers in discussing the various physical symmetries of the problem, and the complications that emerge due to the sign problem in the case of a real quark chemical potential. It also discusses a novel implementation of the background-field method at finite temperature which does not suffer from some of the limitations of the standard approach and should, therefore, provide more robust results.

The material covered in these lecture notes, which derives from my habilitation thesis work defended in June 2019, is intended for any graduate student or confirmed researcher that wishes to deepen their understanding of the confinement/deconfinement transition from an analytical perspective in Yang-Mills theory or related theories. Basic knowledge of these theories at finite temperature is required, although the text is designed in a self-contained manner, with most concepts and tools introduced when needed. In particular, the Cartan-Weyl bases, so useful in the context of background-field applications at finite temperature, are thoroughly reviewed in a dedicated Appendix. The same goes for the various techniques of computation of Matsubara sums at finite temperature. At the end of each chapter, a series of exercises is proposed in support of the material in the main text. Exercices marked with one ★ are direct applications of the material in the main text, those with two ★s are moderately demanding, while those with three ★s require a deeper investment.

I would like to thank Duifje van Egmond for checking the main text and the problems. Also, the new version of the final sections of Chap. 3 derives from enlightening discussions that I had with her during our work in common. My most since gratitude goes also to Lisa Scalone for smooth and efficient interactions during the whole submission process.

Palaiseau, France   Urko Reinosa
May 2022

Acknowledgements

The time that I have spent writing this manuscript strangely reminds me that of the imaginary time formalism with periodic boundary conditions used throughout this work: it does not flow in the usual direction, but rather orthogonal to it, and it is characterized by perpetual, almost periodic, repetitions until the final result is achieved. Now that I am freed from the constraints of this unusual time, I would like to express my most sincere gratitude to all the people that supported me on many different levels, prior, during, and after the writing process. This manuscript would clearly not be the same without them.

First of all, my warmest thanks to the members of the jury, Maxim Chernodub, François Gelis, Antal Jakovac, Jean-Loïc Kneur, Dominique Mouhanna, and Samuel Wallon, who accepted the invitation to read and review the manuscript and to participate in the oral defense. Thanks also to Frédéric Fleuret, Régine Perzynski, Marco Picco, and Nathalie Suirco for their many advises and for accepting to move the deadline by a few months, in order to allow me to cope with some real time, real life imperatives. Thanks finally to Juana Isabel Mallmann for making the defense in Alan Turing's building possible and incredibly smooth.

I want to express my deepest gratitude to the collaborators with whom this adventure started: Marcela Peláez, Julien Serreau, Matthieu Tissier, and Nicolás Wschebor. I consider myself very lucky to have crossed paths with these top-notch researchers, and I am very proud of the work that we have accomplished together. Thanks also to their respective laboratories (and by this I do not mean the walls but the people within them) where a substantial part of the present work was developed: Laboratoire de Physique Théorique de la Matière Condensée (Université Paris Sorbonne), Laboratoire AstroParticule et Cosmologie (Université Paris-Diderot), and Instituto de Física (Universidad de la República, Montevideo). Of course, I have also benefited from interactions with many other collaborators, which have contributed to shaping my knowledge in one way or another and whose insight certainly contributes to the flavor (or I should say the color) of this thesis. In particular, a special thank you goes to Jean-Paul Blaizot, Edmond Iancu, David Dudal, Jan M. Pawlowski, and Zsolt Szép.

This thesis was written partly at the "Centre de Physique Théorique" (CPHT) on Ecole Polytechnique campus (even though many hours were also spent at the central library, amidst the noisy workers that were renewing it). I would like to thank the whole CPHT members for the outstanding working atmosphere that they contribute to create, much above the average I would say. In particular, I want to thank Florence Auger, Fadila Debbou, and Malika Lang for their promptness in dealing with any kind of administrative issue. Thanks also to our director, Jean-René Chazottes, and to Ecole Polytechnique, for accepting to pay my registration at the University.

A special thank you goes of course to past and present members of the Particle Physics group at CPHT, Tran Truong, Georges Grunberg, Tri Nang Pham, Bernard Pire, Claude Roiesnel, Stéphane Munier, Cyrille Marquet, and Cédric Lorcé (plus postdoctoral fellows as well as master's and PhD students) for the enjoyable working atmosphere and for accepting me as a member of the group despite the crazy taste for gluonic mass operators. Hopefully, this thesis will succeed in conveying the message that this is not that crazy after all. An even more special thank you goes to Bernard Pire and Stéphane Munier who have invested a considerable amount of energy for many years now into strengthening the visibility of our group/laboratory. I owe them much since my hiring 12 years ago. They have always been available to listen to my queries and to provide valuable pieces of advice.

Thank you to all the friends and colleagues that, every now and then, inquired about the progress of the thesis, in particular those, Laetitia and Mathieu, Xoana and Daniel, who lent their houses so that I could find a few hours of focus on the writing of the manuscript.

Finally, my most special thank you to my dear wife and kids: Rosa, Aitor, Guillem, and Rosa (did I mention periodic boundary conditions?). Your patience has been infinite. Mine, certainly the inverse.

Contents

Acronyms

BRST	Becchi-Rouet-Stora-Tyutin
CF	Curci-Ferrari
FP	Faddeev-Popov
GZ	Gribov-Zwanziger
LDW	Landau-DeWitt
QCD	Quantum Chromodynamics
QED	Quantum Electrodynamics
YM	Yang-Mills

Introduction: The Many Paths to QCD

Quantum chromodynamics, or QCD for short, is by now well accepted as the fundamental theory governing the strong force. According to this theory, the elementary particles sensible to the strong interaction, known as *quarks* and *anti-quarks*, carry a generalized notion of charge, the *color*, that allows them to exchange quanta, known as *gluons*, in a way similar to the exchange of photons by electrons and positrons in quantum electrodynamics (QED). A crucial difference with this latter theory is, however, that the gluons themselves are carriers of color, allowing them to self-interact. Correspondingly, the SU(3) symmetry group associated with the color charge is non-abelian, in contradistinction with the abelian U(1) group at the basis of QED. The fundamental theory of the strong interaction appears, therefore, as the non-abelian generalization of QED.

Although conceptually quite appealing, this generalization hides, in fact, the long process that lead to the construction of QCD as the fundamental theory of the strong interaction, from the thorough study of the many observed particles that reacted to the strong force and the proposal of the quark model as a way to bring order to complexity [1–3], to the experimental evidence for the existence of quarks [4,5], the proposal of a new type of charge with an associated non-abelian symmetry group [6], and the final formulation of QCD in the form of a non-abelian gauge theory.

The main reason explaining this long process is that, unlike the other theories describing the fundamental forces of nature, the elementary bricks of QCD are not directly observable. Instead, quarks and anti-quarks appear to us in the form of a large fauna of bound states or resonances, the *hadrons,* of which the protons, the neutrons, and the pions are just a few representatives. Moreover, these compound particles come with the added mystery to always appear in a color-neutral form. This property, known as (color) *confinement* [7], as evaded a fully satisfactory theoretical grasp since the advent of QCD, and, even though it is now pretty much accepted to be a mathematical property of the theory [8], its rigorous first-principle derivation is one of the open challenges in theoretical particle physics. The challenge is rooted in the fact that the coupling of the strong interaction is much larger than the

© The Author(s), under exclusive license to Springer Nature Switzerland AG 2022

U. Reinosa, *Perturbative Aspects of the Deconfinement Transition*, Lecture Notes in Physics 1006, https://doi.org/10.1007/978-3-031-11375-8_1

corresponding coupling of the electromagnetic interaction and perturbation theory, so useful in this latter case, is admittedly of no use here.

Confinement characterizes, however, the low-energy regime of the strong interaction. In the opposite, high-energy limit, QCD displays a totally different behavior. Indeed, as any other relativistic field theory, the coupling of the interaction varies, or "runs," with the energy scale relevant to the particular process under scrutiny. In the case of QCD, the running coupling decreases and approaches zero logarithmically for asymptotically large values of the energy. This special property is known as *asymptotic freedom* [9, 10] and turns QCD at high energies into a weakly interacting system of quarks and gluons, thus providing access to some of its properties from first-principle perturbative calculations.[1]

Another exciting property of the high-energy regime can be revealed by imagining coupling the system to a thermostat. Indeed, owing to asymptotic freedom, one expects the interaction to decrease as the temperature is increased, up to the point where quarks and anti-quarks cannot remain bound anymore inside of hadrons. The low-temperature *confined phase* is then expected to evolve into a *deconfined phase*, sometimes dubbed as *quark-gluon plasma*, in which quarks and gluons are liberated and color neutrality constraints do not apply anymore. Similarly, one expects to find a deconfined phase at large matter densities, known as the *color-flavor-locked phase* [11], although it is superconducting in nature and therefore rather different from the deconfined phase at high temperature.

Deconfined phases of matter are believed to be relevant in various physical situations of interest, for instance, during the thermal history of the early universe [12] or in the core of certain ultra-dense stellar objects [13]. The quest for the quark-gluon plasma and its decay into a confining phase (as the system cools down) is also the central motivation for heavy-ion collision experiments, at RHIC (BNL, Brookhaven, USA), LHC (CERN, Geneva, Switzerland), or FAIR (GSI, Darmstadt, Germany). Beyond providing the experimental evidence for the existence of deconfined phases of matter, the ultimate goal of these experiments is to acquire valuable insight into the QCD-phase diagram, not only as a function of the temperature but also as a function of additional external parameters. For instance, the possible presence of a critical end-point [14, 15] terminating a line of first-order phase transitions in the plane defined by temperature and density has been and remains nowadays a hotly debated issue. The classification and analysis of exotic phases along the density axis or as a function of a possible magnetic field are also of topical relevance [16–20].

Parallel to these large experimental programs, an intense theoretical activity has been devoted to extract the properties of the QCD-phase diagram from first-principle

---

[1] In practice, a given process involves both hard and soft scales. If *factorization* applies, one can rewrite the process as a convolution between hard and soft components. The former, because they involve large momentum transfers, can be treated within perturbative QCD. The latter, although non-perturbative, are universal and evolve with the running scale in a way that can again be determined within perturbative QCD.

calculations.[2] As already mentioned, perturbation theory is certainly a valuable tool in inferring the behavior of the system at asymptotically large temperatures or densities. However, for lower values of these parameters, it has to face the counter-effect of asymptotic freedom, namely, that the coupling becomes larger and larger, eventually invalidating the use of any perturbative expansion. As a matter of fact, if one insists in decreasing the energy further, the perturbative running coupling diverges at a finite scale, known as $\Lambda_{QCD} \simeq 200\,\mathrm{MeV}$, that sets an apparently impassable barrier for perturbative methods. Moreover, because this energy scale is not much different from the deconfinement temperature (see below), one concludes that neither the deconfinement transition nor the confined phase are amenable to perturbative methods. In fact, even when combined with hard-thermal-loop resummation techniques at high temperature [26], perturbation theory is believed to give a good insight on the deconfined phase only down to temperatures of the order of two to three times the transition temperature [27, 28]. We will have more to add below on the relevance of perturbative approaches, but the above considerations are usually taken as the starting point for the development of non-perturbative tools.

The most famous of them is certainly lattice QCD for it yields a full numerical solution to the theory within a given spacetime discretization [29, 30]. It is based on the Euclidean functional integral formulation of QCD and the probabilistic interpretation of the latter that allows for the use of importance sampling Monte Carlo techniques. After various decades of improvements in order to solve many practical implementation issues, lattice QCD has now reached the era of precision, providing much physical insight as well as a wealth of valuable data that other approaches can use for benchmarking. In the vacuum, this robust approach gives a very good account of the hadronic spectrum of the theory [31] as well as compelling evidence that confinement is a property of QCD [8]. At finite temperature, it provides a clear evidence for the existence of a transition between confined and deconfined phases at a temperature of around $T_d \simeq 154\,\mathrm{MeV}$ [32]. The transition is not a sharp transition, however, rather a crossover [33, 34] characterized by a rapid but smooth variation of the thermodynamical properties of the system, which can also be accurately evaluated within lattice simulations [35].

One of the main drawbacks of lattice simulations is that they rely crucially on the probabilistic interpretation of the functional integral. Away from this comfort zone, that is, whenever the functional to be integrated is not positive definite, they suffer a tremendous loss of accuracy, known as *sign problem* [36, 37]. The latter prevents the investigation of many interesting quantities such as the QCD-phase diagram for moderate to large densities (and small temperatures) or the evaluation of dynamical quantities such as transport coefficients. Many approaches have been devised to circumvent or at least tame the lattice sign problem, such as Taylor expansions around small densities, re-weightings of the functional integral,

---

[2] We shall not review here the many interesting approaches that are based on low-energy models of QCD. See for instance [21–25] and references therein.

numerical continuations from imaginary chemical potentials, and more recently the use of Lefschetz thimbles [38] or of complex Langevin dynamics [39]. Although valuable progress could be achieved within each of these approaches, no complete solution to the QCD sign problem is available so far.

The second possible class of methods beyond perturbation theory goes under the name of continuum (or functional) non-perturbative methods. To avoid the sign problem, these approaches do not aim at a direct numerical evaluation of the QCD functional integral but rather at the (approximate) resolution of sets of exact equations characterizing the dynamics of the system. These equations can be, for instance, the set of quantum equations of motion, known as Dyson-Schwinger equations [40–44], or the hierarchy of renormalization group equations that one derives from the Wetterich equation [45–47]. Related approaches include the use of n-particle-irreducible effective actions [48, 49], the Hamiltonian formalism and its variational principle [50, 51], or Dyson-Schwinger equations modified through the pinch technique [52].

One common denominator to most continuum approaches is that the elementary quantities they give directly access to are the correlation functions of the system. The latter obey infinite hierarchies of coupled equations that cannot, in general, be solved exactly. As such, there exists in general no simple systematics for improvement, as opposed to the perturbative or lattice approaches. Instead, one usually resorts to truncations of the infinite hierarchy of equations, dictated either by physical intuition or computability criteria. The quality of these truncations needs in any case to be tested *a posteriori*. An added difficulty in the case of gauge theories such as QCD is that the correlation functions are not uniquely defined. Indeed, within any gauge theory, the definition of correlation functions makes sense only within a specified gauge, turning the correlation functions into subtly gauge-dependent quantities, as opposed to the physical, gauge-invariant observables that one can access from lattice simulations. In principle, observables can be reconstructed in terms of the gauge-variant correlation functions, but this requires an accurate determination of the latter through appropriate truncations, the quality of which can again be tested only *a posteriori*. Comparison between various continuum approaches or confrontation to lattice results (when available) is therefore a crucial element in finding the appropriate truncations.

Here, lattice simulations come specially in handy. Indeed, the latter cannot only be formulated in a gauge-invariant setting but also within a specified gauge. In particular, over the past 20 years, an intense activity has been devoted to the evaluation of Landau gauge correlation functions, both from lattice simulations and from continuum non-perturbative approaches. The reason for choosing the Landau gauge is twofold. First, this gauge can be formulated as an extremization problem which is perfectly suited for a lattice implementation [53–62]. Second, and even more importantly, the Landau gauge correlation functions have been thought to provide direct access to the physics of confinement. Indeed, in their seminal work [63], by combining the symmetry properties of the Landau gauge together with the hypothesis of confinement, Kugo and Ojima could predict a very characteristic behavior of the two-point correlation functions for the ghost fields (certain additional degrees of freedom that are introduced when specifying to the Landau gauge). According to them, the ghost two-point function should be strongly enhanced at low momenta, with respect to the two-point function in the absence of interactions. Correspondingly, the gluon propagator should vanish at low momenta. This characteristic low momentum behavior, known nowadays as *scaling*, could be observed in the various continuum non-perturbative approaches mentioned above [43, 44, 64–67].

This solution, however, does not seem to be the one seen in lattice simulations where the gluon two-point function and the ghost dressing function (the ratio of the ghost two-point function to its non-interacting version) both saturate to finite non-zero values at low momentum, defining what is nowadays referred to as a *decoupling* behavior [53, 68–71]. It has eventually been shown that non-perturbative continuum approaches can accommodate both scaling- and decoupling-type solutions [72], with however the price of modifying the boundary conditions of the corresponding infinite hierarchy of equations. Since then, the agreement between lattice and non-perturbative approaches has considerably increased, providing better control on the truncations that are considered and opening the way to a myriad of applications of functional methods to QCD. In particular, many aspects of the QCD-phase diagram can now be addressed with these methods [73–81].

This manuscript reviews the results obtained within yet a third route, as originally proposed in [82, 83]. To understand it better, we need to take a few steps back. We argued above that perturbation theory, although quite relevant at high energies, seems to contain the seeds of its own breakdown as the energy is lowered, since the perturbative running coupling increases and eventually diverges at a finite scale. It is to be noted, however, that, in a gauge theory, the very definition of perturbation theory requires a gauge to be specified. Put it differently, there is not a unique perturbative expansion in QCD, but infinitely many, in fact as many as there are ways to fix the gauge. It is true that the way the coupling runs is universal at high energies.[3] However, both the actual value of the coupling (at all energies) and its running at low energies are not universal. In general, they depend both on the renormalization scheme and on the gauge that one works with. Therefore, the properties of the perturbative expansion, including its range of validity, depend on the precise procedure used to fix the gauge.

Gauge fixing is usually performed by following a standard approach, known as the *Faddeev-Popov (FP) procedure*. The outcome of this procedure is that, in practice, one should not work with the original QCD action but, rather, with a gauge-fixed version of it, known as the *Faddeev-Popov action*. In principle, these two formulations are identical. In practice, however, the Faddeev-Popov construction relies on certain mathematical assumptions which are known not to be realized due to the infamous *Gribov copy problem* or *Gribov ambiguity* [84]. It is generally accepted that this mathematical subtlety can be neglected in the high-energy regime of the theory. However, in the opposite limit, no one really knows

---

[3] The β function that controls this running is shown to be two-loop universal.

how it could impact the gauge-fixed implementation of QCD. Even the breakdown of perturbation theory at low energies could be questioned and some quantities could become amenable to perturbative methods. The thesis to be defended in this manuscript is that certain aspects, in particular the physics of the deconfinement transition, could become akin to perturbative methods once the standard FP gauge-fixing procedure is appropriately extended.

In fact, we know already that this perturbative scenario is too naïve for the strict QCD case [85,86]. There are compelling evidences, however, that the scenario could apply to the gluonic or *pure gauge* sector of QCD. As a matter of fact, lattice simulations in the Landau gauge, which can be seen as one particular approach beyond the FP procedure, show that the pure gauge coupling, rather than diverging at a Landau pole, remains finite at all scales.[4] In addition, the corresponding expansion parameter of pure YM theory never exceeds one,[5] thus opening the interesting possibility of a perturbative approach to some of the properties of pure YM theories. This manuscript aims at exploring this possibility further.[6]

We mention of course that the study of the pure gauge sector is not a purely academic question, disconnected from QCD. The corresponding Yang-Mills (YM) theory remains non-trivial due to the self-interaction of the gluons, and, to some extent, it is believed to capture some of the non-trivial features of QCD, in a simplified setting. In particular, unlike the physical QCD case, the deconfinement transition appears here as a genuine phase transition, associated with the breaking of a symmetry, namely, the *center symmetry* of YM theory at finite temperature, that can be probed with order parameters such as the *Polyakov loop*. Moreover, as discussed in [85,87], a perturbative grasp on the gluon dynamics could open the way to the study of some of the properties of QCD, if not with perturbative methods, at least by means of a systematic expansion scheme, controlled by small parameters.

Roughly speaking, the approaches beyond the standard gauge-fixing procedure can be classified into two categories: semi-constructive approaches on the one hand that aim at resolving the Gribov problem, at least in some approximate form and more phenomenologically inspired approaches on the other hand that aim at constraining the operators that could appear beyond the FP prescription. In this second type of approaches, constraints could come from experimental measurements but also from lattice simulations. In particular, gauge-fixed lattice simulations are a method of choice in constraining whatever model beyond the FP prescription, precisely because they themselves do not rely on it.

---

[4] More precisely, this is the gauge coupling extracted from the ghost-antighost-gluon vertex in the Taylor scheme.

[5] If g denotes the coupling constant and N the number of colors, the expansion parameter is given by $g^2 N/(16\pi^2)$. In terms of the strong coupling constant α_s, this corresponds to $\alpha_s N/(4\pi)$.

[6] If a perturbative approach is possible, one could ask why the standard one fails in exploring the properties of pure gauge theories in the infrared. Once more, the point of view taken in this manuscript is that the starting point of such a standard perturbative expansion (i.e., the FP action), even though well-grounded at high energies, has no solid justification in the infrared and one needs to find a different starting point.

Among the possible phenomenological models beyond the FP action, the Curci-Ferrari (CF) model [88] (also known as Fradkin-Tyutin model [89]) is a particularly interesting one. This model was originally introduced as an alternative to the Higgs mechanism. More recently, it has been proposed as a model beyond the FP procedure. This proposal is grounded on how (surprisingly) well the lattice Landau gauge decoupling-type correlators in the vacuum can be accommodated by the one-loop correlation functions of the model [82, 83]. This applies not only to the two-point correlator functions but also to the three-point correlators [90]. The model rests on the adjustment of one additional free parameter, the Curci-Ferrari mass.[7] However, the model being renormalizable, this is the only additional parameter as compared to the FP action that can be adjusted from the comparisons to lattice results in the vacuum. The success of the model in reproducing the vacuum Landau gauge correlators from high to low momentum scales relies on the existence of *infrared-safe* renormalization group trajectories along which the coupling of the interaction remains moderate.[8]

As a further stringent test of the model, the goal of this manuscript is to review how consistent are its perturbative predictions away from the vacuum, in particular with regard to the deconfinement transition and the corresponding phase structure. The plan is as follows:

- In the next chapter, after reviewing the basic properties and limitations of the standard gauge-fixing procedure, we discuss some of the possible approaches beyond it. We also introduce the approach to be examined in this manuscript, based on the CF model, and review some of the results obtained with this approach in the vacuum, regarding the Landau gauge correlation functions. A first attempt at studying finite temperature effects is also reviewed, based on the hypothesis that the finite temperature Landau gauge correlators could carry some imprint of the deconfinement transition. It turns out that this analysis is inconclusive essentially because the Landau gauge, so useful in the vacuum, does not properly capture the order parameter associated with the deconfinement phase transition. In order to understand the limitations of the Landau gauge, we recall some basic considerations related to center symmetry and the Polyakov loop in Chap. 3. We also explain why these basic properties are difficult to capture not only within the Landau gauge but also in fact within any standard gauge-fixed setting.
- The previous difficulties can be solved by generalizing the gauge fixing in the presence of a background field. Even though part of this is a known material, we dedicate Chaps. 4 and 5 to a comprehensive presentation of the use of background-field methods at finite temperature. In particular, we stress

---

[7] What is really free is the value of the CF mass at a given scale. Interestingly enough, the running CF mass runs to 0 both at high and at low energies.

[8] In particular, no Landau pole is found, as opposed to renormalization group trajectories in the FP model.

the importance of a description of the states of the system that is free of the redundancy associated with gauge invariance, as well as the role of *self-consistent backgrounds* as alternative order parameters for center symmetry. We also discuss other symmetry constraints such as charge conjugation, homogeneity, and isotropy, which are rarely discussed in the literature. As much as possible, we try to critically discuss the various implicit assumptions that are usually made when applying background-field methods at finite temperature, in particular regarding the properties of the gauge-fixed measure.

- The perturbative study of the YM deconfinement transition within the CF model in the presence of a background is given in Chap. 6, at leading order, together with a comparison to other approaches. The convergence properties of the approach are investigated in Chap. 7 where we evaluate the next-to-leading-order corrections. Chapter 8 investigates further the relation between center symmetry and the deconfinement transition by providing a general discussion in the SU(N) case and treating yet another example, that of the SU(4) gauge group.

- We also discuss the perturbative predictions of the model for a theory cousin to YM theory, namely, QCD in the regime where all quarks masses are considered heavy. Although this does not correspond to the physical QCD case, this formal regime of QCD has received a lot of attention lately since it possesses a rich phase structure that can be probed with the same order parameters as in the pure YM case and which can be used as a benchmarking of any method that one plans to extend to the real QCD case. We analyze to which extent the phase structure in this regime can be described using perturbation theory within the CF model. This is done in Chap. 10 after some additional material is provided in Chap. 9 on the use of background-field methods at finite density.

- Finally, in Chap. 11, we revisit once more the background-field method at finite temperature and review a recently proposed novel implementation, which, although equivalent to the one based on self-consistent backgrounds, in principle (i.e., in the absence of approximations or modeling of the gauge fixing in the infrared), differs from it in practice and could potentially yield more robust results. This is also the opportunity to revisit the question of a possible imprint of the deconfinement transition on the two-point correlation functions. In particular, we show that within the novel implementation of the background-field method, the gluon propagator can be used as a probe of the deconfinement transition. We compare these results with similar ones obtained within the self-consistent approach or within the standard Landau gauge.

- The manuscript ends with a conclusion where the main results are summarized and an outlook is proposed.

This manuscript reviews results covered in Refs. [91–100]. Beyond the mere review, we have tried as much as possible to provide a self-contained document, in particular with regard to the used methodology. We have also included some unpublished work, such as the material in Sects. 3.3, 3.4, 4.3, 4.4, 8.4, and 9.3.4; the Appendices of Chaps. 1 and 5 and the related Problems 5.6 and 5.7. The presentation in Chaps. 8 and 9 is also new. Finally, Chap. 11 was considerably rewritten as compared to the original version of this manuscript, so as to include the latest developments. In particular, some of the conjectured results in the earlier version have now been proven and have opened a new way into the study of the deconfinement transition within the background-field method, beyond the one described in Chaps. 4 and 5.

Faddeev-Popov Gauge Fixing and the Curci-Ferrari Model

2

The point of view taken in this manuscript is, first, that tackling the low-energy properties of non-abelian gauge theories in the continuum requires extending the standard, but incomplete, Faddeev-Popov gauge-fixing procedure and, second, that once such an extension is found, a new perturbative scheme could become available in the infrared providing access to some of the low-energy properties. In this first chapter, we introduce the Curci-Ferrari action, as a phenomenological model for an extension of the Faddeev-Popov action in the Landau gauge.

For the sake of completeness, we first review the standard gauge-fixing procedure together with its main properties and limitations. We then discuss some of the approaches that have been devised in order to go beyond it and, finally, particularize to the Curci-Ferrari model. We recall that the latter does a pretty reasonable job in reproducing some of the known low-energy properties of Landau gauge Yang-Mills theory in the vacuum, already at one-loop order. We also discuss the difficulties that appear when using the model at finite temperature, serving as the main motivation for the developments in the rest of the manuscript.

2.1 Standard Gauge Fixing

For simplicity, let us review the standard gauge-fixing procedure in the case of Yang-Mills (YM) theory, the theory obtained from quantum chromodynamics (QCD) after neglecting the dynamics of quarks. Of course, a similar discussion could be carried out in the presence of matter fields.

2.1.1 Gauge Invariance

YM theory describes the dynamics of a non-abelian gauge field $A_\mu^a(x)$. The index a corresponds to an internal degree of freedom (color) and labels the generators it^a

© The Author(s), under exclusive license to Springer Nature Switzerland AG 2022
U. Reinosa, *Perturbative Aspects of the Deconfinement Transition*, Lecture Notes in Physics 1006, https://doi.org/10.1007/978-3-031-11375-8_2

of a non-abelian gauge group, SU(N) in what follows. The non-abelian structure is encoded in the structure constants f^{abc} such that $[t^a, t^b] = i f^{abc} t^c$, and the dynamics is specified by the Yang-Mills action :

$$S_{\text{YM}}[A] = \frac{1}{4g^2} \int d^d x \, F^a_{\mu\nu}(x) \, F^a_{\mu\nu}(x) \,, \tag{2.1}$$

with $F^a_{\mu\nu} \equiv \partial_\mu A^a_\nu - \partial_\nu A^a_\mu + f^{abc} A^b_\mu A^c_\nu$ as the non-abelian generalization of the QED field-strength tensor. Since the applications to be discussed in this manuscript concern the equilibrium properties of the QCD/YM system, we have here chosen the Euclidean version of the action, and, therefore, there is no distinction between covariant and contravariant indices [101, 102]. Moreover, the gauge field should be taken periodic along the Euclidean time direction, with a period equal to the inverse temperature $\beta \equiv 1/T$. Correspondingly, the integration symbol $\int d^d x$ needs to be understood as

$$\int d^d x \equiv \int_0^\beta d\tau \int d^{d-1}x \,, \tag{2.2}$$

and we have $A^a_\mu(\tau + \beta, \mathbf{x}) = A^a_\mu(\tau, \mathbf{x})$. These boundary conditions will play a major role in what follows.[1]

The main feature of the YM action is of course that it is gauge-invariant. This is most easily seen by rewriting the action (2.1) in an intrinsic form that does not depend on the particular coordinate system used to describe the color degrees of freedom.[2] One interprets any colored object X^a as an element of the SU(N) Lie algebra $X \equiv i X^a t^a$. To any two such elements X and Y, one then associates the Killing form[3]

$$(X; Y) \equiv -2 \operatorname{tr} XY \,, \tag{2.3}$$

which allows one to rewrite the Yang-Mills action as

$$S_{\text{YM}}[A] = \frac{1}{4g^2} \int d^d x \, \left(F_{\mu\nu}(x); F_{\mu\nu}(x) \right), \tag{2.4}$$

---

[1] Very schematically, the need for an Euclidean version of the action and the presence of periodic boundary conditions relate to the fact that the partition function at finite temperature is the trace of an evolution operator $e^{-\beta H}$ in imaginary time (see Refs. [101, 102] for details).

[2] Later, this will also facilitate the change from the standard, Cartesian bases to Cartan-Weyl bases. We shall introduce and use these bases in subsequent chapters.

[3] The minus sign is chosen such that $(i t^a; i t^b) = \delta^{ab}$.

with $F_{\mu\nu} = \partial_\mu A_\nu - \partial_\nu A_\mu - [A_\mu, A_\nu]$. In this intrinsic representation, a gauge transformation of the gauge field is defined to be

$$A_\mu^U(x) \equiv U(x) A_\mu(x) U^\dagger(x) - U(x) \partial_\mu U^\dagger(x), \tag{2.5}$$

with $U(x) \in SU(N)$, $\forall x$. It is then easily verified that the field-strength tensor transforms correspondingly as $F_{\mu\nu}^U(x) = U(x) F_{\mu\nu}(x) U^\dagger(x)$. From its definition, the Killing form is trivially invariant under color rotations in the sense $(UXU^\dagger; UYU^\dagger) = (X; Y)$. It follows, as announced, that the action (2.1) is gauge-invariant. In more pompous terms, it is constant along a given *gauge orbit*, defined as the collection of configurations A^U as U spans the possible gauge transformations, for a given A. The notion of gauge orbit, which we will specify further in the case of a finite temperature setting, will play a major role in subsequent discussions.

In what follows, we shall often make use of the adjoint covariant derivative $D_\mu X \equiv \partial_\mu X - [A_\mu, X]$. Upon expansion along a basis $\{it^a\}$, this corresponds to $D_\mu X^a \equiv \partial_\mu X^a + f^{acb} A_\mu^c X^b$.

2.1.2 Observables

The gauge invariance of the YM action is just the expression of a certain arbitrariness in the choice of the gauge-field configuration that describes a given physical situation. Observables cannot depend on this arbitrariness and, therefore, are to be represented by gauge-invariant functionals.[4] To any such observable $O[A]$, one associates an expectation value as

$$\langle O \rangle \equiv \frac{\int \mathcal{D}A\, O[A]\, e^{-S_{YM}[A]}}{\int \mathcal{D}A\, e^{-S_{YM}[A]}}, \tag{2.6}$$

where $S_{YM}[A]$ and $\mathcal{D}A$ are also gauge-invariant.

The usual difficulty with the above definition is that it involves two indefinite integrals. Indeed, since $O[A]$, $S_{YM}[A]$ and $\mathcal{D}A$ are gauge-invariant, the integrals sum redundantly over the orbits of the gauge group and are thus proportional to the (infinite) volume of the latter. Even though these two infinities should formally factorize and cancel between the numerator and the denominator of Eq. (2.6), their presence prevents any expansion of $\langle O \rangle$ in terms of expectation values of gauge-variant functionals, typically products of gauge fields at the same spacetime point, as needed by most continuum approaches.

---

[4] We are deliberately being vague here concerning the type of gauge transformations that should be considered. At finite temperature, where the gluon field is periodic along the Euclidean time direction, the true, unphysical gauge transformations are also periodic. There are more general gauge transformations that preserve the periodicity of the fields, but those are associated with physical transformations in a sense to be clarified in the next chapter.

One possibility to tackle this problem is to find a way to rewrite the definition (2.6) identically as

$$\langle O \rangle = \frac{\int \mathcal{D}_{\text{gf}}[A] \, O[A] \, e^{-S_{\text{YM}}[A]}}{\int \mathcal{D}_{\text{gf}}[A] \, e^{-S_{\text{YM}}[A]}}, \tag{2.7}$$

where the *gauge-fixed measure* $\mathcal{D}_{\text{gf}}[A]$ is restricted to gauge-field configurations obeying a certain *gauge-fixing condition* $F[A] = 0$. Since $\mathcal{D}_{\text{gf}}[A]$ is typically not invariant under gauge transformations, the expression (2.7) for the expectation value extends to gauge-variant functionals and can then be used as a starting point for developing continuum methods.

Of course, the crux of the problem lies in the construction of the gauge-fixed measure $\mathcal{D}_{\text{gf}}[A]$. The usual strategy is the Faddeev-Popov (FP) procedure [103, 104] which we now recall.

2.1.3 Faddeev-Popov Procedure

In the FP approach, under the assumption that the constraint $F[A] = 0$ admits a unique solution on each orbit, one considers a *decomposition of unity* in the form

$$1 = \int \mathcal{D}U \, \Delta[A^U] \delta\big(F[A^U]\big), \tag{2.8}$$

where $\Delta[A^U]$ is the determinant of $\delta F[A^U]/\delta U$, also known as the *Faddeev-Popov operator*.[5]

Plugging the identity (2.8) into the definition (2.6) and using the gauge invariance of $O[A]$, $S_{\text{YM}}[A]$ and $\mathcal{D}A$, one obtains

$$\langle O \rangle_{\text{FP}} = \frac{\int \mathcal{D}U \int \mathcal{D}A^U \, O[A^U] \, \Delta[A^U] \delta(F[A^U]) \, e^{-S_{\text{YM}}[A^U]}}{\int \mathcal{D}U \int \mathcal{D}A^U \, \Delta[A^U] \delta(F[A^U]) \, e^{-S_{\text{YM}}[A^U]}}. \tag{2.9}$$

Changing variables from A^U to A, the volume of the gauge group factorizes and cancels between the numerator and the denominator. One then arrives at the

---

[5] Strictly speaking, it is the absolute value of the determinant that should appear in Eq. (2.8). However, since it is assumed that there is only one solution to $F[A] = 0$ on each orbit, and if one further assumes that this solution changes continuously as one changes the orbit, the sign of the determinant is constant and, therefore, irrelevant in the Faddeev-Popov approach. Moreover, that $\Delta[A^U]$ does not depend independently on A, and U follows from the identity $A^{VU} = (A^U)^V$ which implies $F[A^{(\delta V)U}] = F[(A^U)^{\delta V}]$.

following *gauge-fixed* expression for the expectation value:

$$\langle O \rangle_{\text{FP}} = \frac{\int \mathcal{D}_{\text{FP}}[A] \, O[A] \, e^{-S_{\text{YM}}[A]}}{\int \mathcal{D}_{\text{FP}}[A] \, e^{-S_{\text{YM}}[A]}} \, , \tag{2.10}$$

where the Faddeev-Popov gauge-fixed measure is defined as

$$\mathcal{D}_{\text{FP}}[A] \equiv \mathcal{D}A \, \Delta[A] \, \delta(F[A]) \, . \tag{2.11}$$

A similar analysis applies to the partition function except for an overall volume factor which, however, does not affect the temperature-dependent part of the free-energy density. One finds $Z = \mathcal{V} \times Z_{\text{FP}}$, with

$$Z_{\text{FP}} = \int \mathcal{D}_{\text{FP}}[A] \, e^{-S_{\text{YM}}[A]} \, , \tag{2.12}$$

and $\mathcal{V}$ the volume of the group of gauge transformations, such that $-\ln \mathcal{V} \propto \beta\Omega$ with $\beta\Omega$ the Euclidean spacetime volume.

We mention that the previous derivation relies on a rather strong assumption, namely, that the solution to the gauge-fixing condition $F[A] = 0$ is unique along a given orbit. As pointed out by Gribov [84], this assumption is generally wrong due to the existence, instead of multiple solutions, known as *Gribov copies*. For this reason, we have denoted by $\langle O \rangle_{\text{FP}}$ the gauge-fixed expression for the expectation value of an observable as obtained from the FP approach, which may differ from $\langle O \rangle$ as originally defined in Eq. (2.6). We shall come back to this important point in the next section. For the time being, we continue reviewing the properties of the FP approach.

2.1.4 Faddeev-Popov Action

The previous formulation is not very practical due to the presence of both the determinant $\Delta[A]$ and the functional Dirac distribution $\delta(F[A])$ in the gauge-fixed measure (2.11). As it is well known, one can transform the latter into a standard field theory with the price of introducing additional fields.

First, the factor $\delta(F[A])$ can be treated using a Nakanishi-Lautrup field h^a as

$$\delta(F[A]) = \int \mathcal{D}h \, \exp\left\{ -\int d^d x \, i h^a(x) \, F^a[A](x) \right\} . \tag{2.13}$$

Second, the determinant $\Delta[A]$ can be evaluated by assuming that $F[A] = 0$ due to the presence of the factor $\delta(F[A])$. One finds (see Problem 2.1)

$$\Delta[A] = \det\left(\int d^d z \, \frac{\delta F^a[A](x)}{\delta A_\mu^c(z)} \, D_\mu^{cb} \delta(z-y)\right), \tag{2.14}$$

where $D_\mu^{ab} \equiv \partial_\mu \delta^{ab} + f^{acb} A_\mu^c$ denotes the adjoint covariant derivative (see above). Finally, by introducing (Grassmannian) ghost and antighost fields c^a and $\bar{c}^a$, this rewrites

$$\Delta[A] = \int \mathcal{D}[c, \bar{c}] \, \exp\left\{\int d^d x \int d^d y \, \bar{c}^a(x) \frac{\delta F^a[A](x)}{\delta A_\mu^b(y)} \, D_\mu c^b(y)\right\}. \tag{2.15}$$

Altogether, the gauge-fixed expression for the expectation value of an observable in the FP approach reads

$$\langle O \rangle_{\text{FP}} = \frac{\int \mathcal{D}[A, c, \bar{c}, h] \, O[A] \, e^{-S_{\text{FP}}[A,c,\bar{c},h]}}{\int \mathcal{D}[A, h, c, \bar{c}] \, e^{-S_{\text{FP}}[A,c,\bar{c},h]}}, \tag{2.16}$$

where $S_{\text{FP}}[A, c, \bar{c}, h] \equiv S_{\text{YM}}[A] + \delta S_{\text{FP}}[A, c, \bar{c}, h]$ is known as the Faddeev-Popov action, with

$$\delta S_{\text{FP}} \equiv -\int d^d x \int d^d y \, \bar{c}^a(x) \frac{\delta F^a[A](x)}{\delta A_\mu^b(y)} \, D_\mu c^b(y) + \int d^d x \, i h^a(x) \, F^a[A](x). \tag{2.17}$$

Similarly, the partition function rewrites

$$Z_{\text{FP}} = \int \mathcal{D}[A, c, \bar{c}, h] \, e^{-S_{\text{FP}}[A,c,\bar{c},h]}. \tag{2.18}$$

The Landau gauge to be considered in this manuscript corresponds to the choice $F^a[A](x) = \partial_\mu A_\mu^a(x)$ whose FP operator is $\partial_\mu D_\mu^{ab} \delta(x-y)$. It follows that

$$\delta S_{\text{FP}} = \int d^d x \left\{\partial_\mu \bar{c}^a(x) \, D_\mu c^b(x) + i h^a(x) \, \partial_\mu A_\mu^a(x)\right\}, \tag{2.19}$$

where an integration by parts has been used in the ghost term.

2.1.5 BRST Symmetry

As we now recall, the FP action possesses a very important symmetry, due to Becchi, Rouet, Stora, and Tyutin [105], at the origin of many properties, including its renormalizability [105–107].

Suppose that we perform an infinitesimal gauge transformation of the form $\delta A_\mu^a = \bar{\eta} \, D_\mu c^a \equiv \bar{\eta} \, s A_\mu^a$, with $\bar{\eta}$ as a constant Grassmannian parameter, and let us more generally denote by $\delta \phi = \bar{\eta} s \phi$ the corresponding transformation of any other field ϕ. The Yang-Mills part of the action is of course invariant by construction, so let us focus on the gauge-fixing part δS_{FP}. It is easily shown that $s \delta S_{\text{FP}}$ rewrites as a linear combination of $i h^a - s \bar{c}^a$, $s i h^a$ and $s^2 A_\mu^a$ (see Problem 2.2). Then the gauge-fixing contribution is invariant if we set $s \bar{c}^a = i h^a$ and $s i h^a = 0$ and choose $s c^a$ such that $s^2 A_\mu^a = 0$. A straightforward calculation reveals that this is ensured if $s c^a = -\frac{1}{2} f^{abc} c^b c^c$ (see Problem 2.2).

In summary, the FP action is invariant under

$$s A_\mu^a = D_\mu c^a, \quad s c^a = -\frac{1}{2} f^{abc} c^b c^c, \quad s \bar{c}^a = i h^a, \quad s i h^a = 0, \tag{2.20}$$

which is known as BRST symmetry. It is easily verified that s^2 vanishes not only over A_μ^a but over the whole functional space, and the BRST symmetry s is then said to be nilpotent (see Problem 2.3). In fact, the BRST invariance of the FP action can be understood in terms of the nilpotency of s and the gauge invariance of S_{YM}. Indeed, the action rewrites

$$S_{\text{FP}} = S_{\text{YM}} + s \int d^d x \, \bar{c}^a(x) \, F^a[A](x), \tag{2.21}$$

and therefore $s S_{\text{FP}} = s S_{\text{YM}} = 0$.

2.1.6 Gauge-Fixing Independence

For the gauge-fixing procedure to make sense at all, it should be of course such that the expectation value of a gauge-invariant functional does not depend on the choice of gauge-fixing condition. In the FP approach, this basic property is not guaranteed a priori since, as mentioned above, the procedure relies on an incorrect assumption. It is one of the merits of the BRST symmetry to ensure nonetheless the gauge-fixing independence of the observables in the FP framework. To see how this works, let us first derive the following useful result:

Lemma
Given an infinitesimal symmetry $\varphi \to \varphi + \delta\varphi$ of a theory $S[\varphi]$, the expectation value of any corresponding infinitesimal variation $\delta O[\varphi]$ needs to vanish.

Proof By performing a change of variables in the form of an infinitesimal symmetry under the functional integral definition of the expectation value of $O[\varphi]$, one obtains

$$\int \mathcal{D}\varphi \, O[\varphi] \, e^{S[\varphi]} = \int \mathcal{D}\varphi \, O[\varphi + \delta\varphi] \, e^{S[\varphi+\varphi]}. \qquad (2.22)$$

Using the invariance of the action and subtracting the left-hand side from the right-hand side, one arrives at the desired result.[6]   □

Let us now apply the lemma to prove the gauge-fixing independence of the observables in the FP approach. Consider first the FP partition function Z_{FP}, and let λ denote any parameter that may enter the definition of the gauge fixing. From Eq. (2.21), we find

$$\frac{d}{d\lambda} \ln Z_{\text{FP}} = \left\langle s \int d^d x \left(\bar{c}^a(x) \frac{d F^a[A](x)}{d\lambda} \right) \right\rangle. \qquad (2.23)$$

Being the expectation value of an infinitesimal symmetry transformation, the right-hand side needs to vanish according to the lemma, from which we deduce that $d Z_{\text{FP}}/d\lambda = 0$ and thus that the partition function does not depend on the choice of gauge-fixing functional $F[A]$. This result extends in fact to any observable. Using the previous result together with $sO[A] = 0$, we find indeed

$$\frac{d}{d\lambda} \langle O \rangle_{\text{FP}} = \left\langle s \left(O[A] \int d^d x \, \bar{c}^a(x) \frac{d F^a[A](x)}{d\lambda} \right) \right\rangle, \qquad (2.24)$$

which vanishes for the same reason.

---

[6]It is to be mentioned that this standard derivation does not pay too much attention to how the measure and the integration domain are transformed. This is of course because, in general, they are both invariant. However, in the case of a BRST transformation, the initial integration domain for the gauge-field A is made of purely numerical functions, whereas the transformed domain is a more general space, including still commuting but non-numerical contributions of the form $\bar{\eta} \, Dc$. We argue in the Appendix of this chapter that the above lemma is unaffected by these subtleties.

2.2   Infrared Completion of the Gauge Fixing

As we have already mentioned, the derivation of the FP action (2.17), together with its main properties, relies on a strong assumption, which is known to not always hold true [84]. In particular, it is incorrect in the case of the Landau gauge [108] to which we restrict from now on.

2.2.1   Gribov Copies

The point is that, contrary to what is assumed in the FP construction, the gauge-fixing condition $F[A] = 0$ admits multiple solutions along a given orbit, known as Gribov copies. The existence of the latter invalidates the use of Eq. (2.8) and, therefore, the identification of $\langle O \rangle$ and $\langle O \rangle_{\text{FP}}$, as given, respectively, by Eqs. (2.6) and (2.16). The identification $\langle O \rangle_{\text{FP}} \approx \langle O \rangle$ is nonetheless believed to be legitimate at high energies since only a perturbative region of the space of gauge-field configurations contributes to the functional integral and the Gribov copies can be neglected. In this case, one can evaluate $\langle O \rangle$ using the perturbative expansion of $\langle O \rangle_{\text{FP}}$ which is controlled at high energies thanks to the asymptotic freedom property [10].

In contrast, in the low-energy regime, the situation is much less clear. The perturbative expansion of $\langle O \rangle_{\text{FP}}$ is useless due to the presence of an infrared Landau pole. At the same time, there is no argument anymore in favor of the identification $\langle O \rangle_{\text{FP}} \approx \langle O \rangle$. As a matter of fact, undoing the step from Eq. (2.10) to Eq. (2.9) and integrating over the gauge group by taking into account possible copies, one finds

$$\langle O \rangle_{\text{FP}} = \frac{\int \mathcal{D}A\, O[A]\, e^{-S_{\text{YM}}[A]} \sum_{i(A)} s_{i(A)}}{\int \mathcal{D}A\, e^{-S_{\text{YM}}[A]} \sum_{i(A)} s_{i(A)}}, \tag{2.25}$$

where $i(A)$ labels all the solutions to $F[A] = 0$ along the orbit of A, including the Gribov copies, and $s_{i(A)}$ is the sign of the FP determinant on the solution labeled $i(A)$. If the quantity $\sum_{i(A)} s_{i(A)}$ were non-zero and the same for each orbit, the identification of $\langle O \rangle_{\text{FP}}$ and $\langle O \rangle$ would indeed be correct. Unfortunately, for compact gauge groups, the sum vanishes instead, leading to a $0/0$ indetermination or *Neuberger zero problem* [109,110], which prevents the formal identification of $\langle O \rangle_{\text{FP}}$ and $\langle O \rangle$ beyond the one discussed above in the high-energy, perturbative domain. Similarly, the FP partition function vanishes.

This calls for constructing alternative gauge-fixing procedures that take into account the effect of the Gribov copies, at least in some approximate form.[7] The quest for such an infrared completion of the FP gauge fixing is not only formal. In fact, according to certain scenarios, once such a gauge fixing is found, a new

---

[7] Another interesting approach relies on decomposing the gauge group into a subgroup where the gauge-fixing problem is trivial and a quotient group where the Neuberger zero is absent [111,112].

perturbative expansion could become available in the low-energy regime [82,83,90]. Various strategies have been devised to include the Gribov copies in a more rigorous way. Let us briefly review some of them.

2.2.2 Gribov-Zwanziger Approach

One possibility is to restrict the functional integration over gauge-field configurations to a subdomain such as some (but in general not all) copies are excluded. This is, for instance, achieved in the Gribov-Zwanziger (GZ) approach where *infinitesimal copies* are excluded by restricting to a region such that the FP operator $-\partial_\mu D_\mu$ is positive definite [113, 114].

Just as in the FP approach, the GZ procedure can be formulated as a local and renormalizable theory. The standard BRST is broken.[8] A nilpotent BRST symmetry can be identified, but it is non-local [115]. More recently, a proposal has been made to rewrite the GZ action into an alternative local form that displays a local and nilpotent BRST symmetry [116]. This rewriting requires, however, in a certain sense, neglecting once more the presence of copies.

One very appealing feature of the GZ approach is that a mass scale is dynamically generated and determined only in terms of the YM coupling, a feature that any *bona fide* gauge fixing should possess. However, in its original formulation, the GZ action predicts correlation functions at odds with the ones obtained on the lattice. It has since then been refined by the inclusion of condensates in order to improve the comparison with lattice results [117, 118]. The impact of these condensates on the propagators is usually considered at tree-level. However, because the condensates emerge from a cancellation between tree-level and one-loop terms of the corresponding potential (more precisely its derivative with respect to the condensate), the true impact of the condensates should be tested at least at the level of one-loop correlators (where similar cancelations could occur).

2.2.3 Serreau-Tissier Approach

Another possible strategy, followed, for instance, by Serreau and Tissier [119] (see also Ref. [120] for a similar idea), is to sum formally over all copies, just as in Eq. (2.25) but with a weighting factor $w[A^{U_{i(A)}}]$ that avoids the Neuberger zero problem. To ensure that the result is an identical rewriting of Eq. (2.6), one considers

---

[8] This can be understood from the presence of a boundary in the space of gauge-field configurations and the fact that a BRST transformation changes the original integration domain (see the discussion in the Appendix).

an average rather than a sum. To this purpose, one inserts in Eq. (2.6) the identity

$$1 = \frac{\sum_{i(A)} s_{i(A)} w[A^{U_{i(A)}}]}{\sum_{i(A)} s_{i(A)} w[A^{U_{i(A)}}]} = \frac{\int \mathcal{D}U \, \Delta[A^U] \, \delta(F[A^U]) \, w[A^U]}{\int \mathcal{D}U' \, \Delta[A^{U'}] \, \delta(F[A^{U'}]) \, w[A^{U'}]}. \tag{2.26}$$

Under the integral over A, the volume of the gauge group can be factored out in the integral over U while leaving the integral over U' unaffected thanks to the change of variables $U' \to U'U$. One eventually arrives at

$$\langle O \rangle = \frac{\int \mathcal{D}_{\text{ST}}[A] \, O[A] \, e^{-S_{\text{YM}}[A]}}{\int \mathcal{D}_{\text{ST}}[A] \, e^{-S_{\text{YM}}[A]}}, \tag{2.27}$$

with the Serreau-Tissier measure defined as

$$\mathcal{D}_{\text{ST}}[A] \equiv \mathcal{D}A \, \Delta[A] \, \delta(F[A]) \, e^{-\ln \int \mathcal{D}U' \Delta[A^{U'}] \delta(F[A^{U'}]) w[A^{U'}]}. \tag{2.28}$$

Note that we have not included any labeling of $\langle O \rangle$ since the rewriting is exact at this point.

In principle, the logarithm of the U'-integral in the above expression provides a non-trivial correction to the FP action. In practice, however, it is not possible to evaluate this correction exactly, and the following strategy has been adopted, based on the well-known replica trick of statistical physics [121]. One writes

$$\mathcal{D}_{\text{ST}}[A] = \lim_{z \to 0} \mathcal{D}_{\text{ST}}^z[A], \tag{2.29}$$

with

$$\mathcal{D}_{\text{ST}}^z[A] \equiv \mathcal{D}A \, \Delta[A] \, \delta(F[A]) \left(\int \mathcal{D}U' \Delta[A^{U'}] \, \delta(F[A^{U'}]) \, w[A^{U'}] \right)^{z-1}. \tag{2.30}$$

For any strictly positive integer value of z, the right-hand side can be rewritten as a standard field theory involving z-replicated versions of the ghost, antighosts, and Nakanishi-Lautrup fields, in addition to $z - 1$ replicas of the field U'.

This formulation in terms of replicas can be used for practical calculations. The subtle question is, however, how the results obtained for an integer number of replicas $z \in \mathbb{N}^*$ can be extrapolated to the zero replica limit $z \to 0$, which requires $z \in \mathbb{R}$. To date, it is not clear how this limit should be taken.[9] For this reason we

---

[9] An analytic function $f(z)$ is not uniquely determined from its values for $z \in \mathbb{N}^*$.

shall write the last step of the Serreau-Tissier approach as

$$\langle O \rangle_{\mathrm{ST}} = \lim_{z \to 0} \frac{\int \mathcal{D}_{\mathrm{ST}}^z[A]\, O[A]\, e^{-S_{\mathrm{YM}}[A]}}{\int \mathcal{D}_{\mathrm{ST}}^z[A]\, e^{-S_{\mathrm{YM}}[A]}}\,, \tag{2.31}$$

with a labeled expectation value $\langle O \rangle_{\mathrm{ST}}$ to emphasize that, just as in the case of the FP approach, the identification between $\langle O \rangle_{\mathrm{ST}}$ and $\langle O \rangle$ is still open to debate.

2.2.4   Curci-Ferrari Approach

In addition to these semi-constructive strategies, one can envisage a more phenomenological approach. Indeed, since the standard FP action is meant to be modified from a proper account of the Gribov copies, one can try to propose possible corrections in the form of operators added to the FP action[10] and constrain the corresponding couplings from comparison to experiment and/or numerical simulations. All things considered, the idea is pretty similar to the search of theories beyond the Standard Model, of course at a more pedestrian level and with the important difference that, in the present case, the action for the physical theory is known, be it the YM action or the QCD action. What is searched after is a gauge-fixed version of this physical action, beyond the standard FP recipe.

For the Landau gauge, one proposal which has been explored in recent years is the one based on the Curci-Ferrari (CF) model [88, 89]. It consists in supplementing the FP action (2.19) with a mass term for the gluon field, $S_{\mathrm{CF}}[A, c, \bar{c}, h] \equiv S_{\mathrm{YM}}[A] + \delta S_{\mathrm{CF}}[A, c, \bar{c}, h]$, with

$$\delta S_{\mathrm{CF}} = \int d^d x \left\{ \partial_\mu \bar{c}^a(x)\, D_\mu c^a(x) + i h^a(x)\, \partial_\mu A_\mu^a(x) + \frac{1}{2} m^2 A_\mu^a(x) A_\mu^a(x) \right\}. \tag{2.32}$$

We stress that this is just a model and the question to be asked here is not whether this action represents a *bona fide* gauge fixing but, rather, whether the CF mass term could represent a dominant contribution to the unknown gauge-fixed action beyond the FP terms. The main reason to believe that this could be the case is the indiscutable observation made on the lattice that the gluon propagator follows the decoupling solution (see "Introduction") and saturates to a finite non-zero value at low momenta. Of course, other operators could be dominant, and one way to test this is to confront the predictions of the CF model with those of alternative approaches. The rest of the manuscript will be essentially concerned with this question.

A note on terminology here: the presence of a mass term makes it very tempting to refer to the Curci-Ferrari model as "massive Yang-Mills" theory. We do not deem this a very good choice, however, since this opens the door to confusion. Indeed,

---

[10] While fulfilling of course the basic symmetries compatible with the gauge fixing at hand.

this terminology could also refer to a non-abelian version of Proca theory in which the mass term is directly added to the YM action. Although the two approaches seem quite similar from a distance, they are quite different when studied more closely. On a conceptual level, adding a mass directly to YM theory would constitute an intentional modification of a fundamental theory which is not what we aim at here. On the other hand, the mass term in the CF model is added to a gauge-fixed version of YM theory, and the motivation for it is that the FP gauge-fixed version is incomplete as we have argued above. Of course, most probably, the addition of a mass term to the FP action does not qualify as an exact gauge fixing and accordingly also modifies the fundamental theory under scrutiny. However, since the role of the mass term is to model observations made on the lattice that are missed by the FP procedure, we expect these modifications to be milder and that some features of Landau gauge-fixed YM theory can be described in that way. One of the main purposes of this manuscript is to test these expectations on some concrete examples, giving therefore an incentive to eventually find a first-principle completion of the Landau gauge fixing.

The differences between the non-abelian Proca theory and the CF model are also of practical nature. The former is known to be non-renormalizable which prevents any serious prediction. On the contrary, the CF model is renormalizable [88]. In practice, this means that there is only one additional parameter to be dealt with and that once this parameter is fixed (by comparison to simulations, for instance), the model becomes predictive. The proof of renormalization relies in particular on the fact that the tree-level gluon propagator (see Problem 2.5) is given by $P_{\mu\nu}^{\perp}(Q)G(Q)$ with $P_{\mu\nu}^{\perp}(Q) \equiv \delta_{\mu\nu} - Q_\mu Q_\nu/Q^2$ and

$$G(Q) = \frac{1}{Q^2 + m^2}, \tag{2.33}$$

and thus decreases fast enough in the ultraviolet (yet another difference with Proca theory). The proof and implementation of the renormalization program are also simplified thanks to two non-renormalization theorems [83, 122] that follow from the presence of a BRST-like symmetry s_m defined as (see Problem 2.4)

$$s_m A_\mu^a = D_\mu c^a, \quad s_m c^a = -\frac{1}{2} f^{abc} c^b c^c, \quad s_m \bar{c}^a = ih^a, \quad s_m ih^a = m^2 c^a, \tag{2.34}$$

where s_m differs from s only on its action over the Nakanishi-Lautrup field. One of these theorems is Taylor's non-renormalization theorem which is also present in the FP case (see Problem 2.6 for a proof that does not directly exploits BRST symmetry).

Other properties of the FP action do not survive the presence of the Curci-Ferrari mass term. In particular, the modified BRST symmetry is not nilpotent anymore since $s_m^2 \bar{c}^a = m^2 c^a$. Moreover, the "gauge-fixed" part of the action does not rewrite as a variation under this modified BRST transformation. It should be stressed, however, that it is not clear whether the standard BRST symmetry should be manifest in the infrared, precisely because it appears as a result of neglecting the Gribov copies.[11]

Let us close this section by mentioning that the CF model bears interesting relations to the other approaches to gauge fixing described above. For instance, as we have mentioned, the standard GZ action needs to be refined with the use of condensates that modify the usual GZ propagator as

$$G_{GZ}(Q) = \frac{Q^2}{Q^4 + \gamma^4} \quad \rightarrow \quad G_{RGZ}(Q) = \frac{Q^2 + M^2}{Q^4 + (M^2 + m^2)Q^2 + M^2 m^2 + \gamma^4}.$$
(2.35)

A recent investigation on the dynamical generation of these condensates finds that M^2 is larger than both the Gribov parameter γ^2 and the gluon condensate m^2 [125]. Although the differences are not dramatic, it is interesting to note that for large enough M^2, the refined GZ propagator approaches the CF one, so not only the corresponding fields decouple, but the Gribov parameter disappears from the tree-level propagator.

Another connection exists with the Serreau-Tissier approach. Indeed, by choosing the weighting functional as

$$w[A] \equiv \exp\left\{ -\int d^d x \, \frac{1}{2} \rho \, A_\mu^a(x) A_\mu^a(x) \right\},$$
(2.36)

it could be shown that the perturbative evaluation of any correlation function involving only the original, non-replicated fields is equivalent to the evaluation of the same correlation function within the CF model with $m \equiv z\rho$ [119]. The naïve zero replica limit $z \rightarrow 0$ brings us back to the FP action. However, one could consider the limit in a different way, by first interpreting $z\rho$ as the bare mass of the CF model and by using it to absorb the corresponding divergences, after which the limit $z \rightarrow 0$ would again be trivial but would lead to the CF model instead.[12] Of course, the above argument does not fix the value of the renormalized CF mass in terms of the YM coupling, a necessary condition for the procedure to correspond to a *bona fide* gauge fixing, but it gives some indication that the CF model is maybe not that far from this goal. As a matter of fact, a mechanism for the dynamical

---

[11] Interestingly enough, the complicated landscape of Gribov copies is very similar to the landscape of extrema of the energy functional in disordered systems. The latter are usually treated using the Parisi-Sourlas procedure [123], which takes the same steps as the Faddeev-Popov procedure. In this context, it is known that the associated BRST-like symmetry is spontaneously broken below a critical dimension [124]. One could very well imagine a similar breaking as a function of the energy in the context of YM theories.

[12] The fact that the zero replica limit does not commute with other limits is a well-known fact. In applications to disordered systems, it is generally true that one needs to take the infinite volume limit before the zero replica limit.

generation of such a mass could be identified in a gauge cousin to the Landau gauge [126].

2.2.5 Connection to Other Approaches

In addition to the above connections, the CF model could also be relevant for other continuum approaches including the functional renormalization group [46, 72, 127], Dyson-Schwinger equations [44, 54, 128–130], the pinch technique [52, 131, 132], or the variational approach of Ref. [133].

All these approaches take the FP action as a starting point and aim ideally at circumventing the difficulties of the latter in the infrared by going beyond perturbation theory. However, because it requires introducing a cut-off, the practical implementation of these approaches necessarily leads to an explicit breaking of the BRST symmetry and implies de facto that the starting action is more general than the FP one. In particular, in all present continuum approaches to YM theory, a bare mass term is introduced in one way or another to deal with the quadratic divergences that the use of a cut-off entails. This mass term can take the form of a subtraction in the Dyson-Schwinger and pinch technique approaches [131, 134], an additional parameter m_Λ at the UV scale in the functional renormalization group approach [127], or even an explicit mass counterterm in the variational approach [133].

We stress here that this is not a problem per se. One is always allowed (and sometimes forced) to use a regulator that breaks certain symmetries of the problem. This just means that the theory space that needs to be considered is larger. The relevant question is rather how the additional couplings are fixed in terms of the YM coupling. The answer to this question depends on whether or not BRST symmetry survives in the infrared. In the case it does, one should fix all the additional parameters such that the consequences of BRST symmetry are fulfilled. For instance, in the functional renormalization group approach of Ref. [127], one proposal is to adjust the m_Λ parameter so as to trigger the scaling solution, if one deems the latter a consequence of BRST symmetry (from the Kugo-Ojima scenario [63]).

However, in the case where BRST symmetry does not survive in the infrared, the appropriate adjustment of the additional couplings remains an open question, and, for all practical purposes so far, the bare mass remains a free parameter. For instance, in the functional renormalization group approach, if one wants to reproduce the lattice correlation functions, the parameter m_Λ needs to be adjusted to fit the lattice data. The same is true in the case of the pinch technique with the parameter $\Delta_{\text{reg}}^{-1}(0)$ introduced in Ref. [131]. It could still be that the lattice decoupling solution can be generated within a BRST-invariant framework, but this requires subtle mechanisms such as the irregularities proposed in Ref. [127] or the Schwinger mechanism discussed in Refs. [135–137]. To date, and to our knowledge, none of these mechanisms was seen to be realized. In this type of scenarios, studying the CF model perturbatively can bring an interesting perspective to the rest of continuum approaches, as a way to disentangle what is genuinely non-perturbative from what could become perturbative once a mechanism for the generation of a CF mass has been identified. We shall present some of these comparisons in subsequent chapters.

As an added note to the original version of this manuscript, let us mention some recent and interesting developments that appeared 1 year after the writing of this thesis. In the context of Dyson-Schwinger equations, new truncations seem to indicate that the earlier ambiguities related to the removal of quadratic divergences do not seem to impact much the gluon propagator or its corresponding dressing function [138]. Whether the full arbitrariness that the subtraction of quadratic divergences entails has been tested in Ref. [138] as well as how the observed insensitivity to this subtraction depends on the specifics of the truncation and how it can be implemented in other non-perturbative continuum approaches remain open questions. For a related analysis, see also Ref. [139] and for recent advances on the front of the Schwinger mechanism, see [140].

2.3   Review of Results

Let us conclude this first chapter by reviewing some of the tests that the CF model has already passed. We shall also discuss some cases where it is not fully conclusive, as a motivation for the developments to be presented in the rest of the manuscript.

The more direct way to test the CF model is to compare its predictions for Landau gauge-fixed correlation functions with the corresponding predictions obtained from the lattice: on the one hand, gauge-fixed correlation functions are the best quantities to test the specificities of a given gauge; on the other hand, gauge-fixed lattice simulations are less sensitive to the Gribov ambiguity since gauge fixing is done (statistically) by selecting one copy per orbit. Differences are seen depending on which copy is selected, but they seem to concern only the deep infrared region.

2.3.1   Zero Temperature

In a series of works [82, 83, 90], the lattice results for the vacuum two-point correlation functions in the Landau gauge [55–62], as well as for the three-point vertices, have been systematically compared to the one-loop perturbative predictions of the CF model.

The agreement for the two-point functions is rather impressive (see Fig. 2.1). A recent two-loop calculation confirms (and even improves) these results showing that the very good agreement at one-loop was not accidental [141]. The agreement is less impressive for the three-point vertices but still qualitatively very good, given the simplicity of the one-loop approach and the uncertainties of the lattice simulations. In particular, the CF model predicts a zero crossing of the three-gluon vertex, as also seen in other approaches [142]. The comparison to other continuum approaches has been more systematically discussed in Ref. [97]. For instance, the qualitative dependence of the results on the value of the CF mass has been found to be similar

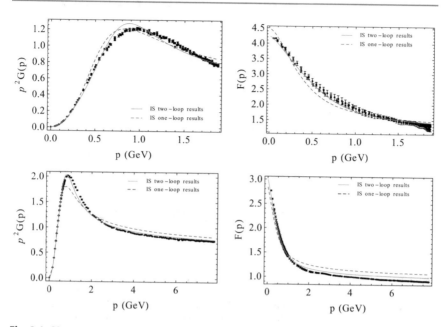

Fig. 2.1 Vacuum gluon dressing function (left) and ghost dressing function (right) as computed in the CF model at one-loop order (dashed) and two-loop order (plain), compared to Landau gauge-fixed lattice simulations of the same quantities, in the SU(2) case (top) and the SU(3) case (bottom). The parameters are defined in the infrared-safe scheme of [83]. Reprinted from Ref. [141]

to the dependence of the functional renormalization group results with respect to the value of the initial parameter m_Λ alluded to above.

It is also to be mentioned that the flow structure of the CF model has been studied at one-loop order [83, 97, 119] (and also, more recently, at two-loop order [141]) (see Fig. 2.2). Interestingly enough, it is found that, in addition to renormalization group trajectories displaying an infrared Landau pole, reminiscent from the one in the FP approach, there exists a family of trajectories which can be defined over all energy scales. In particular, the one trajectory that best fits the lattice results remains always in the perturbative regime (see Fig. 2.3), supporting the idea that the perturbative expansion can be used at all scales in the CF model.

2.3.2 Finite Temperature

Gauge-fixed lattice simulations also provide results for the YM two-point correlation functions at finite temperature [78, 144–153], as a further testing ground for the various continuum approaches. One particularly scrutinized quantity has been the longitudinal susceptibility given by the zero momentum value of the longitudinal

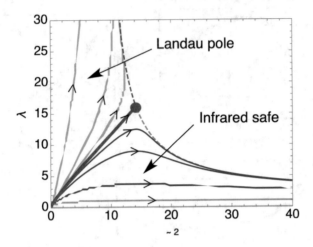

Fig. 2.2 The one-loop flow structure of the CF model in the plane $(\tilde{m}^2, \lambda)$, where $\tilde{m}^2(\mu) \equiv m^2/\mu^2$ and $\lambda(\mu) \equiv g^2(\mu)N/(16\pi^2)$. The plot represents two sets of trajectories, on each side of a separatrix terminating at a critical point (dot). The trajectories labeled *Landau pole* are only defined down to a certain scale in the infrared, while those labeled *infrared safe* are defined at all scales. Reprinted from Ref. [97]

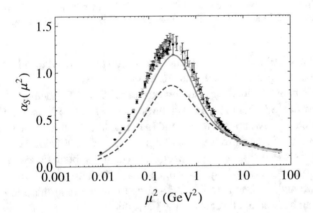

Fig. 2.3 The running of α_S as computed from the CF model in the infrared scheme and compared (after scheme translation) to the lattice Taylor's scheme [143]. The dashed and plain lines correspond to the one-loop and two-loop CF results in the IRS scheme, respectively. In both cases, a normalization factor is applied so that the value $\alpha_S(\mu_0)$ agrees with the lattice result at $\mu_0 = 10$ GeV (see Ref. [141] for details). When multiplied by $N/(4\pi)$, this particular α_S turns out to be the relevant expansion parameter in the CF model, which does not exceed 0.4 over the all range. Reprinted from Ref. [141]

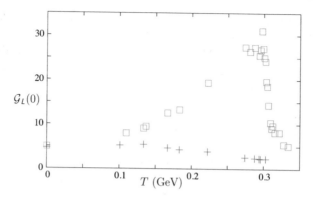

Fig. 2.4 Zero momentum value of the longitudinal gluon propagator (susceptibility), as obtained from the CF model at one-loop (crosses) and compared to the results of Landau gauge-fixed lattice simulations (squares). Reprinted from Ref. [91]

gluon propagator,[13] $\chi = \mathcal{G}_L(p^2 = 0)$. Early lattice results showed a rather sharp variation of this quantity around the YM deconfinement transition [149, 152] and opened the way to speculations about the possibility of accessing the physical transition from the study of gauge-variant quantities [154–156].

The fit of the lattice data to the one-loop CF two-point functions at finite temperature is globally satisfactory in the ghost and transversal gluonic sectors (see Ref. [91] for more details). However, the temperature dependence of the longitudinal susceptibility differs substantially from the one obtained in lattice simulations (see Fig. 2.4). This, however, does not necessarily point out to a limitation of the CF approach since the other continuum approaches also fail to reproduce this particular feature of the lattice results.[14]

Rather, it has been suggested in Ref. [154] that the discrepancy between continuum and lattice results may be due to the fact that the order parameter associated with the deconfinement transition is not properly accounted for in the Landau gauge (and in fact in most standard gauge fixings). As put forward in Ref. [73], one way to tackle this problem is to extend the standard Landau gauge into the Landau-DeWitt gauge that allows for the possibility of a background gauge-field configuration. Following this idea, we shall extend the CF model in the presence of a background and investigate the perturbative predictions of the model at finite temperature, in particular with regard to the QCD/YM theory phase diagrams and the confinement/deconfinement transition. In Chap. 11, we shall come

---

[13] In the Landau gauge, the gluon propagator is (4d) transverse. At finite temperature, however, this transverse component decomposes into (3d) longitudinal and transverse components.

[14] It should be mentioned that the longitudinal gluon propagator is very sensitive to the details of the lattice simulation. Improved lattice results show a much less pronounced sensitivity around the transition. It would be interesting to update the comparisons between the continuum and the lattice approaches in light of these more recent results.

back to the question of the identification of the transition using the longitudinal susceptibility. Before embarking on this adventure, in the next three chapters, we shall make a technical digression on the application of background-field methods at finite temperature, explaining in detail why and how they allow to circumvent the limitations of standard gauge fixings.

Appendix: BRST Transformations Under the Functional Integral

It may happen that one needs to consider a change of variables under a functional integral in the form of a BRST transformation. We have seen one particular application in the proof of the gauge-fixing independence of expectation values of gauge-invariant functionals in the FP approach. Another well-known example is the derivation of the Zinn-Justin equation associated with BRST symmetry or the related Slavnov-Taylor's identities.

A peculiarity of such changes of variables is that they modify the integration domain a priori since numerical gauge fields are transformed into still commuting but non-numerical fields. Here we show that this can lead to an additional (boundary) contribution in the formula for the change of variables. We argue that this term can be ignored in the case where no boundary is considered on the space of gauge-field configurations, as it is the case in particular when discussing the FP or the CF model. It could however have consequences in the presence of restrictions such as the ones imposed in the GZ model and connect directly to the breaking of the standard BRST symmetry in this framework.

A Simple Case as a Toy Example

Let us first illustrate these considerations using the case of a one-dimensional numerical integral

$$I = \int_a^b dx\, f(x)\,, \tag{2.37}$$

defined over a certain interval $[a, b]$ of the real axis.

Let us embed the real axis in a Grassmannian algebra generated by 1, θ, and $\bar{\theta}$, and let us consider a change of variables of the form

$$x = \psi(x') + \bar{\theta}\theta\, \varphi(x')\,, \tag{2.38}$$

where ψ and φ denote two numerical functions, with ψ invertible of inverse ψ^{-1}. These functions can be extended over the Grassmannian algebra using a Taylor's expansion. In particular, reshuffling the terms in Eq. (2.38) and applying ψ^{-1} on both sides, we find that

$$x' = \psi^{-1}\left(x - \bar{\theta}\theta\,\varphi(x')\right)$$

$$= \psi^{-1}(x) - \bar{\theta}\theta\,\varphi(x')\frac{d\psi^{-1}}{dx}$$

$$= \psi^{-1}(x) - \bar{\theta}\theta\,\varphi(\psi^{-1}(x))\frac{d\psi^{-1}}{dx}. \tag{2.39}$$

This shows that, when $\varphi \neq 0$, the new variable x' cannot be numerical, and it is a priori not clear how the change of variables should be applied under the integral.

We now argue that, if the following condition is satisfied

$$\varphi(\psi^{-1}(b))f(b) - \varphi(\psi^{-1}(a))f(a) = 0, \tag{2.40}$$

one can proceed as if x' were real and write

$$I = \int_{\psi^{-1}(a)}^{\psi^{-1}(b)} dx'\left(\frac{d\psi}{dx'} + \bar{\theta}\theta\,\frac{d\varphi}{dx'}\right)f(\psi(x') + \bar{\theta}\theta\,\varphi(x')). \tag{2.41}$$

To see this, let us start from the announced result and write

$$\int_{\psi^{-1}(a)}^{\psi^{-1}(b)} dx'\left(\frac{d\psi}{dx'} + \bar{\theta}\theta\,\frac{d\varphi}{dx'}\right)f(\psi(x') + \bar{\theta}\theta\,\varphi(x'))$$

$$= \int_{\psi^{-1}(a)}^{\psi^{-1}(b)} dx'\left(\frac{d\psi}{dx'} + \bar{\theta}\theta\,\frac{d\varphi}{dx'}\right)\left(f(\psi(x')) + \bar{\theta}\theta\,\varphi(x')\frac{df}{dx}\bigg|_{x=\psi(x')}\right)$$

$$= \int_{\psi^{-1}(a)}^{\psi^{-1}(b)} dx'\frac{d\psi}{dx'}f(\psi(x')) + \bar{\theta}\theta\int_{\psi^{-1}(a)}^{\psi^{-1}(b)} dx'\frac{d}{dx'}\left(\varphi(x')f(\psi(x'))\right)$$

$$= \int_a^b dx\,f(x), \tag{2.42}$$

where we have used the condition (2.40). On the contrary, if Eq. (2.40) is not fulfilled, the formula for the change of variables features an extra (boundary term)

$$\int_a^b dx\,f(x) = \int_{\psi^{-1}(a)}^{\psi^{-1}(b)} dx'\left(\frac{d\psi}{dx'} + \bar{\theta}\theta\,\frac{d\varphi}{dx'}\right)f(\psi(x') + \bar{\theta}\theta\,\varphi(x'))$$

$$- \bar{\theta}\theta\left[\varphi(\psi^{-1}(b))f(b) - \varphi(\psi^{-1}(a))f(a)\right]. \tag{2.43}$$

The Case of a BRST Transformation

When a BRST transformation $\varphi = \varphi' + \bar{\theta} s\varphi'$ is applied under the functional integral, it is usually assumed that ("sdet" stands for the super-determinant)

$$\int \mathcal{D}\varphi\, I[\varphi] = \int \mathcal{D}\varphi'\, \mathrm{sdet}\left(\mathbb{1} + \frac{\delta\bar{\theta}s\varphi'}{\delta\varphi'}\right) I[\varphi' + \bar{\theta}s\varphi']\,, \qquad (2.44)$$

where the fields that are originally numerical are assumed to remain numerical, despite the fact that, for such fields, $\varphi - \varphi' = \bar{\theta}s\varphi$ is non-numerical. Let us now analyze under which conditions the above identity applies. To this purpose, we shall start from the RHS of Eq. (2.44) and try to go back to the LHS.

One has ($q_m = 0$ for any bosonic component φ_m and $q_m = 1$ for any fermionic component)

$$\int \mathcal{D}\varphi'\, \mathrm{sdet}\left(\mathbb{1} + \frac{\delta\bar{\theta}s\varphi'}{\delta\varphi'}\right) I[\varphi' + \bar{\theta}s\varphi']$$

$$= \int \mathcal{D}\varphi' \left(1 + \underbrace{\int_x (-1)^{q_m} \frac{\delta\bar{\theta}s\varphi'_m(x)}{\delta\varphi'_m(x)}}_{\text{supertrace}}\right) \left(I[\varphi'] + \int_x \bar{\theta}s\varphi'_m(x) \frac{\delta I}{\delta\varphi'_m(x)}\right)$$

$$= \int \mathcal{D}\varphi'\, I[\varphi'] + \int \mathcal{D}\varphi' \left(\int_x (-1)^{q_m} \frac{\delta\bar{\theta}s\varphi'_m(x)}{\delta\varphi'_m(x)} I[\varphi'] + \int_x \bar{\theta}s\varphi'_m(x) \frac{\delta I}{\delta\varphi'_m(x)}\right)$$

$$= \int \mathcal{D}\varphi'\, I[\varphi'] + \bar{\theta} \int \mathcal{D}\varphi' \int_x \frac{\delta}{\delta\varphi'_m(x)}\left(s\varphi'_m(x)\, I[\varphi']\right)\,. \qquad (2.45)$$

The identity (2.44) is then valid if

$$0 = \int \mathcal{D}\varphi' \int_x \frac{\delta}{\delta\varphi'_m(x)}\left(s\varphi'_m(x)\, I[\varphi']\right)\,, \qquad (2.46)$$

which states the vanishing of the (field) volume integral of a super-divergence (over a space made of numerical and Grassmannian fields). Otherwise, the formula for the change of variables gets corrected by a boundary term:

$$\int \mathcal{D}\varphi'\, I[\varphi'] = \int \mathcal{D}\varphi'\, \mathrm{sdet}\left(\mathbb{1} + \frac{\delta\bar{\theta}s\varphi'}{\delta\varphi'}\right) I[\varphi' + \bar{\theta}s\varphi']$$

$$- \bar{\theta} \int \mathcal{D}\varphi' \int_x \frac{\delta}{\delta\varphi'_m(x)}\left(s\varphi'_m(x)\, I[\varphi']\right)\,. \qquad (2.47)$$

Using Stokes' theorem, we should be able to relate this to the flux of a certain current through the boundary, which would be the generalization of Eq. (2.40). To see how this works in the case where the volume contains Grassmannian directions, let us consider a simple example with four variables $x, y, \theta, \bar{\theta}$. The integral over the superspace of the super-divergence of four component field writes

$$\int dx\,dy\,d\theta\,d\bar{\theta} \left(\frac{\partial f_x}{\partial x} + \frac{\partial f_y}{\partial y} + \frac{\partial f_\theta}{\partial \theta} + \frac{\partial f_{\bar{\theta}}}{\partial \bar{\theta}} \right) = \int dx\,dy \left(\frac{\partial^3 f_x}{\partial x \partial \theta \partial \bar{\theta}} + \frac{\partial^3 f_y}{\partial y \partial \theta \partial \bar{\theta}} \right),$$

(2.48)

where we used $\int d\theta \partial/\partial\theta = 0$. We obtain a standard volume integral of the divergence of a current field with components $\partial^2 f_x/\partial\theta\partial\bar{\theta}$ et $\partial^2 f_y/\partial\theta\partial\bar{\theta}$. The final result is then the flux of such vector field through the boundary of the standard volume. Coming back to Eq. (2.47), the boundary to be considered is that of the configuration space for A and h,[15] while the current whose flux needs to be integrated over the boundary is made out of the A and h components of the super-current $s\varphi'_m(x)\,I[\varphi']$, integrated over the fields c and $\bar{c}$. This current has no components in the h directions (because $sih = 0$). Finally, whatever restriction we might consider on the configuration space, it better not affect the h direction; otherwise we would depart from the Landau gauge condition $\partial_\mu A^a_\mu = 0$. It follows that the flux through the boundary of the (A, h) configuration space writes as the integral over h of the flux through the boundary $\partial\mathcal{V}$ of the A configuration space. All in all, we arrive at the following corrected formula for the BRST change of variables

$$\int \mathcal{D}\varphi\, I[\varphi] = \int \mathcal{D}\varphi\, I[\varphi + \bar{\theta}s\varphi] - \bar{\theta} \int_{\partial\mathcal{V}} d\sigma_\mu \int \mathcal{D}[c, \bar{c}, h]\, D_\mu c\, I[\varphi],$$

(2.49)

where we have used that the Jacobian is equal to 1 and we have renamed the dummy integration variable φ' as φ.

In particular, when discussing the gauge-fixing independence of observables within the FP approach, we applied a BRST change of variables to the integral

$$\int \mathcal{D}[A, c, \bar{c}, h]\, O[A]\, \bar{c}\, \frac{\delta F[A]^a(x)}{\delta\lambda} e^{-S_{\mathrm{FP}}[A, c, \bar{c}, h]}.$$

(2.50)

In order for this change of variables to be justified, we need to show the flux of

$$O[A] \frac{\delta F[A]^a(x)}{\delta\lambda} \int \mathcal{D}[c, \bar{c}, h]\, D_\mu c\, \bar{c}\, e^{-S_{\mathrm{FP}}[A, c, \bar{c}, h]}$$

(2.51)

---

[15] This is a welcomed feature since the boundary along the other directions is not really defined.

vanishes through the boundary of the integration domain for the gauge field. In fact this current is exponentially suppressed at the boundary due to the factor $e^{-S_{YM}[A]}$.[16] Similar considerations would apply to the derivation of the Zinn-Justin equation associated with BRST symmetry, with however a different current.

The previous discussion is more subtle in the presence of restrictions on the space of gauge-field configurations (as would be required in order to avoid the presence of Gribov copies). In the case of the Zinn-Justin equation, the flux of the corresponding current is not zero a priori and could lead to a modification of the Slavnov-Taylor's identities, connecting to the breaking of the standard BRST symmetry in this framework. This question deserves a separate investigation beyond the scope of the present notes (see [157]).

Problems

2.1 Faddeev-Popov Determinant (⋆)

(a) Consider an infinitesimal gauge transformation $U = 1 + \theta$ with $\theta = i\theta^a t^a$ depending on x. Express the corresponding variation of the gauge field $\delta A_\mu \equiv A_\mu^U - A_\mu$ in terms of the covariant derivative $D_\mu^{ab} \equiv \partial_\mu \delta^{ab} + f^{acb} A_\mu^c$.

(b) Assuming that A obeys the gauge-fixing condition $F[A] = 0$, show that the Faddeev-Popov determinant $\Delta[A]$ defined in Eq. (2.8) is given by

$$\Delta[A] = \det\left(\int d^d z \, \frac{\delta F^a[A](x)}{\delta A_\mu^c(z)} D_\mu^{cb} \delta(z - y) \right).$$

Tip: use the chain rule together with the result obtained in (a).

2.2 BRST Symmetry (⋆⋆)

(a) Given a transformation $\delta\phi = \bar{\eta} s\phi$ of the various fields at hand, $\phi = A_\mu$, c, $\bar{c}$ and h, with $\bar{\eta}$ a Grassmann parameter, show that $s\delta S_{FP}$ rewrites as a linear combination of $ih - s\bar{c}$, sih and $s^2 A_\mu$ and, therefore, that S_{FP} is invariant if one sets $s\bar{c} = ih$, $sih = 0$ and arranges for $s^2 A_\mu = 0$. Tip: use that $\delta^2 F^a[A](x)/\delta A_\mu^b(y)\delta A_\nu^c(z)$ is symmetric under $(b, y) \leftrightarrow (c, z)$ and thus vanishes when contracted with a tensor antisymmetric under the same permutation.

(b) Choosing $s A_\mu = D_\mu c$, show that $s^2 A_\mu = 0$ is ensured if $sc^a = -\frac{1}{2} f^{abc} c^b c^c$. Tip: use the Jacobi identity $f^{abd} f^{cde} + f^{bcd} f^{ade} + f^{cad} f^{bde} = 0$.

---

[16] There are no flat directions due to the gauge fixing.

2.3 Nilpotency (★★)

(a) Show that s^2 not only vanishes on A_μ but also on ih, $\bar{c}$, and c.
(b) Show that $s^2(\phi\chi) = (s^2\phi)\chi + \phi(s^2\chi)$. *Tip:* start by evaluating $s(\phi\chi)$.
(c) Deduce that $s^2 = 0$ over all possible field combinations.

2.4 Curci-Ferrari Model and Modified BRST Symmetry (★)

(a) Define $s_m A_\mu = s A_\mu$ and show that the Curci-Ferrari mass term transforms as

$$s_m \int d^d x \, \frac{1}{2} m^2 A_\mu^a(x) A_\mu^a(x) = - \int d^d x \, m^2 c^a(x) \, \partial_\mu A_\mu^a(x).$$

(b) Revisit the derivation in Problem 2.2 and show that the Curci-Ferrari action is invariant under the modified BRST transformation $s_m A_\mu = s A_\mu$, $s_m c = sc$, $s_m \bar{c} = s\bar{c}$, and $s_m ih = m^2 c$.

2.5 Curci-Ferrari Model and Gluon Propagator (★★★)

(a) Write the quadratic part of the Curci-Ferrari action $S_{YM} + \delta S_{CF}$ in the (A_μ, ih)-sector.
(b) Invert this quadratic form, and deduce that the tree-level gluon propagator writes:

$$G_{0,\mu\nu}(Q) = \frac{1}{Q^2 + m^2} \left[\delta_{\mu\nu} - \frac{Q_\mu Q_\nu}{Q^2} \right].$$

In particular, it is transverse. How could this property be anticipated?

2.6 Curci-Ferrari Model and Taylor's Non-renormalization Theorem (★★★)

(a) Derive the Feynman rule for the ghost-antighost-gluon tree-level vertex, and show that it is proportional to the antighost momentum.
(b) Consider a loop correction to the full ghost-antighost-gluon vertex. By looking at that part of the Feynman graph that involves the tree-level vertex attached to the external ghost leg, show that the loop correction vanishes in the limit where the external ghost momentum vanishes. *Tip:* use (a) and the result of Problem 2.5.
(c) If we denote by Z_A, Z_c, and Z_g, the renormalization factors associated with the gluon field, ghost field, and coupling constant, deduce that the combination $Z_A^{1/2} Z_c Z_g$ is necessarily finite. *Tip:* evaluate the renormalized full vertex in the limit of vanishing external ghost momentum.

Deconfinement Transition and Center Symmetry

3

The goal of this chapter is to motivate why standard gauge fixings do not provide the appropriate framework in order to study the confinement/deconfinement transition in the continuum. To this purpose, we shall start reviewing some basic knowledge of finite temperature QCD and the associated phase transition in a non-gauge-fixed setting and then touch on the difficulties that can arise when fixing the gauge.

In fact, the physical QCD transition does not correspond to a sharp transition between two distinct phases of matter but, rather, to a smooth crossover of the various thermodynamical observables characterizing the system [34]. To some extent, this obscures the theoretical understanding of the transition itself. It proves convenient, therefore, to consider particular limits for which the transition becomes a sharp one, testable with robust order parameters. One may then hope to access the physical case more simply from these limiting cases. For the most part, this manuscript will be concerned with the case of pure YM theory which can be seen as the limit of QCD with infinitely heavy quarks: $S_{\mathrm{QCD}} \to S_{\mathrm{YM}}$. In this simplified setting, the transition can be probed with the help of an observable known as the Polyakov loop. Moreover, it can be interpreted as the spontaneous breaking of a special symmetry known as center symmetry. This chapter is mainly devoted to a review of these concepts for they play a major role in the rest of the manuscript, even when quarks are included back in the analysis.

Although the applications to be covered in the present manuscript will concern mainly the SU(2) and SU(3) gauge groups, we shall introduce the various relevant concepts and later derive the various relevant formulas within the general case of SU(N). Some general discussion and an application to the SU(4) gauge group can be found in Chap. 8.

3.1   The Polyakov Loop

Let us start recalling the definition of the Polyakov loop and how it relates to the confinement/deconfinement transition [158].

3.1.1   Definition

For a given gauge-field configuration $A_\mu(x) = i A_\mu^a(x) t^a$ on the Lie algebra, we define the Wilson line wrapped along the Euclidean time direction as

$$L_A(\mathbf{x}) \equiv \mathcal{P} \exp\left\{ \int_0^\beta d\tau \, A_0(\tau, \mathbf{x}) \right\}.$$ (3.1)

Here, $\mathcal{P}$ denotes the path-ordering along the compact time direction that arranges Lie algebra elements $A_0(\tau, \mathbf{x})$ from left to right according to the decreasing value of the argument τ. One central property of $L_A(\mathbf{x})$ that will be needed below is that it transforms very simply under a gauge transformation (2.5):

$$L_{A^U}(\mathbf{x}) = U(\beta, \mathbf{x}) \, L_A(\mathbf{x}) \, U^\dagger(0, \mathbf{x}).$$ (3.2)

More details are provided in Problem 3.1.

We also define the normalized trace of $L_A(\mathbf{x})$ as $\Phi_A(\mathbf{x}) \equiv \operatorname{tr} L_A(\mathbf{x}) / N$. Finally, the *Polyakov loop* ℓ is defined as the expectation value of $\Phi_A(\mathbf{x})$:

$$\ell \equiv \langle \Phi_A(\mathbf{x}) \rangle \equiv \frac{\int \mathcal{D}A \, \Phi_A(\mathbf{x}) \, e^{-S_{\text{YM}}[A]}}{\int \mathcal{D}A \, e^{-S_{\text{YM}}[A]}}.$$ (3.3)

We mention that the Wilson line $L_A(\mathbf{x})$ is a unitary matrix and, then, the modulus of its normalized trace $\Phi_A(\mathbf{x})$ is not larger than unity. Moreover, since the averaging weight in Eq. (3.3) is positive, it follows that $|\ell| \leq 1$.[1]

3.1.2   Order Parameter Interpretation

It can be shown that ℓ is related to the free-energy cost ΔF for bringing a static test quark into the thermal bath of gluons as [158, 159]

$$\ell = e^{-\beta \Delta F}.$$ (3.4)

---

[1] At this level of the discussion, we work within a bare regularized setting. The question of renormalization of the Polyakov loop will be touched upon in subsequent chapters, when necessary.

The interpretation of the Polyakov loop as an order parameter for the confinement/deconfinement transition follows immediately. Indeed, if we assume that the system is in the confined phase, no free quarks are allowed to roam around in the thermal bath ($\Delta F = \infty$) and then $\ell = 0$. In contrast, in the deconfined phase, quarks can be introduced in the system provided some finite amount of energy is paid ($\Delta F < \infty$) and then $\ell \neq 0$.

The behavior of the Polyakov loop as a function of the temperature has been studied on the lattice [159–163] where it yields a confined phase at low temperatures and, as the temperature is increased, a second order type *deconfinement transition* in the case of the SU(2) gauge group and a first-order type transition for SU(N) with $N \geq 3$. It is one of the purposes of this manuscript to review how these seemingly non-perturbative results can be accessed from perturbative methods, in the framework of the CF model.

3.2 Center Symmetry

The previous considerations relate the confinement/deconfinement transition to the behavior of the Polyakov loop but do not explain why the latter can vanish in the low-temperature phase. As we now recall, this is related to a symmetry of the Euclidean YM action (2.1) at finite temperature known as center symmetry [7, 159, 164].

3.2.1 The Role of Boundary Conditions

The YM action is invariant under the gauge transformations $U(x)$ defined in Eq. (2.5). At finite temperature, the only, however important, constraint on the field $U(x)$ is that it should preserve the periodic boundary conditions of the gauge fields along the Euclidean time direction, that is, it should be such that

$$A_\mu^U(\tau + \beta, \mathbf{x}) = A_\mu^U(\tau, \mathbf{x}), \tag{3.5}$$

for all $A_\mu(\tau, \mathbf{x})$ that are themselves β-periodic.

Beyond the obvious case of β-periodic gauge transformations, one can find more general solutions to the constraint (3.5). Indeed, consider a gauge transformation obeying the boundary condition:

$$U(\tau + \beta, \mathbf{x}) = U(\tau, \mathbf{x}) Z, \quad \forall \tau, \forall \mathbf{x}, \tag{3.6}$$

where Z is an element of the center of SU(N). Since Z commutes with any other element of SU(N), we have

$$A_\mu^U(\tau + \beta, \mathbf{x}) = U(\tau + \beta, \mathbf{x}) A_\mu(\tau + \beta, \mathbf{x}) U^\dagger(\tau + \beta, \mathbf{x}) - U(\tau + \beta, \mathbf{x}) \partial_\mu U^\dagger(\tau + \beta, \mathbf{x})$$

$$= U(\tau, \mathbf{x}) Z A_\mu(\tau, \mathbf{x}) Z^\dagger U^\dagger(\tau, \mathbf{x}) - U(\tau, \mathbf{x}) Z \partial_\mu (Z^\dagger U^\dagger(\tau, \mathbf{x}))$$

$$= U(\tau, \mathbf{x}) A_\mu(\tau, \mathbf{x}) Z Z^\dagger U^\dagger(\tau, \mathbf{x}) - U(\tau, \mathbf{x}) Z Z^\dagger \partial_\mu U^\dagger(\tau, \mathbf{x})$$

$$= A_\mu^U(\tau, \mathbf{x}) . \tag{3.7}$$

In fact, the condition (3.6) defines the most general gauge transformations preserving the β-periodicity of the gluon field (see Problem 3.2 for more details). Since the center elements form a subgroup of SU(N), the gauge transformations obeying Eq. (3.6) define a group which we refer to in what follows as the group of *twisted gauge transformations* and which we denote $\mathcal{G}$.

It is important not to identify the group $\mathcal{G}$ with the center of SU(N). These groups are not isomorphic. In fact, the former is infinite, while the second is seen to correspond to the finite group $Z_N \equiv \{e^{i\frac{2\pi}{N}k} \mathbb{1} | k = 0, \ldots, N - 1\}^2$ (see Problem 3.3). The connection between $\mathcal{G}$ and the center of SU(N) will be clarified below.

3.2.2   Relation to the Deconfinement Transition

To unveil the relevance of the twisted gauge transformations for the deconfinement transition, let us see how the Polyakov loop transforms under the action of $\mathcal{G}$. First, owing to Eq. (3.2), we find the transformation rule

$$\Phi_{A^U}(\mathbf{x}) = \frac{1}{N} \operatorname{tr} U(\beta, \mathbf{x}) L_A(\mathbf{x}) U^\dagger(0, \mathbf{x})$$

$$= \frac{1}{N} \operatorname{tr} U^\dagger(0, \mathbf{x}) U(\beta, \mathbf{x}) L_A(\mathbf{x}) = e^{i\frac{2\pi}{N}k} \Phi_A(\mathbf{x}) , \tag{3.8}$$

where we have used the cyclicality of the trace as well as Eq. (3.6).

In order to derive the corresponding transformation rule for ℓ, we note that, in general, the definition (3.3) makes sense only in the presence of an infinitesimal, symmetry-breaking source term added to the action: $S_{\mathrm{YM}} \to S_{\mathrm{YM}} - 0^+ e^{i\alpha} \int d\mathbf{x}\, \Phi_A(\mathbf{x})$. To account for the presence of this external source of modulus 0^+ and direction $e^{i\alpha}$, we denote the Polyakov loop more rigorously as ℓ_α. With these words of caution, we can now perform a change of variables in

---

^{2} Z_N is isomorphic to $\mathbb{Z}_N \equiv \mathbb{Z}/N\mathbb{Z}$, the group of relative integers modulo N.

Eq. (3.3) in the form of a twisted gauge transformation:

$$
\ell_\alpha = \frac{\int \mathcal{D}A^U \, \Phi_{A^U}(\mathbf{x}) \, e^{-S_{\mathrm{YM}}[A^U]+0^+ e^{i\alpha} \int d\mathbf{x} \, \Phi_{A^U}(\mathbf{x})}}{\int \mathcal{D}A^U \, e^{-S_{\mathrm{YM}}[A^U]+0^+ e^{i\alpha} \int d\mathbf{x} \, \Phi_{A^U}(\mathbf{x})}}
$$

$$
= \frac{\int \mathcal{D}A \, \Phi_{A^U}(\mathbf{x}) \, e^{-S_{\mathrm{YM}}[A]+0^+ e^{i\alpha} \int d\mathbf{x} \, \Phi_{A^U}(\mathbf{x})}}{\int \mathcal{D}A \, e^{-S_{\mathrm{YM}}[A]+0^+ e^{i\alpha} \int d\mathbf{x} \, \Phi_{A^U}(\mathbf{x})}} = e^{i\frac{2\pi}{N}k}\ell_{\alpha+\frac{2\pi}{N}k} \, , \forall k . \qquad (3.9)
$$

In the last steps, we have used the transformation rule (3.8), the invariance of both $\mathcal{D}A$ and $S_{\mathrm{YM}}[A]$ under twisted gauge transformations and the fact that the periodic boundary conditions of the gauge field are preserved.

We have thus found that, under a phase multiplication of the external source by $e^{-i\frac{2\pi}{N}k}$, the Polyakov loop transforms as

$$
\ell_{\alpha-\frac{2\pi}{N}k} = e^{i\frac{2\pi}{N}k}\ell_\alpha . \qquad (3.10)
$$

From here, two scenarios are possible:

- If the symmetry G is realized in the Wigner-Weyl sense, then the definition (3.3) makes sense without the need of the symmetry-breaking source. The transformation rule (3.10) becomes a constraint equation for the Polyakov loop implying that the latter has to vanish.[3]
- In contrast, in the case where the symmetry is spontaneously broken in the Nambu-Goldstone sense, one expects degenerate ground states with different values of ℓ connected to each other by Eq. (3.10). The Polyakov loop does not need to be zero anymore and the system is in the deconfined phase.[4]

We have thus related the existence of a confined/deconfined phase to the explicit/broken realization of the symmetry group G. This connection offers a promising way for studying the confinement/deconfinement transition which we shall exploit in subsequent chapters.

3.2.3 The Center Symmetry Group

For later considerations, it will be important to realize that not all the transformations of G are physical. To see this, consider the subgroup G_0 of *periodic gauge transformations,* corresponding to twisted gauge transformations with $k = 0$. The operator $\Phi_A(\mathbf{x})$ is an observable (in the sense that its expectation value ℓ measures a physical quantity) which is left invariant by the elements of G_0. It is therefore legitimate to assume that the elements of G_0 correspond to true, and therefore unphysical, gauge transformations.

---

[3] Strictly speaking, the constraint could also imply $\ell = \infty$. However, we have shown above that $|\ell| \leq 1$, so this possibility is excluded.

[4] The statement is strictly correct for the SU(2) and SU(3) groups. The general discussion in the case of SU(N) is slightly more involved, as we discuss in Chap. 8.

In contrast, even though they equally look like gauge transformations, the elements of $\mathcal{G}$ that do not belong to $\mathcal{G}_0$ correspond to physical transformations because they lead to explicit modifications of the observable $\Phi_A(\mathbf{x})$. In fact, two transformations $U_1, U_2 \in \mathcal{G}$ that are related to each other by $U_1 = U_2 U_0$, with $U_0 \in \mathcal{G}_0$, lead to the same modification of $\Phi_A(\mathbf{x})$ and, therefore, have the same physical meaning. It follows that the true physical content of $\mathcal{G}$ lies within the quotient set $\mathcal{G}/\mathcal{G}_0$. Moreover, since the subgroup $\mathcal{G}_0$ is normal within $\mathcal{G}$ (see Problem 3.3), this quotient set $\mathcal{G}/\mathcal{G}_0$ inherits a group structure from $\mathcal{G}$, isomorphic to the center of SU(N). This is known as the *center symmetry group*.

In summary, two transformations of $\mathcal{G}$ that belong to the same class in $\mathcal{G}/\mathcal{G}_0$ should be seen as redundant versions of the same, physical center transformation. In particular, the elements of $\mathcal{G}_0$ do not correspond to physical transformations since they define the neutral element of $\mathcal{G}/\mathcal{G}_0$. These simple remarks will play a major role in the next two chapters, as well as in the final chapter.

3.3   Center Symmetry and Gauge Fixing

The results that we have recalled so far rely on the gauge-invariant formulation of the theory. In practice, however, continuum approaches to the deconfinement transition require working in a given gauge. It is therefore important to assess whether and how the previous results extend to a gauge-fixed setting.

At first sight, it may seem that gauge fixing, which has to do with the treatment of an unphysical symmetry (incarnated in the group $\mathcal{G}_0$), can in no way interfere with a physical symmetry such as center symmetry (as described by the quotient group $\mathcal{G}/\mathcal{G}_0$). This should definitely be true at an exact level of treatment. Most continuum approaches, however, require working within a truncation/approximation scheme. The latter can negatively affect the way a physical symmetry appears in the formalism,[5] and, depending on the chosen gauge, it might be more or less difficult to devise approximation schemes where the symmetry is explicit.[6] Similarly, as we have reviewed in the previous chapter, going beyond the Faddeev-Popov gauge-fixing procedure requires, in practice, some type of modeling when extending it to the infrared. There again, depending on the chosen gauge, it might be more or less difficult to construct models that keep the physical symmetries of the problem explicit.

In the rest of this chapter, we would like to illustrate these difficulties in a more precise way, for they may take diverse manifestations. Various fixes will be detailed in Chaps. 4 and 5, as well as in Chap. 11.

---

[5] We shall see in Chaps. 4 and 5 that the discussion around center symmetry provides in fact a good prototype of how physical symmetries should be treated within gauge-field theories, the physical symmetry group being naturally enlarged into a group $\mathcal{G}$ of physical transformations modulo gauge transformations, whose quotient $\mathcal{G}/\mathcal{G}_0$ is the actual physical symmetry group. Then, the present discussion regarding the impact of gauge fixing on symmetries applies to any type of physical symmetry defined in those terms.

[6] By the term explicit, we put here no prejudice on whether or not the symmetry is spontaneously broken eventually.

3.3.1 Gauge-Fixed Measure

Gauge fixing can be interpreted as the replacement of the gauge-invariant measure $\mathcal{D}A$ that enters the evaluation of any observable by a gauge-variant measure $\mathcal{D}_{\mathrm{gf}}[A]$, without affecting the value of the observable. For instance, in the case of the Polyakov loop, one writes

$$\ell = \frac{\int \mathcal{D}A\, \Phi_A(\mathbf{x})\, e^{-S_{\mathrm{YM}}[A]}}{\int \mathcal{D}A\, e^{-S_{\mathrm{YM}}[A]}} = \frac{\int \mathcal{D}_{\mathrm{gf}}[A]\, \Phi_A(\mathbf{x})\, e^{-S_{\mathrm{YM}}[A]}}{\int \mathcal{D}_{\mathrm{gf}}[A]\, e^{-S_{\mathrm{YM}}[A]}}, \tag{3.11}$$

where the appropriate infinitesimal symmetry-breaking source (see the discussion above) has been left implicit. The subtleties related to the construction of bona fide gauge-fixed measures that realize the above type of identities have been discussed in the previous chapter, and we shall not repeat them here.[7] However, in order to discuss the difficulties regarding symmetries that might occur in the presence of gauge fixing, it is important to consider a gauge-fixing procedure that is free of these subtleties, at least at a formal level. We shall then consider a rather general gauge-fixing procedure that can be seen as a generalization of the Serreau-Tissier gauge fixing reviewed in the previous chapter. Although formal (and not necessarily practical), it deals properly with the question of Gribov copies.

Consider a functional $\rho[A]$, and assume that it integrates to a finite non-zero value along each $\mathcal{G}_0$-orbit:

$$0 < \int_{\mathcal{G}_0} \mathcal{D}U_0\, \rho[A^{U_0}] < \infty. \tag{3.12}$$

We can then introduce the functional

$$z[A] \equiv \frac{\rho[A]}{\int_{\mathcal{G}_0} \mathcal{D}U_0\, \rho[A^{U_0}]}, \tag{3.13}$$

which has the special property of integrating to 1 along each $\mathcal{G}_0$-orbit:

$$\int_{\mathcal{G}_0} \mathcal{D}U\, z[A^{U_0}] = 1, \tag{3.14}$$

(see Problem 3.4). This decomposition of unity allows one to factorize the volume of the gauge group $\mathcal{G}_0$ and to realize the gauge fixing in the form

$$\langle O \rangle = \frac{\int \mathcal{D}A\, O[A]\, e^{-S_{\mathrm{YM}}[A]}}{\int \mathcal{D}A\, e^{-S_{\mathrm{YM}}[A]}}$$

$$= \frac{\int_{\mathcal{G}_0} \mathcal{D}U \int \mathcal{D}A\, z[A^U]\, O[A]\, e^{-S_{\mathrm{YM}}[A]}}{\int_{\mathcal{G}_0} \mathcal{D}U \int \mathcal{D}A\, z[A^U]\, e^{-S_{\mathrm{YM}}[A]}}$$

---

[7] For instance, the gauge-fixed measure $D_{gf}[A]$ can be non-local. Its localization requires one to introduce integrations over auxiliary fields, such as the Faddeev-Popov ghost and antighost fields or the Nakanishi-Lautrup field. Conveniently, however, the formal discussion to be presented in this and the next two chapters can be done without introducing these fields.

$$= \frac{\int_{\mathcal{G}_0} \mathcal{D}U \int \mathcal{D}A^U z[A^U] O[A^U] e^{-S_{\mathrm{YM}}[A^U]}}{\int_{\mathcal{G}_0} \mathcal{D}U \int \mathcal{D}A^U z[A^U] e^{-S_{\mathrm{YM}}[A^U]}}$$

$$= \frac{\int_{\mathcal{G}_0} \mathcal{D}U \times \int \mathcal{D}A \, z[A] O[A] e^{-S_{\mathrm{YM}}[A]}}{\int_{\mathcal{G}_0} \mathcal{D}U \times \int \mathcal{D}A \, z[A] e^{-S_{\mathrm{YM}}[A]}}$$

$$= \frac{\int \mathcal{D}A \, z[A] O[A] e^{-S_{\mathrm{YM}}[A]}}{\int \mathcal{D}A \, z[A] e^{-S_{\mathrm{YM}}[A]}}, \tag{3.15}$$

that is, with a gauge-fixed measure $\mathcal{D}_{\mathrm{gf}}[A] = \mathcal{D}A \, z[A]$.

The formula (3.15) has a relatively simple interpretation: by weighting the contribution of the gluon field along a given orbit by the functional $z[A]$, one replaces the redundant counting of configurations along the orbit in the numerator and denominator of the left-hand side of Eq. (3.15) by a non-redundant counting in the right-hand side, where, although each configuration along the orbit is taken into consideration, it contributes only partially (i.e., in the proportion $z[A]$) to the final value of the observable. For this reason, the expression in the last line of Eq. (3.15) can be extended to the case of gauge-variant functional such as the ones that enter the definition of correlation functions. Of course, in the case of gauge-invariant functionals, this expression is independent of the choice of functional $z[A]$ provided the latter is a decomposition of unity along each $\mathcal{G}_0$-orbit. We note, however, that this requires one, in general, to take into account the contribution from all the gauge-field configurations within each orbit with $z[A] \neq 0$, so that $z[A]$ integrates formally to 1. We shall come back to this point below.

Let us mention that the previous gauge fixing, which we could refer to as *gauge fixing on average* is more general than the usual *conditional gauge fixing*. Indeed, one does not try here to restrict to configurations obeying a certain condition $F[A] = 0$, but, rather, one integrates over all configurations, with a certain weight $z[A]$ that lifts the degeneracy of the contributions along a given $\mathcal{G}_0$-orbit while not affecting the observables. On the other hand, conditional gauge fixings can be seen as particular cases of gauge fixing on average, with the choice $\rho[A] = \delta(F[A])$ and the requirement (3.12) rewriting

$$0 < \sum_i \left| \Delta[A^{U_0^{(i)}[A]}] \right|^{-1} < \infty, \tag{3.16}$$

where the functional Δ was defined in Eq. (2.8) and the $U_0^{(i)}[A]$ are the gauge transformations that take the configuration A into the solutions of the equation $F[A] = 0$ along the $\mathcal{G}_0$-orbit of A. Any good conditional gauge fixing should intersect each $\mathcal{G}_0$-orbit at least once for, otherwise, relevant orbits to some of the observables could be missed. In this case, the left inequality in Eq. (3.16) is automatically fulfilled. On the other hand, there could be Gribov copies, that is, multiple solutions to $F[A] = 0$ along each $\mathcal{G}_0$-orbit. These are properly taken into account in the present approach because the corresponding functional $z[A]$ reads

$$z[A] = \frac{\delta(F[A])}{\sum_i \left| \Delta[A^{U_0^{(i)}[A]}] \right|^{-1}}, \tag{3.17}$$

rather than the generally inexact $\Delta[A]\,\delta(F[A])$. The only constraint is that the right inequality in Eq. (3.16) should be satisfied, which is not necessarily the case for all possible choices of $F[A]$.

We note that, in the presence of Gribov copies, there is no unique way of implementing conditional gauge fixings as gauge fixings on average since one could also choose $\rho[A] = w[A]\,\delta(F[A])$. In this case, the sum over copies in Eqs. (3.16) and (3.17) needs to be supplemented with a factor $w[A_0^{U_0^{(i)}}[A]]$. Also, by redefining $w[A] \to \Delta[A]w[A]$, one recovers the Serreau-Tissier-type gauge fixings. Only in the particular, ideal case, where there are no Gribov copies, all these implementations lead to $z[A] = \Delta[A]\,\delta(F[A])$.

3.3.2 Constraints on Observables

The principal way by which symmetries act on a system is through the simple, in general linear, transformation of some of the observables. Under the assumption that the symmetry is not broken, these simple transformation rules turn into constraints for these observables in the symmetric phase. The prototype for these considerations is given by the Polyakov loop whose constraint from center symmetry we have already analyzed in the absence of gauge fixing (see the discussion around Eq. (3.9)). Let us now repeat this discussion in the presence of a gauge-fixing functional $z[A]$. For simplicity, we assume that the system is in the center-symmetric phase, so there is no need for introducing an external, infinitesimal source. We can then write

$$\frac{\int \mathcal{D}A\, z[A]\, \Phi_A(\mathbf{x})\, e^{-S_{\mathrm{YM}}[A]}}{\int \mathcal{D}A\, z[A]\, e^{-S_{\mathrm{YM}}[A]}}$$

$$= \frac{\int \mathcal{D}A^U\, z[A^U]\, \Phi_{A^U}(\mathbf{x})\, e^{-S_{\mathrm{YM}}[A^U]}}{\int \mathcal{D}A^U\, z[A^U]\, e^{-S_{\mathrm{YM}}[A^U]}}$$

$$= e^{i\frac{2\pi}{N}k}\frac{\int \mathcal{D}A\, z[A^U]\, \Phi_A(\mathbf{x})\, e^{-S_{\mathrm{YM}}[A]}}{\int \mathcal{D}A\, z[A^U]\, e^{-S_{\mathrm{YM}}[A]}}. \tag{3.18}$$

As opposed to the derivation in the absence of gauge fixing, we cannot here immediately read this equation as a constraint on the Polyakov loop since the functional integral expression in the left-hand side is not exactly the same than the corresponding functional integral expression in the right-hand side, due to the presence of the a priori different functionals $z[A]$ and $z[A^U]$. Indeed, the only constraint on $z[A]$ is to provide a decomposition of unity along each $\mathcal{G}_0$-orbit. There is however no reason why $z[A]$ should be invariant under the transformations $U \in \mathcal{G}$. Even worst, if $z[A]$ were invariant under $\mathcal{G}$, it would also be invariant under $\mathcal{G}_0$, thus contradicting the fact that it provides a decomposition of unity along each $\mathcal{G}_0$-orbit.

The looked-after constraint for the Polyakov loop originates instead in the fact that $z_U[A] \equiv z[A^U]$, although different from $z[A]$, is also a decomposition of unity along each $\mathcal{G}_0$ (see Problem 3.4). This means that $z_U[A]$ qualifies as a gauge-fixing functional. Then, it can be replaced by $z[A]$ in the evaluation of any observable, and in particular in the last line of Eq. (3.18), thus allowing one to finalize the argument and retrieve the constraint on the Polyakov loop.

3.3.3 Approximations and/or Modeling

Yet, as we have already mentioned, the independence of observables with respect to the choice of $z[A]$ relies on taking into account all the gauge-field configurations along each $\mathcal{G}_0$-orbit with $z[A] \neq 0$, in such a way that $z[A]$ integrates to 1. This of course is very difficult to maintain within a given approximation scheme that usually looses contact with the notion of orbits and, in any case, treats the contribution of each orbit approximately. The same applies in the presence of modeling for the gauge-fixing procedure in the infrared. As a consequence, the argumentation of the previous section does not go through in practice, in the presence of approximations or modeling.

We mention that the problem identified above applies in particular to conditional gauge fixings in the presence of Gribov copies, as we have considered them in Eq. (3.17): only when one sums over all copies along a given orbit does one reconstruct the unity that allows the argument to go through in the absence of any other particular property of the gauge-fixing condition. On the other hand, in the ideal case of conditional gauge fixings without Gribov copies, one may wonder whether the problem could be avoided. Indeed, there is only one configuration per orbit contributing to the observables and reconstructing the unity in this case just amounts to taking this configuration into account. We discuss this particular case in Problem 3.5 and show that one can indeed identify a certain, orbit-dependent symmetry that connects, for each A, that one configuration obeying the gauge-fixing condition on the orbit of A to that one configuration obeying the gauge-fixing condition on the orbit of A^U and that implements the center invariance of the functional integral measure $\Delta[A]\,\delta(F[A])$. However, whether this is an actual symmetry under the functional integral in the continuum is far from obvious due to the orbit-dependent nature of this transformation, potentially leading to a non-trivial Jacobian. Moreover, after localization (see Eq. (2.17)), this transformation could correspond to a very complicated symmetry in terms of the Nakanishi-Lautrup and ghost/antighost fields and thus become fragile to approximations/modeling.[8]

The above discussion shows that, in the presence of approximations/modeling, it might be convenient to develop the formalism within a gauge-fixed set-up that makes center symmetry explicit in some sense, without having to rely on the gauge-fixing independence of the observables.

3.3.4 Strategies to Be Followed Next

In the next two chapters, we show how the above shortcomings can be cured by upgrading standard gauge fixings into background-extended gauge fixings. More precisely, a given gauge fixing, characterized by a gauge-fixing functional $z[A]$, will be upgraded into a background gauge fixing, characterized by a functional $z_{\bar{A}}[A]$ that depends on a background field configuration $\bar{A}$. The latter should be seen as an infinite collection of gauge-fixing parameters characterizing the choice of gauge and which allows one to implement an

---

[8] Interestingly, that same orbit-dependent symmetry is the one that allows one to maintain center symmetry in the lattice formulation of the gauge fixing, even in the case where Gribov copies are present. In this case, the two continuum problems identified here, namely, the non-trivial Jacobian and localization, are not present. For remarks on how the discussion extends on the lattice, see Sect. 3.5.

invariance under twisted gauge transformations in the form $z_{\bar{A}^U}[A^U] = z_{\bar{A}}[A]$, $\forall U \in \mathcal{G}$. This formal invariance, which one should more legitimately refer to as covariance, will not be the end of the story, however, since it is achieved to the price of transforming the background and thus changing the gauge. This makes it difficult, if not impossible, to identify the symmetric states of the system, a minimum pre-requisite to study the spontaneous breaking of the symmetry. We will then see how the by now standard notion of self-consistent backgrounds allows one to circumvent the problem and identify the invariant states. This will also allow us to understand how the center symmetry constraint on the Polyakov loop is retrieved.

In Chap. 11, we will discuss yet another proposal, also based on background-extended gauge fixings, but not on self-consistent backgrounds. It is based on the observation that, even though $z[A]$ can in no way be invariant under $\mathcal{G}$, for the reasons given above, it is possible to construct certain functionals $z[A]$ that are invariant, not under $\mathcal{G}$ but rather under $\mathcal{G}$ modulo $\mathcal{G}_0$[9], or equivalently, under certain representatives U_k (with $k = 1, \ldots, N-1$) of the non-trivial center transformations within each class of $\mathcal{G}/\mathcal{G}_0$. These particular transformations, although they do not cover the whole group $\mathcal{G}$, allow one to deduce all the consequences of the physical center symmetry on the relevant observables. We will see that, for some specific choices $\bar{A} = \bar{A}_c$ of background, one can achieve the invariance of the gauge-fixed functional in the form $z_{\bar{A}_c}[A^{U_k}] = z_{\bar{A}_c}[A]$. Since this approach is very recent and its consequences are yet to be fully explored, we have decided to discuss it at the end of the manuscript and to dedicate the next two chapters to the more accepted approach based on self-consistent backgrounds. We believe, however, that the novel proposal, because it avoids some of the problems related to the use of the self-consistent backgrounds (also discussed in the next chapter), should lead to even more robust results in the future.

3.4 Effective Action

So far, we have focused our analysis on observables. In non-gauged systems, another quantity that usually reflects the physical symmetries of the problem is the effective action. This functional is actually very useful in the study of the spontaneous symmetry breaking because its minima play the role of order parameters. The same is true in a gauge theory for the effective action of gauge-invariant observables such as the Polyakov loop.

It is very often convenient, however, to work in terms of the effective action for the gluon field which is not gauge-invariant. This effective action $\Gamma_z[A]$ is defined only after a gauge fixing z has been specified and it is not always the case that it makes the symmetry explicit, even at an exact level of treatment. As a consequence, the minima of $\Gamma_z[A]$ in a given gauge are not necessarily order parameters, even though the same physics is obtained from these minima in any gauge. Let us see more precisely how this comes about.

---

[9] In Problem 3.5, we discuss under which conditions a gauge-fixing functional $F[A]$ leads to a functional $z[A]$ invariant under $\mathcal{G}$ modulo $\mathcal{G}_0$.

3.4.1   Definitions

For a given choice of functional $z[A]$, one defines the generating functional

$$W_z[J] \equiv \ln \int \mathcal{D}A \, z[A] \, e^{-S_{YM}[A]+(J_\mu; A_\mu)} \,, \tag{3.19}$$

where $(\,;\,)$ is the Killing form introduced in Eq. (2.3). Being the expectation value of a gauge-variant functional, $W_z[J]$ depends a priori on z and so even at an exact level of treatment. On the other hand, its zero-source limit $W_z[0]$ should be independent of z:

$$0 = \frac{\delta}{\delta z} W_z[0] \,, \tag{3.20}$$

as it corresponds to a physical quantity: the free energy of the system.

The effective action $\Gamma_z[A]$ is defined as the Legendre transform of $W_z[J]$ with respect to J:

$$\Gamma_z[A] \equiv -W_z[J_z[A]] + (J_{z,\mu}[A]; A_\mu) \,, \tag{3.21}$$

with $J_z[A]$ the inverse of[10]

$$A_z[J] \equiv -\frac{1}{2} \frac{\delta W_z}{\delta J^t} \,. \tag{3.22}$$

As usual, the functional inverse of $A_z[J]$ is obtained as

$$J_z[A] = -\frac{1}{2} \frac{\delta \Gamma_z[A]}{\delta A^t} \,. \tag{3.23}$$

As for $W_z[J]$, the functional $\Gamma_z[A]$ depends a priori on z. However, corresponding to Eq. (3.20), we have that

$$0 = \frac{\delta}{\delta z} \Gamma_z[A_{\min}[z]] \,, \tag{3.24}$$

where $A_{\min}[z]$ denotes the absolute minimum of $\Gamma_z[A]$ with respect to A.[11] This is just the statement that the same physics should be reached from any gauge. Despite this indisputable fact, $W_z[J]$ and $\Gamma_z[A]$ do not necessarily reflect center symmetry as we now argue.

---

[10] The factor $-1/2$ and the transposition originate in our choice of convention when defining the Killing form (see Eq. (2.3)).

[11] That the limit of zero sources corresponds to an extremum of $\Gamma_z[A]$ follows from Eq. (3.23). That it corresponds to an absolute minimum follows from the convexity of $W_z[J]$ which is guaranteed if $z[A]$ is not too negative.

3.4.2 Center Symmetry and Order Parameters

To see this, let us make use of the identity

$$(J_\mu; A^U) = (J_\mu^{U^\dagger}; A_\mu) - (J_\mu; U\partial_\mu U^\dagger), \tag{3.25}$$

with $J^U \equiv UJU^\dagger$ (see Problem 3.6) to repeat the argument around Eq. (3.18). The only difference is that, because $W_z[J]$ is not the expectation value of an observable, we cannot replace $z_U[A]$ by $z[A]$. We thus arrive at

$$W_z[J] = W_{z_U}[J^{U^\dagger}] - (J_\mu; U\partial_\mu U^\dagger). \tag{3.26}$$

Rather than a symmetry constraint on the generating functional of a given gauge fixing, this is a relation between the generating functionals in two different gauges, corresponding to $z[A]$ and $z_U[A]$, respectively. When translated in terms of the effective action, this relation reads

$$\Gamma_z[A] = \Gamma_{z_U}[A^{U^\dagger}], \tag{3.27}$$

(see Problem 3.6). Once again, because $\Gamma_z[A]$ depends on z, this is in general not a symmetry constraint on the effective action in a given gauge but rather a relation between the effective actions in two different gauges corresponding to $z[A]$ and $z_U[A]$. Thus, the effective action in a given gauge does not necessarily reflect the center symmetry.

It also follows from Eq. (3.27) that $A_{\min}[z]$ is in general not an order parameter for center symmetry. Indeed, if we apply a transformation $U \in G$ on this minimum, from Eq. (3.27), we can write

$$\Gamma_{z_{U^\dagger}}[A^U_{\min}[z]] = \Gamma_z[A_{\min}[z]] \le \Gamma_z[A^{U^\dagger}] = \Gamma_{z_{U^\dagger}}[A]. \tag{3.28}$$

Thus, $A^U_{\min}[z]$ is the minimum of the effective action in another gauge. This change of gauge as one transforms the minimum makes it difficult, if not impossible, to decide whether the minimum is explicitly symmetric under the symmetry. This, in turn, prevents the interpretation of the minimum as an order parameter for the symmetry. Similarly, the correlation functions in the gauge defined by $z[A]$ will typically not reflect the symmetry and, in particular, will not be specially sensitive to its spontaneous breaking.

Let us stress that, an exact level of treatment, the fact that the symmetry is not necessarily explicit at the level of the effective action (or the corresponding minimum) is not really a problem. Indeed because of Eq. (3.24), the breaking of the symmetry, which reflects itself at the level of the free energy, should be visible in any gauge. On the contrary, in the presence of approximations or the modeling of the gauge-fixing procedure in the infrared, this identity might be violated.[12] This implies that not all gauges are equivalent in practice for the study of symmetry breaking, not even from the point of view of the free energy. This is yet another reason for looking for gauges where the symmetry is explicit at the level of the effective action.

In the next two chapters, this will be realized not directly in terms of an effective action (in the sense of a Legendre transform) but of a related construct known as the background effective action which connects directly to the notion of self-consistent backgrounds. In Chap. 11, we will construct a center-symmetric genuine effective action that avoids some of the problems of the background effective action.

3.5   Lattice Implementations of the Gauge Fixing

Before closing this chapter, and in order to appreciate the problem differently, let us examine how the above discussion would appear on the lattice. We will remain at a very informal level though and see the lattice as a procedure that selects gauge-field configurations according to some rule. Of course, a more rigorous discussion would require introducing the link variables.

Let us first imagine implementing the gauge fixing on average on the lattice. In order to evaluate a correlation function in the gauge associated with a generic partition of unity $z[A]$, one would first generate relevant configurations A_i for the non-gauge-fixed weight $e^{-S_{YM}}$, and then, from these configurations, one would generate new configurations $A_i^{U_{0,j}}$ along the corresponding $\mathcal{G}_0$-orbits. The correlation functions in the considered gauge would then be evaluated as

$$\frac{\sum_{i,j} C[A_i^{U_{0,j}}] z[A_i^{U_{0,j}}] e^{-S_{YM}[A_i^{U_{0,j}}]}}{\sum_{i,j} z[A_i^{U_{0,j}}] e^{-S_{YM}[A_i^{U_{0,j}}]}}, \tag{3.29}$$

where $C[A]$ is a gauge-variant functional characteristic of the considered correlation function. With the same implementation, an observable would be computed as

$$\frac{\sum_{i,j} O[A_i^{U_{0,j}}] z[A_i^{U_{0,j}}] e^{-S_{YM}[A_i^{U_{0,j}}]}}{\sum_{i,j} z[A_i^{U_{0,j}}] e^{-S_{YM}[A_i^{U_{0,j}}]}}, \tag{3.30}$$

but this time, if the number of generated configurations along each orbit is not enough, the unity is not reconstructed, and the result depends in practice on the choice of $z[A]$. For the same reason, the computed Polyakov is not necessarily 0.

In contrast, in the case of a particular gauge-fixing functional $z[A]$ that is invariant under certain U_ks as described above, and provided that for each configuration in the statistical

---

[12] Within a strict perturbative expansion, in a scheme that does not depend on z, Eq. (3.24) is true order by order. However, it is very often the case that perturbation theory is not applied to $\Gamma_z[A_{\min}[z]]$ but rather to $\Gamma_z[A]$ in which case $\Gamma_z[A_{\min}[z]]$ contains all orders, some of them only partially, and then Eq. (3.24) applies only approximately. This is also most probably the case of most truncations of non-perturbative functional equations.

ensemble, one associates $A_i^{U_k U_{0,j}}$ where U_k is a representative of each center transformation that leaves $z[A]$ invariant, we have, for the Polyakov loop

$$\frac{\sum_{i,j,k} \ell[A_i^{U_k U_{0,j}}] z[A_i^{U_k U_{0,j}}] e^{-S_{YM}[A_i^{U_k U_{0,j}}]}}{\sum_{i,j,k} z[A_i^{U_k U_{0,j}}] e^{-S_{YM}[A_i^{U_k U_{0,j}}]}}$$

$$= \frac{\sum_{i,j,k} e^{i2\pi k/N} \ell[A_i^{U_{0,j}}] z[A_i^{U_{0,j}}] e^{-S_{YM}[A_i^{U_{0,j}}]}}{\sum_{i,j,k} z[A_i^{U_{0,j}}] e^{-S_{YM}[A_i^{U_{0,j}}]}}$$

$$= \frac{\sum_k e^{i2\pi k/N}}{\sum_k 1} \times \frac{\sum_{i,j} z[A_i^{U_{0,j}}] e^{-S_{YM}[A_i^{U_{0,j}}]}}{\sum_{i,j} z[A_i^{U_{0,j}}] e^{-S_{YM}[A_i^{U_{0,j}}]}} = 0, \qquad (3.31)$$

so that the symmetry is explicit.

On the other hand, and contrary to the continuum setting, the case of conditional gauge fixings is peculiar. Indeed, even in the presence of Gribov copies, the lattice can proceed differently to Eq. (3.17) by selecting one copy per orbit (we assume that there is statistically little chance to generate two copies on the same orbit). In that case, similar to the discussion we had in Problem 3.5 in the absence of copies, one can identify an orbit-dependent transformation that connects the selected copies between themselves and allows one to realize $\ell = 0$. In this case, the two obstacles that we have identified in the continuum, namely, the Jacobian and the localization in terms of auxiliary fields, are not present. It is however worth noting that this argument applies to observables only. In the case of the generating functional $W_z[J]$, this orbit-dependent transformation does not allow one to derive a simple symmetry identity, and neither there is a simple symmetry identity for $\Gamma_z[A]$. So, even on the lattice, and even though ℓ is a good order parameter, $\langle A \rangle_z$ is only a good order parameter for very specific choices of gauge fixing. Similarly, we should expect sharp signatures of the transition via the correlation functions only for those specific choices of gauge fixing.

Problems

3.1 Gauge Transformation of the Wilson Line (⋆)
Consider the Wilson line

$$L_A(\sigma_f, \sigma_i, \mathbf{x}) \equiv \mathcal{P} \exp\left\{ \int_{\sigma_i}^{\sigma_f} d\tau \, A_0(\tau, \mathbf{x}) \right\},$$

with $0 < \sigma_i < \sigma_f < \beta$ and where $\mathcal{P}$ was defined in Eq. (3.1).

(a) Show that $L_A(\sigma_f, \sigma_i, \mathbf{x})$ is entirely determined as the solution of a first-order differential equation with respect to σ_f obeying a particular initial condition at $\sigma_f = \sigma_i$.

(b) Given a gauge transformation $U(\tau, \mathbf{x})$, show that $U(\sigma_f, \mathbf{x}) L_A(\sigma_f, \sigma_i, \mathbf{x}) U^\dagger(\sigma_i, \mathbf{x})$ and $L_{A^U}(\sigma_f, \sigma_i, \mathbf{x})$ obey the same differential equation and the same initial condition. Deduce that

$$L_{A^U}(\sigma_f, \sigma_i, \mathbf{x}) = U(\sigma_f, \mathbf{x}) L_A(\sigma_f, \sigma_i, \mathbf{x}) U^\dagger(\sigma_i, \mathbf{x}).$$

In particular, this yields Eq. (3.2) in the case $\sigma_i = 0$ and $\sigma_f = \beta$.

3.2 Periodic Boundary Conditions and Gauge Transformations (★★)

We look for gauge transformations U that preserve the periodic boundary condition (3.5) of any gauge field at finite temperature.

(a) Show that the condition (3.5) rewrites

$$\partial_\mu Z(\tau, \mathbf{x}) - [A_\mu(x), Z(\tau, \mathbf{x})] = 0,$$

where $Z(\tau, \mathbf{x}) \equiv U^\dagger(\tau, \mathbf{x}) U(\tau + \beta, \mathbf{x})$. *Tip:* express the gauge transformations on each side of the condition (3.5), and reshuffle the resulting expression.

(b) Deduce that $Z(\tau, \mathbf{x})$ should be a constant SU(N) matrix Z commuting with all the generators of the su(N) Lie algebra and, therefore, with all the elements of the SU(N) group.

3.3 Center Symmetry Group (★)

(a) Show that the center of SU(N) is made of the special unitary matrices $e^{i \frac{2\pi}{N} k} \mathbb{1}$, with $k = 0, \ldots, N - 1$. *Tip:* use that any element in the center commutes in particular with "infinitesimal" special unitary matrices and thus with elements of the corresponding Lie algebra.

(b) Show that the subgroup $\mathcal{G}_0$ is normal within $\mathcal{G}$, that is

$$\forall U \in \mathcal{G}, \ \forall U_0 \in \mathcal{G}_0, \ U U_0 U^\dagger \in \mathcal{G}_0.$$

Deduce that $\mathcal{G}/\mathcal{G}_0$ is a group isomorphic to the center of SU(N).

3.4 Decomposition of Unity (★)

(a) Show that $z[A]$ as defined in Eq. (3.13) integrates to 1 along each $\mathcal{G}_0$-orbit. *Tip:* use the invariance of the Haar measure on the underlying gauge group under multiplication on the right.

(b) Assuming that the Haar measure is invariant under both left and right multiplication (this is the case for a compact group such as the group SU(N) considered in this manuscript), show that, for any $U \in \mathcal{G}$, $z_U[A] \equiv z[A^U]$ also integrates to 1 along each $\mathcal{G}_0$-orbit.

3.5 Conditional Gauge Fixing (★★★)

In this problem, we would like to study more precisely the implementation of conditional gauge fixings $F[A] = 0$ as gauge fixings on average, corresponding to the choice $\rho[A] = \delta(F[A])$ that leads to the gauge-fixing functional

$$z[A] = \frac{\delta(F[A])}{\int_{\mathcal{G}_0} \mathcal{D}U_0 \, \delta(F[A^{U_0}])} .$$

In particular, we would like to investigate under which conditions on $F[A]$, the functional $z[A]$ is invariant under $\mathcal{G}$ modulo $\mathcal{G}_0$, that is invariant under certain representatives U_k in each class of $\mathcal{G}/\mathcal{G}_0$. In Chap. 11, we shall construct explicit realization of such conditional gauge fixings (see in particular Prob. 11.2).

(a) Show first that $z[A^{U_k}] = z[A]$ iff $\delta(F[A^{U_k}]) = \alpha[\mathcal{A}] \, \delta(F[A])$ where $\alpha[\mathcal{A}]$ depends only on the $\mathcal{G}_0$-orbit $\mathcal{A}$ to which A belongs.

(b) Show that a necessary condition for $\delta(F[A^{U_k}]) = \alpha[\mathcal{A}] \, \delta(F[A])$ to hold true is that the set of configurations $A^{U_0^{(i)}[A]}$ such that $F[A] = 0$ is invariant under the U^k. *Tip:* examine the support of functional Dirac distributions.

(c) Show that another necessary condition is that the following ratio

$$\frac{\left| \det \left. \frac{\delta F[(A^{U_0})^{U_k}]}{\delta U_0} \right|_{U_0^{(i)}(A)} \right|^{-1}}{\left| \det \left. \frac{\delta F[A^{U_0}]}{\delta U_0} \right|_{U_0^{(i)}(A)} \right|^{-1}}$$

does not depend on the considered solution $U_0^{(i)}[A]$. *Tip:* express the assumed relation between the functional Dirac distributions along a given orbit $\{A^{U_0} \mid U_0 \in \mathcal{G}_0\}$.

(d) Show that the two conditions identified in (b) and (c) are in fact sufficient for $\delta(F[A^{U_k}]) = \alpha[\mathcal{A}] \, \delta(F[A])$ to hold true.

(e) Consider now the case where there are no copies. Explain why the condition identified in (c) is always fulfilled. Argue why the condition identified in (b) is not fulfilled a priori but that it can be fulfilled provided one allows for orbit-dependent transformations U_k.

(f) Show that, for these orbit-dependent transformations, the relation $\delta(F[A^{U_k}]) = \alpha[\mathcal{A}] \, \delta(F[A])$ still holds true and, therefore, that $\Delta[A]\delta(F[A])$ is invariant under these peculiar transformations. As we explain in the main text, however, it is not obvious that this symmetry is an actual symmetry under the functional integral in the continuum and that it easily survives approximations or modeling. On the other hand, those are precisely the transformation that allows one to maintain the center symmetry explicit in the lattice set-up.

3.6 Effective Action (★★★)

(a) Derive Eq. (3.25).
(b) Use it to deduce the transformation law Eq. (3.26) for the functional $W_z[J]$.
(c) From the transformation law obtained in (b), show that $A_z[J] = A_{zU}^U[J^{U^\dagger}]$.
(d) Upon inversion of $A_z[J]$, derive $J_z[A] = J_{zU}^U[A^{U^\dagger}]$.
(c) Finally, deduce the transformation law (3.27) for the effective action $\Gamma_z[A]$.

Background-Field Gauges: States and Symmetries

4

In this chapter, we recall how the shortcomings of standard gauge fixings concerning the description of the confinement/deconfinement transition in the continuum can be cured by means of a self-consistent background field. We put a special emphasis on the justification of some of the steps that are usually taken when implementing self-consistent backgrounds at finite temperature. In particular, we discuss the underlying hypotheses concerning the gauge-fixed measure, and we stress the importance of finding a good, redundancy-free description of the states of the system that then allows for the analysis of center symmetry breaking.

We try to follow a semi-deductive approach: we start recalling the basic properties of background-field gauges, argue why these properties are a priori not sufficient to solve the problem identified at the end of the previous chapter, and, finally, show that one possible solution to the problem is via the notion of self-consistent backgrounds. The main point of this chapter is that such backgrounds play the role of order parameters for center symmetry. We give a formal but relatively compact proof of this known result and postpone a more explicit illustration to the next chapter.[1]

Self-consistent backgrounds offer an alternative route to the study of the confinement/deconfinement transition [73] that we shall pursue in subsequent chapters within the framework of the Curci–Ferrari model. The present chapter is also the opportunity to introduce various related constructs, such as the background-field effective action or the background-dependent Polyakov loop, that shall ubiquitously appear in the rest of the manuscript. For later use, we also extend the considerations on center symmetry to other symmetries, in particular to charge conjugation.

---

[1] In Chap. 11, we shall propose yet another order parameter.

© The Author(s), under exclusive license to Springer Nature Switzerland AG 2022
U. Reinosa, *Perturbative Aspects of the Deconfinement Transition*, Lecture Notes in Physics 1006, https://doi.org/10.1007/978-3-031-11375-8_4

4.1   The Role of the Background Field

In the previous chapter, we claimed that the limitations of standard gauge fixings regarding the continuum description of the deconfinement transition could be cured with the help of background-field gauges [165, 166]. Let us now review some of the properties of this type of gauges that shall eventually help reaching this conclusion. We stress however that these properties, although helpful, will not be sufficient. The ultimate solution to the problem will be presented in Sect. 4.2. An alternative, more recent solution will be discussed in Chap. 11.

4.1.1   Center-Covariant Gauge-Fixed Measure

As we saw in the previous chapter, the problem with standard gauge fixings is that the gauge-fixing conditions and, in turn, the associated gauge-fixed measures $\mathcal{D}_{gf}[A]$ are in general not invariant under twisted gauge transformations. This makes it difficult to ensure the appropriate transformation rule of the Polyakov loop under center transformations (and, therefore, its order parameter interpretation), unless the Polyakov loop is computed exactly. The idea behind introducing a background is to upgrade the gauge fixing in such a way that the corresponding gauge-fixed measure becomes invariant (in a sense to be specified below), with the hope that the transformation rule of the Polyakov loop applies even in the presence of approximations (such as perturbative expansions, truncations of Dyson-Schwinger equations, flow equations, ...).

For the sake of illustrating the procedure, let us consider the case of the Landau gauge. Given a gauge-field configuration $A_\mu(x)$ such that $\partial_\mu A_\mu(x) = 0$ and a twisted gauge transformation $U \in G$, one has typically $\partial_\mu A_\mu^U(x) \neq 0$ and, therefore, $\mathcal{D}_{Landau}[A^U] \neq \mathcal{D}_{Landau}[A]$. To cure this obviously inconvenient feature, one introduces an arbitrary (periodic) background-field configuration $\bar{A}_\mu(x)$ and replaces the Landau gauge-fixing operator $\partial_\mu A_\mu(x)$ by a covariant version of it, namely, $\bar{D}_\mu(A_\mu(x) - \bar{A}_\mu(x))$, where $\bar{D}_\mu \equiv \partial_\mu \_ - [\bar{A}_\mu(x), \_]$ denotes the adjoint covariant derivative in the presence of the background $\bar{A}_\mu(x)$. The condition

$$\bar{D}_\mu(A_\mu(x) - \bar{A}_\mu(x)) = 0 \,, \tag{4.1}$$

defines the Landau–DeWitt (LDW) gauge and is invariant under simultaneous twisted gauge transformations of $A_\mu(x)$ and $\bar{A}_\mu(x)$:

$$A_\mu^U(x) = U(x)A_\mu(x)\,U^\dagger(x) - U(x)\,\partial_\mu U^\dagger(x)\,, \tag{4.2}$$

$$\bar{A}_\mu^U(x) = U(x)\bar{A}_\mu(x)\,U^\dagger(x) - U(x)\,\partial_\mu U^\dagger(x)\,. \tag{4.3}$$

The corresponding gauge-fixed measure is then such that

$$\mathcal{D}_{LDW}[A^U; \bar{A}^U] = \mathcal{D}_{LDW}[A; \bar{A}]\,, \quad \forall U \in G\,. \tag{4.4}$$

It is in this sense that the gauge-fixed measure associated with the background extension of the Landau gauge is invariant under twisted gauge transformations. The procedure can be easily extended to any other gauge fixing, and one generally arrives at a gauge-fixed measure such that

$$\mathcal{D}_{\text{gf}}[A^U; \bar{A}^U] = \mathcal{D}_{\text{gf}}[A; \bar{A}], \quad \forall U \in \mathcal{G}. \tag{4.5}$$

We stress that the background should be interpreted as an infinite collection of gauge-fixing parameters. Of course, any *bona fide* background gauge fixing should lead to observables that do not depend on the chosen background (at least at an exact level of treatment). On the other hand, this means that an identity such that (4.5) qualifies more as center covariance rather than center invariance. Indeed, Eq. (4.5) imposes no constraint on the gauge-fixed measure of a specific gauge but rather connects the gauge-fixed measures in two different gauges. This apparently simple remark will be important for the following discussion.

4.1.2 Background Gauge Fixing and the Polyakov Loop

Indeed, let us now turn back to our original problem of maintaining the transformation rule of the Polyakov loop in the presence of approximations (see the discussion in Sect. 3.3.2, and see if the just derived properties help in that matter).

If we denote by $\ell[\bar{A}]$ the Polyakov loop computed in some approximation that preserves Eq. (4.5), within the LDW gauge with background $\bar{A}$,[2] then, following similar steps as in the previous chapter, we can derive the transformation rule

$$\ell_{\alpha - \frac{2\pi}{N}k}[\bar{A}^U] = e^{i\frac{2\pi}{N}k}\ell_\alpha[\bar{A}], \quad \forall U \in \mathcal{G}, \tag{4.6}$$

where $e^{i\frac{2\pi}{N}k}\mathbb{1} = U^\dagger(0, \mathbf{x})U(\beta, \mathbf{x})$ and $e^{i\alpha}$ is the direction of the necessary, infinitesimal, symmetry-breaking source. In the Wigner–Weyl realization of the symmetry, the Polyakov loop does not depend on α, and then

$$\ell[\bar{A}^U] = e^{i\frac{2\pi}{N}k}\ell[\bar{A}], \quad \forall U \in \mathcal{G}. \tag{4.7}$$

As interesting as these relations might be, they do not allow us, however, to solve our original problem. The reason is that Eq. (4.7) is not really a constraint but rather a relation between two different approximated evaluations of the Polyakov loop. Only when the latter is computed exactly does $\ell[\bar{A}]$ become independent of $\bar{A}$ (seen as an infinite collection of gauge-fixing parameters), and only then Eq. (4.7) implies that the Polyakov loop has to vanish in the center-symmetric phase. The discussion is here just a particular case of the discussion around Eq. (3.18). We are

---

[2] This is easily achieved in practice because the symmetry is realized linearly.

thus back to our original problem: even within a background-extended gauge fixing, the transformation properties of the Polyakov loop seem difficult to maintain in the presence of approximations.

4.1.3 Center-Covariant Effective Action

A similar discussion, which mimics the one in Sect. 3.4, can be done at the level of the effective action $\Gamma_{\bar{A}}[A]$ for the gauge field in the gauge with background $\bar{A}$. Indeed, it is easily shown that $\Gamma_{\bar{A}}[A]$ preserves the covariance property (4.5).

To check this more explicitly, one first introduces the generating functional

$$\hat{W}_{\bar{A}}[J] \equiv \ln \int \mathcal{D}_{\mathrm{gf}}[A; \bar{A}] \exp\left\{-S_{YM}[A] + \int d^d x \ (J_\mu; a_\mu)\right\}, \qquad (4.8)$$

where the source J_μ is coupled to the field $a_\mu(x) \equiv A_\mu(x) - \bar{A}_\mu(x)$. Being the difference of two gauge fields, it transforms under (4.2)–(4.3) as $a_\mu^U(x) = U(x)\, a_\mu(x)\, U^\dagger(x)$. It is pretty immediate to check (see Problem 4.1) that

$$\hat{W}_{\bar{A}^U}[J^U] = \hat{W}_{\bar{A}}[J], \ \forall U \in \mathcal{G}, \qquad (4.9)$$

with $J_\mu^U(x) \equiv U(x)\, J_\mu(x)\, U^\dagger(x)$. Next, one constructs the Legendre transform of $\hat{W}_{\bar{A}}[J]$ with respect to the source:

$$\hat{\Gamma}_{\bar{A}}[a] \equiv -\hat{W}_{\bar{A}}[J_{\bar{A}}[a]; \bar{A}] + \int d^d x \ (J_{\bar{A},\mu}[a]; a_\mu), \qquad (4.10)$$

where $J_{\bar{A}}[a]$ is obtained by inverting the relation

$$a_{\bar{A}}[J](x) = -\frac{1}{2}\frac{\delta \hat{W}_{\bar{A}}[J]}{\delta J^{\mathrm{t}}(x)}, \qquad (4.11)$$

for a fixed background. It can be shown that

$$J_{\bar{A}^U}[a^U] = J_{\bar{A}}^U[a], \ \forall U \in \mathcal{G}, \qquad (4.12)$$

(see Problem 4.1), which together with Eq. (4.9) implies

$$\hat{\Gamma}_{\bar{A}^U}[a^U] = \hat{\Gamma}_{\bar{A}}[a], \ \forall U \in \mathcal{G}, \qquad (4.13)$$

showing that the covariance property (4.5) is indeed preserved by fluctuations, as anticipated.

In what follows, we shall use a source term of the form $(J_\mu; A_\mu)$, which means that we work instead with the functional $W_{\bar{A}}[J] \equiv \hat{W}_{\bar{A}}[J] + \int d^d x \ (J_\mu; \bar{A}_\mu)$. It is easily seen that the corresponding effective action is related to the previous one by $\Gamma_{\bar{A}}[A] = \hat{\Gamma}_{\bar{A}}[A - \bar{A}]$ and obeys, therefore, the symmetry identity [167]

$$\Gamma_{\bar{A}^U}[A^U] = \Gamma_{\bar{A}}[A], \quad \forall U \in \mathcal{G}. \tag{4.14}$$

Yet, this is in general not a symmetry constraint on the effective action in a given gauge, but rather a relation between the effective actions in two different gauges. In other words, within a given gauge defined by the choice of a generic background $\bar{A}$, the effective action $\Gamma_{\bar{A}}[A]$, seen as a functional of A, does not reflect the center symmetry of the problem. As compared to the discussion regarding the Polyakov loop, this is true even at an exact level of treatment because we cannot invoke the gauge-fixing (and thus background) independence of the effective action (which is not an observable).

From these considerations, it follows that the minimum $A_{\min}[\bar{A}]$ of the effective action, defined by the condition

$$\Gamma_{\bar{A}}[A_{\min}[\bar{A}]] \leq \Gamma_{\bar{A}}[A], \quad \forall A, \tag{4.15}$$

does not qualify in general as an order parameter for center symmetry:[3] it is not possible from the actual state of the system as given by $A_{\min}[\bar{A}]$ to determine whether the system is in a symmetric or in a broken phase. Indeed, suppose we wanted to assess whether such a state $A_{\min}[\bar{A}]$ is center-symmetric or center-breaking. To this purpose, we would apply a twisted gauge transformation U to $A_{\min}[\bar{A}]$. The invariance property (4.14) means that the transformed configuration $A_{\min}^U[\bar{A}]$ is also a minimum but of the functional $\Gamma_{\bar{A}^U}[A]$ instead. Indeed, we have

$$\forall A, \quad \Gamma_{\bar{A}^U}[A] = \Gamma_{\bar{A}}[A^{U^\dagger}] \geq \Gamma_{\bar{A}}[A_{\min}[\bar{A}]] = \Gamma_{\bar{A}^U}[A_{\min}^U[\bar{A}]], \tag{4.16}$$

where we have successively used Eqs. (4.14), (4.15), and again (4.14). It follows that a state of the system, as described in the presence of the background $\bar{A}$, is transformed into another state but described in the presence of a different background $\bar{A}^U$. This change of description, or change of gauge, as one transforms the state of the system makes it difficult to identify the states that are invariant and, therefore, to draw any conclusion on the possible breaking of center symmetry at some temperature.

The situation is similar to that in Eq. (4.7) where the background is changed as a center transformation is applied. This is of course reminiscent of the fact that the invariance property (4.14) requires the background to be transformed as well. In the next section, we show nonetheless that the problem can be circumvented by restricting the discussion to what are known as self-consistent backgrounds.

---

[3] In Chap. 11, we discuss one particular case, where it does.

4.2   Self-Consistent Backgrounds

Self-consistent backgrounds are specific background configurations $\bar{A}_s$ defined by the property

$$\bar{A}_s = A_{\min}[\bar{A}_s]. \tag{4.17}$$

As it is well known, the minimum of the effective action coincides with the one-point function, and, therefore, the self-consistency condition also rewrites $\bar{A}_s = \langle A \rangle_{\bar{A}_s}$, where $\langle \ldots \rangle_{\bar{A}_s}$ denotes the expectation value in the presence of the gauge-fixed measure $\mathcal{D}_{gf}[A; \bar{A}_s]$. So, in a sense, the strategy of restricting to self-consistent backgrounds corresponds to constantly adapting the gauge fixing (through the choice of the background configuration) in such a way that the one-point function is always known exactly.[4]

Beyond the obvious practical interest of this choice, the main benefit of self-consistent backgrounds lies in that they allow for the identification of center-symmetric states. In fact, as we now explain, self-consistent backgrounds are order parameters for center symmetry and can be used, therefore, in place of the Polyakov loop. In order to reach this conclusion, we first introduce the background-field effective action that leads both to a simple characterization of the self-consistent backgrounds and to their interpretation as the actual states of the system.

4.2.1   Background-Field Effective Action

The *background-field effective action* is nothing but the effective action $\Gamma_{\bar{A}}[A]$ evaluated along the subspace $A = \bar{A}$:

$$\tilde{\Gamma}[\bar{A}] \equiv \Gamma_{\bar{A}}[A = \bar{A}]. \tag{4.18}$$

It obeys the invariance property

$$\tilde{\Gamma}[\bar{A}^U] = \tilde{\Gamma}[\bar{A}], \quad \forall U \in \mathcal{G}, \tag{4.19}$$

as it can easily be shown using Eqs. (4.14) and (4.18).

Let us now see how it allows for a characterization of the self-consistent backgrounds. One important ingredient to find such characterization will be that the value of the effective action at its minimum does not depend on the chosen background:

$$\Gamma_{\bar{A}}[A_{\min}[\bar{A}]] = \Gamma_{\bar{A}'}[A_{\min}[\bar{A}']], \quad \forall \bar{A}'. \tag{4.20}$$

---

[4] In particular, it follows that self-consistent backgrounds depend on the temperature, as opposed to the fixed backgrounds considered in the previous section.

This is because minimizing the effective action with respect to A corresponds to taking the zero-source limit in Eq. (3.21), that is, to evaluating the free energy of the system up to a factor β. As any other physical observable, the free energy cannot depend on the background $\bar{A}$. This is exactly what is encoded in Eq. (4.20).

With this property in mind, let us derive the promised characterization of self-consistent backgrounds. We first show that a self-consistent background $\bar{A}_s$ is necessarily an absolute minimum of $\tilde{\Gamma}[\bar{A}]$. To this purpose, we write the following chain of relations

$$\tilde{\Gamma}[\bar{A}_s] = \Gamma_{\bar{A}_s}[\bar{A}_s]$$

$$= \Gamma_{\bar{A}_s}[A_{\min}[\bar{A}_s]]$$

$$= \Gamma_{\bar{A}}[A_{\min}[\bar{A}]], \ \forall \bar{A}$$

$$\leq \Gamma_{\bar{A}}[\bar{A}], \ \forall \bar{A}$$

$$\leq \tilde{\Gamma}[\bar{A}], \ \forall \bar{A}, \tag{4.21}$$

where we have successively used Eqs. (4.18), (4.17), (4.20), (4.15), and again (4.18). Thus, as announced, a self-consistent background $\bar{A}_s$ is an absolute minimum of the functional $\tilde{\Gamma}[\bar{A}]$.

Next, we prove the following version of the reciprocal property: assuming that there exists at least one self-consistent background $\bar{A}_s$, then any absolute minimum $\bar{A}_m$ of $\tilde{\Gamma}[\bar{A}]$ is also a self-consistent background. To this purpose, we note that

$$\Gamma_{\bar{A}_m}[\bar{A}_m] = \tilde{\Gamma}[\bar{A}_m] \leq \tilde{\Gamma}[\bar{A}], \forall \bar{A}, \tag{4.22}$$

and write

$$\Gamma_{\bar{A}_m}[\bar{A}_m] \leq \tilde{\Gamma}[\bar{A}_s]$$

$$\leq \Gamma_{\bar{A}_s}[\bar{A}_s]$$

$$\leq \Gamma_{\bar{A}_s}[A_{\min}[\bar{A}_s]]$$

$$\leq \Gamma_{\bar{A}_m}[A_{\min}[\bar{A}_m]]. \tag{4.23}$$

This time, we have successively used Eqs. (4.22), (4.18), (4.17), and (4.20). This implies that $\bar{A}_m = A_{\min}[\bar{A}_m]$ and, therefore, that $\bar{A}_m$ is self-consistent, as announced.

Finally, if we restrict from now on to the background Landau gauges, we note that $\bar{A} = 0$ is always self-consistent. Indeed, $\bar{A} = 0$ corresponds to the standard Landau gauge for which there is no preferred color direction. Since we do not expect global color rotations to break, we have $\langle A \rangle_{\bar{A}=0} = 0 = \bar{A}$, and thus $\bar{A} = 0$ is self-consistent as announced. Putting all the pieces together, we deduce that the self-consistent backgrounds are exactly given by the absolute minima of the functional $\tilde{\Gamma}[\bar{A}]$.[5]

4.2.2 Background Description of the States

Now that we have characterized the self-consistent backgrounds as the absolute minima of the background-field effective action, let us see how they provide a new description of the states of the system that does not suffer from the limitations identified at the end of Sect. 4.1.3.

The point is that, when working with self-consistent backgrounds, all the relevant information can be obtained from the functional $\tilde{\Gamma}[\bar{A}]$ and the (self-consistent) background configurations $\bar{A}_s$ that minimize it, without any reference to the functional $\Gamma_{\bar{A}}[A]$ or to the field configuration $A_{\min}[\bar{A}]$. In particular, up to a trivial factor, the free energy of the system is given by

$$\Gamma_{\bar{A}_s}[A_{\min}[\bar{A}_s]] = \Gamma_{\bar{A}_s}[\bar{A}_s] = \tilde{\Gamma}[\bar{A}_s], \tag{4.24}$$

that is, it is given by the background-field effective action $\tilde{\Gamma}[\bar{A}]$ evaluated at any self-consistent background $\bar{A}_s$. Therefore, in a certain sense, one can interpret the space of background-field configurations $\bar{A}$ over which $\tilde{\Gamma}[\bar{A}]$ is varied as the the space of all potentially available states, the actual state of the system corresponding to those particular (self-consistent) backgrounds that minimize $\tilde{\Gamma}[\bar{A}]$.

Now, the invariance property (4.19) means that, after applying a twisted gauge transformation $U \in \mathcal{G}$ to a given state $\bar{A}_s$, the newly obtained configuration $\bar{A}_s^U$ is also a minimum of the functional $\tilde{\Gamma}[\bar{A}]$ and, therefore, another acceptable state of the system. The important difference with respect to the discussion in Sect. 4.1.3 is that there is no change of description as one transforms the state since the functional that needs to be minimized remains the same. As a consequence, there is no obstacle anymore to the identification of center-symmetric states.

4.2.3 Orbit Description of the States

Before we proceed to this identification, however, it is important to realize that the previous description of the states carries some redundancy. This relates to the fact that the symmetry identity (4.19) applies in particular to the periodic gauge transformations $U \in \mathcal{G}_0$. As we have seen in the previous chapter, these transformations need to be considered as true, and therefore unphysical, gauge transformations. Their presence only reflects the redundancy of the description in

---

[5] The present derivation assumes the unicity of the field configuration $A_{\min}[\bar{A}]$ for each background configuration $\bar{A}$. Similar results can be obtained under the assumption of degenerate minima.

terms of gauge fields, and, consequently, two background configurations that belong to the same orbit under the action of $\mathcal{G}_0$ should be considered as describing the same physical state. In other words, the physical states of the system correspond to the various possible $\mathcal{G}_0$-orbits in the space of background configurations, and a given $\mathcal{G}_0$-orbit (a given physical state) admits various equivalent representations in terms of background configurations.

The redundancy in the description of the physical states by means of background configurations is similar to the redundancy in the description of physical center transformations by means of twisted gauge transformations that we discussed in the previous chapter. And just as it was possible to remove the redundancy inherent to $\mathcal{G}$ by considering the group $\mathcal{G}/\mathcal{G}_0$, it is possible to remove the redundancy inherent to the use of backgrounds by working instead with the corresponding $\mathcal{G}_0$-orbits.

To see how this is achieved in practice, we first notice that the background-field effective action can be defined directly on the $\mathcal{G}_0$-orbits since $\tilde{\Gamma}[\bar{A}]$ depends only on the orbit $\bar{\mathcal{A}}$ the background $\bar{A}$ belongs to. We can then define $\tilde{\Gamma}[\bar{\mathcal{A}}] \equiv \tilde{\Gamma}[\bar{A}]$, where $\bar{A} \in \bar{\mathcal{A}}$ is any background in the orbit $\bar{\mathcal{A}}$, and the physical state is obtained by minimizing $\tilde{\Gamma}[\bar{\mathcal{A}}]$ over the space of $\mathcal{G}_0$-orbits. Second, it is possible to define the action of a center transformation $\mathcal{U} \in \mathcal{G}/\mathcal{G}_0$ directly on the $\mathcal{G}_0$-orbits. Indeed, owing to the property $\forall U \in \mathcal{G}$, $\forall U_0 \in \mathcal{G}_0$, $U U_0 U^\dagger \in \mathcal{G}_0$, all the backgrounds in a given $\mathcal{G}_0$-orbit are transformed under a twisted gauge transformation into backgrounds belonging to one and the same orbit, which means that one can directly define the action of a twisted gauge transformation on $\mathcal{G}_0$-orbits. Moreover, this action depends only on the class of $\mathcal{G}/\mathcal{G}_0$ the twisted gauge transformation belongs to, thereby defining the action of $\mathcal{G}/\mathcal{G}_0$ directly on $\mathcal{G}_0$-orbits. Finally, in this redundancy-free formulation, the symmetry identity (4.19) reads

$$\tilde{\Gamma}[\bar{\mathcal{A}}^{\mathcal{U}}] = \tilde{\Gamma}[\bar{\mathcal{A}}], \quad \forall \mathcal{U} \in \mathcal{G}/\mathcal{G}_0. \tag{4.25}$$

From all these considerations, it follows, as announced, that one can work exclusively in terms of $\mathcal{G}_0$-orbits and center transformations.

4.2.4 Center-Symmetric States

It should be clear by now that the center-symmetric states of the system correspond to the center-invariant $\mathcal{G}_0$-orbits, that is, orbits $\bar{\mathcal{A}}$ such that

$$\forall \mathcal{U} \in \mathcal{G}/\mathcal{G}_0, \quad \bar{\mathcal{A}}^{\mathcal{U}} = \bar{\mathcal{A}}. \tag{4.26}$$

In terms of background configurations, this reads

$$\forall U \in \mathcal{G}, \quad \exists U_0 \in \mathcal{G}_0, \quad \bar{A}_\mu^U(x) = \bar{A}_\mu^{U_0}(x), \tag{4.27}$$

that is, the center-symmetric states correspond to backgrounds that are invariant under twisted gauge transformations modulo periodic gauge transformations.

In the next chapter, we will see how to practically access the center-invariant $\mathcal{G}_0$-orbits using either Eq. (4.26) or Eq. (4.27). For the time being, what we can say is that we have classified the physical states of the system into center-invariant and center non-invariant $\mathcal{G}_0$-orbits. Therefore, the $\mathcal{G}_0$-orbits and, by extension, the background configurations, play the role of order parameters for the confinement/deconfinement transition, as announced earlier. This is the central result of this chapter, at the basis of the applications to be presented in subsequent chapters. For other proofs of this result in the case of the SU(2) gauge group, we refer to [73, 74].

4.3   Other Symmetries

The previous considerations apply not only to center symmetry but also, as we now explain, to any physical symmetry of the quantum action. After introducing some generalities, we discuss in particular the case of charge conjugation symmetry which plays an important role in subsequent applications.

4.3.1   Generalities

Consider a physical transformation $\mathcal{T}$ of the state of the system. In the space of background configurations, it writes $\bar{A} \rightarrow \bar{A}^{\mathcal{T}}$,[6] where $\bar{A}^{\mathcal{T}}$ should not be mistaken with the notation $\bar{A}^U$. In fact, $\mathcal{T}$ is a formal group transformation, not necessarily an element of SU(N).

The transformation $\mathcal{T}$ being physical, the application of $\mathcal{T}$ followed by $\mathcal{T}^{-1}$ to a background representing a given $\mathcal{G}_0$-orbit (a given physical state) should lead to a background of the same $\mathcal{G}_0$-orbit, even when some additional true gauge transformations are inserted between $\mathcal{T}$ and $\mathcal{T}^{-1}$. Mathematically, this writes as

$$\forall U \in \mathcal{G}_0, \ \exists U' \in \mathcal{G}_0, \ ((\bar{A}_\mu^{\mathcal{T}})^U)^{\mathcal{T}^{-1}}(x) = \bar{A}_\mu^{U'}(x). \tag{4.28}$$

From this, it follows immediately that the physical transformation $\mathcal{T}$ can be defined directly on the $\mathcal{G}_0$-orbits, as it should, of course, for a physical transformation.

Moreover, just as with physical center transformations, the group of physical transformations $\mathcal{T}$ can be extended into a group of transformations T defined

---

[6] We are implicitly assuming here that the background-field effective action $\tilde{\Gamma}[\bar{A}]$ is invariant under $\mathcal{T}$, which happens in particular if the gauge-fixed measure $\mathcal{D}_{\mathrm{gf}}[A; \bar{A}]$ is invariant under $(A, \bar{A}) \rightarrow (A^{\mathcal{T}}, \bar{A}^{\mathcal{T}})$ and if the symmetry is realized linearly.

as the transformations $\mathcal{T}$ modulo elements of $\mathcal{G}_0$.[7] Owing to Eq. (4.28), $\mathcal{G}_0$ is a normal subgroup of this extended group, and the corresponding quotient is isomorphic to the group of physical transformations. The reason for introducing such an extended group is that, in the space of background configurations, any transformation of the extended group is a valid representation of the corresponding physical transformation.

The states of the system that are invariant under the physical transformation $\mathcal{T}$ correspond to the $\mathcal{T}$-invariant $\mathcal{G}_0$-orbits. Mathematically, this writes as

$$\forall \mathcal{T}, \quad \bar{\mathcal{A}}^{\mathcal{T}} = \bar{\mathcal{A}}, \tag{4.29}$$

or, in terms of background configurations, as

$$\forall T, \ \exists U_0 \in \mathcal{G}_0, \ \bar{A}_\mu^T(x) = \bar{A}_\mu^{U_0}(x). \tag{4.30}$$

In particular, in terms of background configurations, physical invariance needs to be understood as invariance modulo true gauge transformations. The characterization (4.29) of invariant states is useful when one has a simple description of the orbits, such as the description in terms of Weyl chambers that we review in the next chapter. The characterization (4.30) is used when such a description is not available, for which we also give an example in the next chapter.

4.3.2 Charge Conjugation

Let us now illustrate these general considerations with the important case of charge conjugation. On the gauge field, this transformation reads

$$A_\mu^C(x) \equiv -A_\mu^t(x), \tag{4.31}$$

and changes a given representation into the corresponding contragredient representation. It is easily checked that this indeed corresponds to a physical symmetry in the sense of the condition (4.28).

In order to see the effect of this symmetry on the Polyakov loop, it is convenient to introduce the normalized trace of the Wilson line associated with the anti-quark $\bar{\Phi}_A(\mathbf{x}) \equiv \Phi_{-A^t}(\mathbf{x})$ which transforms according to the contragredient representation.[8] The corresponding Polyakov loop

$$\bar{\ell} \equiv \langle \bar{\Phi}_A \rangle \tag{4.32}$$

---

[7] We do not aim at giving a rigorous definition of this group here. Our intention is just to stress the similarities with the discussion regarding center symmetry.

[8] In terms of the anti-path-ordering, this rewrites $\bar{\Phi}_A(\mathbf{x}) = \mathrm{tr}\,\bar{\mathcal{P}} \exp\left\{-\int_0^\beta d\tau\, A_0(\tau, \mathbf{x})\right\}$.

is referred to as the *anti-Polyakov loop*. From charge conjugation invariance, it is easily checked that[9]

$$\bar{\ell} = \ell \,, \tag{4.33}$$

(see Problem 4.2). This relation is to be expected since charge conjugation symmetry is an unbroken symmetry of the YM system and, therefore, it should cost the same energy to bring a quark or an anti-quark into the thermal bath of gluons.

On the other hand, because $A_\mu = i A_\mu^a t^a$ is anti-Hermitian, we have $-A_\mu^t(x) = A_\mu^*(x)$ and thus $\bar{\Phi}_A(\mathbf{x}) = \Phi_A^*(\mathbf{x})$. This implies

$$\bar{\ell} = \ell^* \,. \tag{4.34}$$

We mention that this result relies crucially on the fact that the averaging measure under the functional integral is real. Combining it with the previous result, we deduce that the Polyakov and anti-Polyakov loops are real, which is a necessary condition for their interpretation as $e^{-\beta \Delta F}$. In Chap. 9, we shall investigate how these properties change in the presence of quarks with a finite chemical potential.

As for the background gauge-fixed theory, assuming that the gauge-fixed measure remains invariant under charge conjugation and because the symmetry is realized linearly, it is not difficult to argue that

$$\tilde{\Gamma}[\bar{A}^C] = \tilde{\Gamma}[\bar{A}] \,, \tag{4.35}$$

for any transformation C equal to C modulo elements of $\mathcal{G}_0$. In terms of $\mathcal{G}_0$-orbits, this reads

$$\tilde{\Gamma}[\bar{\mathcal{A}}^C] = \tilde{\Gamma}[\bar{\mathcal{A}}] \,, \tag{4.36}$$

which expresses charge conjugation invariance at the quantum level. Charge conjugation invariant states are such that $\bar{\mathcal{A}}^C = \bar{\mathcal{A}}$. We shall characterize them more precisely in the next chapter.

4.4   Additional Remarks

Let us conclude this chapter with some critical remarks on the rationale behind the previous background-field gauge construction.

---

[9] Strictly speaking, this identity holds as long as there is no spontaneous breaking of center symmetry. When the symmetry is broken, the identity applies to one of the possible states. The other states are invariant under a combination of charge conjugation and a center transformation.

4.4.1 Back to the Polyakov Loop

The most important feature of the previous construction is that the interpretation of the $\mathcal{G}_0$-orbits as order parameters for the deconfinement transition remains valid in the presence of approximations, provided the latter maintain the symmetry identity (4.25). This is easily achieved in practice because the symmetry is realized linearly.[10] Moreover, since $\mathcal{G}_0$-orbits represent an alternative order parameter, we do not need to be concerned anymore with the Polyakov loop and the question of whether it is possible to maintain its transformation rule in the presence of approximations. It is interesting to note, however, that the same framework allows one to answer this question positively.

To see this, let us allow for a non-zero source J coupled to a which can be seen as a functional $J_{\bar{A}}[a]$ of a and $\bar{A}$ (see the discussion of the effective action $\hat{\Gamma}_{\bar{A}}[a]$ in Sect. 4.1.3). Suppose then that the source is chosen such that $a \equiv A - \bar{A} = 0$, that is, the background $\bar{A}$ is forced to be self-consistent with the help of the source $J[\bar{A}] \equiv J_{\bar{A}}[0]$. We can compute the Polyakov loop in the presence of this source. Because it depends on the background $\bar{A}$, in what follows, we refer to it as the *background-dependent Polyakov loop,* and denote it by $\ell[\bar{A}]$.

Now, from Eq. (4.12), it follows that

$$J[\bar{A}^U] = J^U[\bar{A}], \quad \forall U \in \mathcal{G}, \tag{4.37}$$

from which it is easily deduced that

$$\ell[\bar{A}^U] = e^{i\frac{2\pi}{N}k}\ell[\bar{A}], \quad \forall U \in \mathcal{G}. \tag{4.38}$$

In particular, since $\ell[\bar{A}^U] = \ell[\bar{A}]$ for $U \in \mathcal{G}_0$, $\ell[\bar{A}]$ can be defined directly on $\mathcal{G}_0$-orbits, $\ell[\bar{\mathcal{A}}] \equiv \ell[\bar{A}]$, and the transformation rule (4.38) rewrites

$$\ell[\bar{\mathcal{A}}^{\mathcal{U}}] = e^{i\frac{2\pi}{N}k}\ell[\bar{\mathcal{A}}], \quad \forall \mathcal{U} \in \mathcal{G}/\mathcal{G}_0. \tag{4.39}$$

Now, if the orbit is center-invariant, it follows that

$$\ell[\bar{\mathcal{A}}] = e^{i\frac{2\pi}{N}k}\ell[\bar{\mathcal{A}}], \quad \forall k \in \{0, 1, \ldots, N - 1\}, \tag{4.40}$$

and, therefore, $\ell[\bar{\mathcal{A}}] = 0$. This property relies only on the identity (4.37) which is again easily satisfied in the presence of approximations.

Similarly, under charge conjugation

$$J[\bar{A}^C] = J^C[\bar{A}]. \tag{4.41}$$

---

[10] For instance, the perturbative expansion to be used in the following chapters will fulfil this property.

From this, it is easily deduced that

$$\ell[\bar{A}^C] = \bar{\ell}[\bar{A}] = \ell^*[\bar{A}]. \tag{4.42}$$

In terms of $\mathcal{G}_0$-orbits, this reads

$$\ell[\bar{\mathcal{A}}^C] = \bar{\ell}[\bar{\mathcal{A}}] = \ell^*[\bar{\mathcal{A}}], \tag{4.43}$$

and for a charge conjugation invariant orbit (which should represent the actual state of the YM system), we recover the above results for the physical Polyakov and anti-Polyakov loops.

It should be mentioned that the functional $\ell[\bar{A}]$ defined here is different from the one that we introduced in Sect. 4.1.2. The latter was just an approximated version of the physical Polyakov loop whose $\bar{A}$-dependence stems precisely from the use of approximations. Here, instead, the functional $\ell[\bar{A}]$ does depend on $\bar{A}$, even in the absence of approximations, because it is not the physical Polyakov loop but, rather, the value of the Polyakov loop as the system is forced into a state corresponding to the background $\bar{A}$ by means of a non-zero source $J[\bar{A}]$. The physical Polyakov loop is retrieved when evaluating $\ell[\bar{A}]$ for a self-consistent background $\bar{A}_s$ which is precisely such that $J[\bar{A}_s] = 0.$[11] If it so happens that this self-consistent background belongs to a center-invariant $\mathcal{G}_0$-orbit, it follows from the above that $\ell[\bar{A}_s] = 0$ and so even in the presence of approximations.

4.4.2 Hypothesis on the Gauge-Fixed Measure

The justification of all the previous results relies on two natural but strong assumptions on the generating functional $W_{\bar{A}}[J]$ which in turn imply constraints on the gauge-fixed measure $\mathcal{D}_{gf}[A; \bar{A}]$. These assumptions are, first, that $W_{\bar{A}}[J]$ should be convex with respect to J and, second, that $W_{\bar{A}}[0]$ should not depend on the background $\bar{A}$. It is important to keep these basic assumptions in mind since they are not necessarily implemented exactly in the practical realizations of the gauge fixing and/or in the presence of approximations and could therefore lead to some artifacts. Here, we analyze the basic assumptions in more detail. One of the benefits of the novel approach that we present in Chap. 11 is that it does not require some of these assumptions.

---

[11] One may wonder why we do not need an infinitesimal, symmetry-breaking source here, similar to the one that we introduced in the previous chapter. In the case where the physical state of the system is represented by a background $\bar{A}_s$ belonging to a center-invariant orbit, the system is in the symmetric phase, and the symmetry-breaking source is indeed not needed. For any other physical state, there will typically exist various orbits connected to each other by center transformations. The values of $\ell[\bar{A}]$ on each of these orbits represent the various vacua in the Nambu-Goldstone realization of the symmetry, as they would equivalently be reached from an infinitesimal, symmetry-breaking source.

Convexity of $W_{\bar{A}}[J]$

The convexity of $W_{\bar{A}}[J]$ enters crucially in the identification of the self-consistent backgrounds as the absolute minima of the functional $\tilde{\Gamma}[\bar{A}]$. The reason is that the proof of the standard fact that the limit of zero sources corresponds to the minimization of the functional $\Gamma_{\bar{A}}[A]$ with respect to A requires, in its simpler form, the convexity of $W_{\bar{A}}[J]$. The latter implies indeed that

$$\Gamma_{\bar{A}}\left[\frac{\delta W_{\bar{A}}}{\delta J}\right] \leq \Gamma_{\bar{A}}\left[\frac{\delta W_{\bar{A}}}{\delta J}\bigg|_{J_0}\right] + \int d^d x \left(\frac{\delta W_{\bar{A}}}{\delta J(x)} - \frac{\delta W_{\bar{A}}}{\delta J(x)}\bigg|_{J_0} ; J(x)\right), \quad (4.44)$$

for any sources J and J_0 (see Problem 4.3). Taking the limit $J \to 0$, this implies

$$\Gamma_{\bar{A}}\left[\frac{\delta W_{\bar{A}}}{\delta J}\bigg|_{J \to 0}\right] \leq \Gamma_{\bar{A}}\left[\frac{\delta W_{\bar{A}}}{\delta J}\bigg|_{J_0}\right], \quad \forall J_0, \quad (4.45)$$

which justifies the minimization principle referred to above. The convexity of $W_{\bar{A}}[J]$ is a natural assumption to be made since it is a property of an "ideal gauge fixing" where one configuration per orbit is selected in the space of gauge-field configurations. The gauge-fixed measure associated with this ideal gauge fixing is positive (since the original measure $\mathcal{D}A$ is positive) in which case $W_{\bar{A}}[J]$ should be convex (see Problem 4.4).

In many practical implementations of the gauge fixing, however, the positivity of the gauge-fixed measure is not satisfied. For instance, in the FP approach, positivity is violated due to the presence of the associated determinant which is known not to have a definite sign over the space of gauge-field configurations. The same is true a priori for the Serreau-Tissier approach due to the averaging over Gribov copies with an alternating sign, although it could be that the positivity violation remains moderate if certain copies dominate with respect to others. The situation seems more under control in the GZ approach where the functional integral is, in principle at least, restricted to the first Gribov region defined by the positivity of the FP operator. However, the positivity of the measure is not necessarily guaranteed in practical implementations of the GZ approach, although the violation could again remain small.

In fact, the positivity of the gauge-fixed measure is just a sufficient condition for identifying an order parameter. The order parameter interpretation remains correct provided the positivity violation remains moderate to ensure the convexity of $W_{\bar{A}}[J]$. Since a dynamical mass scale is expected to be generated (and is actually generated in some cases) that damps large enough values of the gauge field, we expect the positivity violation to remain indeed moderate. In the CF approach, this damping factor is there from the beginning, so we expect the positivity requirement to be fulfilled if the CF mass is large enough.

We mention finally that the non-positivity of the integration measure can also have a physical origin, as in the presence of dynamical quarks at finite baryonic density. In this case as well, the very foundation of the background-field method needs to be critically revisited. We shall deal with this matter in Chap. 9.

Background Independence

The other crucial property that enters the justification of the background-field approach is that of the background independence of the free energy. In the FP approach, this is ensured from the general properties derived in Chap. 1. In other approaches, due to the necessary approximations or the degree of modeling of the gauge fixing in the infrared, this property may be violated, thus introducing artifacts in the evaluation of observables that may be tamed only when refining the level of approximation or modeling. For the CF approach, this is discussed more thoroughly in Chap. 6 together with a possible fix in the framework of self-consistent backgrounds. In Chap. 11, we propose a more general fix (which should apply to all continuum methods) based on a novel implementation of the background-field method at finite temperature.

4.4.3   Vanishing Background

We have seen that $\bar{A} = 0$ is a self-consistent background (in the Landau gauge) at any temperature and, thus, a priori a minimum of $\tilde{\Gamma}[\bar{A}]$. At first sight, this may seem in conflict with the ability of $\tilde{\Gamma}[\bar{A}]$ of probing the transition. Indeed, since there should be a minimum of $\tilde{\Gamma}[\bar{A}]$ at the non-center-symmetric configuration $\bar{A} = 0$ at all temperatures, one may wonder how it is possible to conclude to the presence of a center-symmetric phase at low temperatures. In fact, what matters is that, in this range of temperatures, there is another minimum $\bar{A}_s$ located at the center-symmetric configuration $\bar{A}_c$. That $\bar{A} = 0$ is always a minimum of $\tilde{\Gamma}[\bar{A}]$ is expected since, at least in the absence of approximations or modeling, the free energy is background independent and should equally be computable from the Landau gauge. This leads precisely to the fact that $\tilde{\Gamma}[\bar{A} = 0]$, which is nothing but the free energy $\Gamma[A_{\min}[\bar{A} = 0], \bar{A} = 0]$ as computed in the Landau gauge, needs to have the same depth than any other minimum of $\tilde{\Gamma}[\bar{A}]$. As we will see below, in the presence of approximations (which typically break (4.5)) and even though $\bar{A} = 0$ remains a self-consistent background, it is not anymore a minimum of $\tilde{\Gamma}[\bar{A}]$, which, in a sense, lifts any sort of ambiguity regarding the choice of minimum of $\tilde{\Gamma}[\bar{A}]$.

Problems

4.1 Background Gauge Symmetry (★★)
This problem mimics the last problem of the previous chapter in the particular case of background gauge fixings.

(a) Show that the generating functional $\hat{W}_{\bar{A}}[J]$ as defined in Eq. (4.8) obeys the identity

$$\hat{W}_{\bar{A}^U}[J^U] = \hat{W}_{\bar{A}}[J], \quad \forall U \in \mathcal{G}.$$

Tip: use an appropriate change of variables under the functional integral, together with the identity (4.5) and the invariance of both the YM action and the source term. What is the relevance of restricting to gauge transformations that preserve the periodic boundary conditions of the gauge field?

(b) Define $a_{\bar{A}}[J] \equiv -(1/2)\delta \hat{W}_{\bar{A}}[J]/\delta J^t$ and deduce that

$$a_{\bar{A}^U}[J^U] = a_{\bar{A}}^U[J], \quad \forall U \in \mathcal{G}.$$

(c) Define $J_{\bar{A}}[a]$ as the functional inverse of $a_{\bar{A}}[J]$, and provide a clear justification for

$$J_{\bar{A}^U}[a^U] = J_{\bar{A}}^U[a], \quad \forall U \in \mathcal{G}.$$

4.2 Charge Conjugation Invariance and Polyakov Loop (⋆)

Show that in YM theory, $\bar{\ell}$ as defined in Eq. (4.32) coincides with ℓ. *Tip:* use Eq. (4.31) in the definition of the Polyakov loop and the charge conjugation invariance of the YM action.

4.3 Convexity and Effective Action (⋆⋆)

Assume that the generating functional $\hat{W}_{\bar{A}}[J]$ defined in Eq. (4.8) is convex.

(a) Express the fact $\hat{W}_{\bar{A}}[J]$ lies always above any of its tangential planes.
(b) Deduce the inequality (4.44).

4.4 Positivity of the Functional Integral Measure and Convexity (⋆⋆)

Consider two positive real numbers p and q such that $1/p + 1/q = 1$ and two functions f and g such that f^p et g^q are integrable. Then, the function fg is also integrable, and we have the following inequality:

$$\int |fg| \leq \left(\int |f|^p \right)^{1/p} \left(\int |g|^q \right)^{1/q},$$

known as *Hölder's inequality*.

(a) Consider a positive function h, and assume that hf^p and hg^q are integrable. Apply Hölder's inequality to show that

$$\int h|fg| \leq \left(\int h|f|^p \right)^{1/p} \left(\int h|g|^q \right)^{1/q}.$$

We assume in what follows that this result extends to functional integrals with positive measures $\mathcal{D}_+ A$:

$$\int \mathcal{D}_+ A \, | f[A] \, g[A]| \leq \left(\int \mathcal{D}_+ A \, |f[A]|^p \right)^{1/p} \left(\int \mathcal{D}_+ A \, |g[A]|^q \right)^{1/q}.$$

(b) Consider

$$e^{W_{\bar{A}}[J]} \equiv \int \mathcal{D}_{gf} [A; \bar{A}] \, e^{-S_{YM}[A]+J \cdot A},$$

where we assume the gauge-fixed measure $\mathcal{D}_{gf} [A; \bar{A}]$ to be positive and we have introduced the short-hand notation $J \cdot A \equiv \int d^d x \, (J_\mu; A_\mu)$. Given two positive real numbers α_1 and α_2 such that $\alpha_1 + \alpha_2 = 1$, show that

$$W_{\bar{A}}[\alpha_1 J_1 + \alpha_2 J_2] \leq \alpha_1 W_{\bar{A}}[J_1] + \alpha_2 W_{\bar{A}}[J_2],$$

and thus that $W_{\bar{A}}[J]$ is convex with respect to J.

Background-Field Gauges: Weyl Chambers

5

As we have seen previously, within any background-field gauge with self-consistent backgrounds, the physical states of the Yang–Mills system can be interpreted in terms of $\mathcal{G}_0$-orbits in the space of background configurations, where $\mathcal{G}_0$ is the subgroup of periodic gauge transformations within the group $\mathcal{G}$ of twisted gauge transformations. These considerations are crucial to the discussion of the confinement/deconfinement transition since the physical states that are center-symmetric, and thus confining, correspond to the invariant $\mathcal{G}_0$-orbits under center transformations. Preparing the ground for subsequent applications, it is the purpose of the present chapter to recall how these invariant orbits are identified in practice.

We shall do so using a simplifying assumption on the properties of the self-consistent background based on the homogeneity and isotropy of the Yang–Mills system at finite temperature: we shall restrict to constant temporal backgrounds in the diagonal part of the algebra. As a consequence, our analysis of center symmetry will require only a subgroup $\tilde{\mathcal{G}}$ of the twisted gauge transformations, those that preserve this specific form of the background. We shall start by deriving the general form of these transformations in terms of Weyl transformations and what we refer to as the winding transformations. The analysis of these two types of transformations is considerably simplified if one operates a change of color basis from the conventional Cartesian bases to canonical or Cartan–Weyl bases. Since these bases play a major role in subsequent calculations, we spend some time recalling their definition and general properties.

Of particular interest for the analysis of the confinement/deconfinement transition is the subgroup $\tilde{\mathcal{G}}_0$ of periodic transformations within $\tilde{\mathcal{G}}$. It will allow us to interpret the physical states of the system as the $\tilde{\mathcal{G}}_0$-orbits in the restricted space of constant, temporal, and diagonal backgrounds and the center-symmetric states as the center-invariant $\tilde{\mathcal{G}}_0$-orbits. The identification of the latter will then be achieved using the notion of Weyl chambers. In view of its importance when quarks are included back in the analysis, we shall also discuss charge conjugation symmetry using again the Weyl chambers.

© The Author(s), under exclusive license to Springer Nature Switzerland AG 2022
U. Reinosa, *Perturbative Aspects of the Deconfinement Transition*, Lecture Notes in Physics 1006, https://doi.org/10.1007/978-3-031-11375-8_5

Finally, we shall return to the original assumption on the form of the background and critically discuss the homogeneity and isotropy constraints on the possible background configurations.

5.1   Constant Temporal Backgrounds

As we have seen, the actual state of the system is obtained by minimizing the background field effective action $\tilde{\Gamma}[\bar{A}]$ in the space of background-field configurations, or, in a redundancy-free description, by minimizing the background-field effective action $\tilde{\Gamma}[\mathscr{A}]$ in the space of G_0-orbits. As usual, the search for minima can be simplified using the symmetries of the system: any symmetry that is known not to be spontaneously broken should be manifest at the level of the state (Wigner–Weyl realization) and therefore at the level of the minimizing background/orbit.

5.1.1   Homogeneity and Isotropy

In particular, the YM system at non-zero temperature is homogenous and isotropic, that is, it is invariant under both temporal and spatial translations and also under spatial rotations. Because these symmetries are expected not to be broken spontaneously, it makes sense to restrict the search for minima of $\tilde{\Gamma}[\bar{A}]$ to constant temporal backgrounds of the form $\bar{A}_\mu(x) = \bar{A}_0\,\delta_{\mu 0}$.

The discussion is in fact a little bit more subtle due to the general symmetry considerations of the previous chapter. In particular, the homogeneity and isotropy of the state translates into the invariance of the corresponding G_0-orbit but not necessarily into the invariance of the various background configurations that represent the orbit. Thus, even though the orbit associated with a constant temporal background is both homogenous and isotropic, there could be other invariant orbits such that none of the representing backgrounds are invariant, but only invariant modulo a non-trivial element of G_0. We shall return to this interesting possibility at the end of this chapter and assume for the moment (as usually done in the literature) that we can indeed restrict to G_0-orbits that contain constant temporal backgrounds or, more simply, that we can work exclusively with constant temporal backgrounds of the form $\bar{A}_\mu(x) = \bar{A}_0\,\delta_{\mu 0}$.

Moreover, since color rotations are elements of G_0, we can always assume that the constant Lie algebra element $\bar{A}_0$ has been color rotated to the diagonal part of the su(N) algebra.[1] Denoting by $\{iH_j\}$ a basis of the latter, with $[H_j, H_k] = 0$ (see Appendix A), the most general background that we shall consider is then of the form

$$\beta\bar{A}_\mu(x) = ir_j H_j\,\delta_{\mu 0}\,, \tag{5.1}$$

---

[1] This is because the elements of su(N) are anti-Hermitian matrices and are thus diagonalizable by a unitary change of basis.

where the factor $\beta \equiv 1/T$ has been introduced to make the components r_j dimensionless. In fact, the discussion from here on applies more generally to any semi-simple Lie algebra. Such algebras admit maximally commuting sub-algebras, known as Cartan sub-algebras, which any element of the algebra can be rotated into. In what follows, unless specifically stated, we assume that we work in this general context. Denoting by d_C the dimension of the Cartan sub-algebra, the possible background configurations are then represented by a vector $r \in \mathbb{R}^{d_C}$, with $d_C = N - 1$ in the case of SU(N). We refer to this copy of $\mathbb{R}^{d_C}$ as the *restricted background space* and to the vector r as the *restricted background*.

Correspondingly, the analysis of the background effective action $\tilde{\Gamma}[\bar{A}]$ reduces to that of a *background-field effective potential* $\tilde{V}(r)$, and the background-dependent Polyakov loop $\ell[\bar{A}]$ introduced in the previous chapter becomes a function $\ell(r)$ of the restricted background. Similarly, the symmetry transformations should also be restricted to those that preserve the form (5.1) of the background and which we now characterize.

5.1.2 Restricted Twisted Gauge Transformations

The twisted gauge transformations $U(x)$ that preserve the form (5.1) of the background are such that $U(x)\, i r_j H_j \delta_{\mu 0}\, U^\dagger(x) - U(x)\, \beta\, \partial_\mu U^\dagger(x)$ is constant, temporal, and diagonal, for any choice of the restricted background r. We refer to these transformations as *restricted twisted gauge transformations*. They form a subgroup of $\mathcal{G}$, denoted $\tilde{\mathcal{G}}$ in the following, and can be shown to take the form

$$U(\tau) = e^{i \frac{\tau}{\beta} s_j H_j} W\,, \tag{5.2}$$

with W a color rotation that leaves the Cartan sub-algebra globally invariant (see Problem 5.1). Such type of color rotations are known as *Weyl transformations*. The transformations of the form $e^{i \frac{\tau}{\beta} s_j H_j}$ will be referred to as *winding transformations*.

5.1.3 Charge Conjugation

Even though it will play a secondary role for the moment, let us also discuss charge conjugation, which we recall is defined as $A_\mu^C(x) \equiv -A_\mu^t(x)$. In the case of SU(N), since $H_j^t = H_j$, we have $A_\mu^j(x) \to -A_\mu^j(x)$, and charge conjugation leaves the Cartan sub-algebra globally invariant.

We can therefore consider its action directly on the restricted background space, where it corresponds to the geometrical transformation $r \to -r$, that is, the reflection with respect to the origin. The invariance of the potential under charge conjugation reads

$$\tilde{V}(-r) = \tilde{V}(r)\,, \tag{5.3}$$

and the constraint on the background-dependent Polyakov loops is

$$\ell(-r) = \bar{\ell}(r) . \tag{5.4}$$

We shall now study to which geometrical transformations in the restricted background space the Weyl and the winding transformations correspond to.

5.2   Winding and Weyl Transformations

The winding transformation $e^{i\frac{\tau}{\beta}s_j H_j}$ acts on a given gauge-field configuration $A_\mu(x)$ as

$$\beta A_\mu(x) \rightarrow e^{i\frac{\tau}{\beta}s_j H_j} \beta A_\mu(x) e^{-i\frac{\tau}{\beta}s_j H_j} + i s_j H_j \delta_{\mu 0}$$

$$= e^{i\frac{\tau}{\beta}s_j \, \mathrm{ad}_{H_j}} \beta A_\mu(x) + i s_j H_j \delta_{\mu 0} . \tag{5.5}$$

Here, we have introduced $\mathrm{ad}_\theta \equiv [\theta, \_]$, and we have used that $e^\theta X e^{-\theta} = e^{\mathrm{ad}_\theta} X$. The operators ad_{H_j} commute with each other since $[\mathrm{ad}_{H_j}, \mathrm{ad}_{H_k}] = \mathrm{ad}_{[H_j, H_k]} = 0$. To obtain a more explicit representation of the winding transformations, we can try, therefore, to diagonalize simultaneously the action of the operators ad_{H_j} on the Lie algebra. This is precisely the role of the Cartan–Weyl bases. These bases will also be used below for the characterization of the Weyl transformations.[2]

5.2.1   Cartan–Weyl Bases

A Cartan–Weyl basis is a basis $\{i H_j, i E_\alpha\}$ of the Lie algebra which extends the basis $\{i H_j\}$ of the Cartan sub-algebra and which diagonalizes simultaneously the adjoint action of all the elements of the Cartan sub-algebra [168]:

$$\mathrm{ad}_{H_j} H_k \equiv [H_j, H_k] = 0 , \tag{5.6}$$

$$\mathrm{ad}_{H_j} E_\alpha \equiv [H_j, E_\alpha] = \alpha_j E_\alpha . \tag{5.7}$$

The labels α, known as roots, are vectors of $\mathbb{R}^{dc}$ whose components are the α_j. This copy of $\mathbb{R}^{dc}$ is however not the same as the restricted background space defined above. For reasons that will become clear shortly, we refer to it as the *dual (restricted) background space*. It can be shown that roots always appear by pairs $(\alpha, -\alpha)$. They form what is known as the root diagram of the corresponding algebra.

To take a few examples, consider the SU(2) and SU(3) cases. For SU(2), the Cartan sub-algebra is generated by the diagonal Pauli matrix $H_3 = \sigma_3/2$. From the commutation relations $[\sigma_i/2, \sigma_j/2] = i\epsilon_{ijk}\sigma_k/2$, it is easily seen that one possible

---

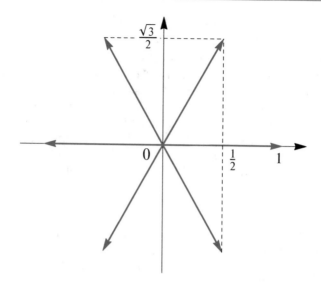

Fig. 5.1 The SU(3) root diagram. The root diagram of SU(2) is also visible if one restricts to the horizontal axis. The other pairs of roots can be interpreted as corresponding to other SU(2) subgroups of SU(3)

Cartan–Weyl basis is given by $\{i\sigma_3/2, i\sigma_+/2, i\sigma_-/2\}$ with $\sigma_\pm = (\sigma_1 \pm i\sigma_2)/\sqrt{2}$ and roots ± 1. In the case of $SU(3)$, the Cartan sub-algebra is generated by the diagonal Gell-Mann matrices $H_3 = \lambda_3/2$ and $H_8 = \lambda_8/2$. Using the expressions for the Gell-Mann matrices, it is easily checked that the roots are $\pm(1, 0)$, $\pm(1/2, \sqrt{3}/2)$, and $\pm(1/2, -\sqrt{3}/2)$ (see Problem 5.2). The corresponding root diagrams are represented in Fig. 5.1. The root diagram of SU(N) is constructed in Appendix A where we show in particular that all roots are of norm unity.

We mention that Cartan–Weyl bases are not bases of the original, real-valued Lie algebra but, rather, of the complexified Lie algebra. Therefore, in such bases, the components of an element X of the original Lie algebra do not need to be real. However, one can always choose the basis elements E_α such that $X = i(X_j H_j + X_\alpha E_\alpha)$, with $X_j \in \mathbb{R}$ and $X_\alpha^* = X_{-\alpha}$ (see Appendix A) for a more detailed discussion.

5.2.2 Winding Transformations

Decomposing the gauge field $A_\mu(x)$ into a Cartan–Weyl basis, it is easily checked that the winding transformation (5.5) writes

$$\beta A_\mu^j(x) \rightarrow \beta A_\mu^j(x) + s_j, \tag{5.8}$$

$$\beta A_\mu^\alpha(x) \rightarrow \beta A_\mu^\alpha(x)\, e^{i\frac{\tau}{\beta} s\cdot\alpha}, \tag{5.9}$$

where $s \cdot \alpha \equiv s_j \alpha_j$ (see Problem 5.3). Then, the components of $\beta A_\mu(x)$ along the Cartan sub-algebra are just shifted by the corresponding components of s, while the orthogonal components[3] are multiplied by the phase $e^{i\frac{\tau}{\beta}s \cdot \alpha}$.

So far, however, we did not implement the fact that winding transformations, like any other twisted gauge transformation, should preserve the β-periodicity of the gauge field along the temporal direction. This implies some constraints on the vector s defining the winding transformation. Since the components of the gauge field along the Cartan sub-algebra are just shifted by a constant, the constraints originate only from the orthogonal components. For these components to remain β-periodic, we need to require the following set of conditions:

$$s \cdot \alpha \in 2\pi \, \mathbb{Z}, \ \forall \alpha. \tag{5.10}$$

In fact, it is sufficient that these conditions be satisfied for a set of roots $\{\alpha^{(j)}\}$ that generate any other root as a linear combination with integer coefficients. In this case, the conditions (5.10) define a lattice, dual to the one generated by the roots. If we introduce a basis $\{\bar{\alpha}^{(k)}\}$ of this dual lattice, such that

$$\alpha^{(j)} \cdot \bar{\alpha}^{(k)} = 2\pi \delta_{jk}, \tag{5.11}$$

the general solution to (5.10) takes then the form

$$s = n_k \bar{\alpha}^{(k)}, \tag{5.12}$$

where $n_k \in \mathbb{Z}$ and a summation over k is implied.

As far as the restricted background r is concerned, because it is diagonal by assumption, it transforms as (5.8), that is, as

$$r \to r + n_k \bar{\alpha}^{(k)}. \tag{5.13}$$

We conclude that, in the restricted background space, the winding transformations are generated by translations along the vectors $\bar{\alpha}^{(k)}$. We can then assume that these vectors belong to the restricted background space. Since the $\alpha^{(j)}$ are in a sense dual to the $\bar{\alpha}^{(k)}$, this explains why we dubbed the copy of $\mathbb{R}^{d_c}$ containing the roots (and in particular the $\alpha^{(j)}$) as the dual (restricted) background space.

In the SU(2) case, the Cartan sub-algebra is one-dimensional, and the restricted background is given by a single real number r. We can take, for instance, $\alpha^{(1)} = 1$ and, therefore, $\bar{\alpha}^{(1)} = 2\pi$. It follows that winding transformations are generated by the shift $r \to r + 2\pi$. In the SU(3) case, the Cartan sub-algebra is two-dimensional, and the restricted background is given by a vector $r = (r_3, r_8)$. We can choose $\alpha^{(1)} = (1/2, -\sqrt{3}/2)$ and $\alpha^{(2)} = (1/2, \sqrt{3}/2)$. Then, winding transformations

---

[3] The reason why we dub these components "orthogonal" is given in Appendix A.

are generated by translations along the vectors $\bar{\alpha}^{(1)} = (2\pi, -2\pi/\sqrt{3})$ and $\bar{\alpha}^{(2)} = (2\pi, 2\pi/\sqrt{3})$.

5.2.3 Weyl Transformations

Let us now consider the Weyl transformations defined as the color rotations that leave the diagonal part of the algebra globally invariant. It is easily seen that these transformations cannot be infinitesimal and, therefore, we need to consider finite transformations of the form $W = e^{\theta}$. The action of such a color transformation on any element X of the algebra is

$$e^{\theta} X e^{-\theta} = e^{\mathrm{ad}_{\theta}} X = \sum_{n=0}^{\infty} \frac{1}{n!} \mathrm{ad}_{\theta}^{n} X \, . \tag{5.14}$$

Suppose now that we take θ of the form $\theta = i(\theta_{\alpha} E_{\alpha} + \theta_{-\alpha} E_{-\alpha}) \equiv w_{\alpha}$, with $\theta_{-\alpha} = \theta_{\alpha}^{*}$ and where no summation over α is implied, hence the notation w_{α}. It can be shown that

$$e^{\mathrm{ad}_{w_{\alpha}}} H_{j} = H_{j} + \left(\cos\left(\sqrt{2\alpha^{2}} |\theta_{\alpha}|\right) - 1 \right) \frac{\alpha_{k} H_{k}}{\alpha^{2}} \alpha_{j}$$
$$- \frac{i\alpha_{j}}{\sqrt{2\alpha^{2}} |\theta_{\alpha}|} \sin\left(\sqrt{2\alpha^{2}} |\theta_{\alpha}|\right) (\theta_{\alpha} E_{\alpha} - \theta_{-\alpha} E_{-\alpha}) \, , \tag{5.15}$$

(see Problem 5.4). Choosing $|\theta_{\alpha}| = \pi/\sqrt{2\alpha^{2}}$, we can enforce the transformation to leave the diagonal part of the algebra globally invariant:

$$H_{j} \to H_{j} - 2 \frac{\alpha_{k} H_{k}}{\alpha^{2}} \alpha_{j} \, . \tag{5.16}$$

Considering this as a passive transformation, the restricted background r transforms as

$$r \to r - 2 \frac{\alpha \cdot r}{\alpha^{2}} \alpha \, , \tag{5.17}$$

which is indeed a Weyl transformation as defined in [168]. In the restricted background space $\mathbb{R}^{dc}$, this corresponds to a reflection with respect to a hyperplane orthogonal to the root α and containing the origin.[4]

---

[4] In Eq. (5.17), it seems that we are shifting a vector of the restricted space (r) by a vector of the dual space (α). However, the root α enters this formula only through its direction $\alpha/\|\alpha\|$, the magnitude of the shift being proportional to the norm of r.

For completeness, we mention that there exist other color rotations that leave the Cartan sub-algebra invariant, in a trivial way. Those correspond to rotations $e^{i\theta_j H_j}$ generated by the elements of the Cartan sub-algebra itself. Those are of no interest for the present discussion since they do not lead to geometrical transformations of the restricted background. They will however play an important role in Chap. 7 as we derive the Feynman rules using Cartan–Weyl bases.

5.2.4 Summary

From the above considerations, we derive the following invariance properties of the background-field effective potential:

$$\tilde{V}(r + \bar{\alpha}^{(k)}) = \tilde{V}(r), \quad \forall \bar{\alpha}^{(k)}, \tag{5.18}$$

as well as

$$\tilde{V}\left(r - 2\frac{\alpha \cdot r}{\alpha^2}\alpha\right) = \tilde{V}(r), \quad \forall \alpha. \tag{5.19}$$

Correspondingly,

$$\ell(r + \bar{\alpha}^{(k)}) = e^{i\varphi(k)}\ell(r), \quad \forall \bar{\alpha}^{(k)}, \tag{5.20}$$

for some $\varphi(k)$ to be fixed in the next section, and

$$\ell\left(r - 2\frac{\alpha \cdot r}{\alpha^2}\alpha\right) = \ell(r), \quad \forall \alpha. \tag{5.21}$$

5.3   Weyl Chambers and Symmetries

In the previous section, we could interpret the various elements of $\tilde{G}$ as geometrical transformations in the restricted background space $\mathbb{R}^{d_c}$. Of particular importance is the subgroup $\tilde{G}_0$ of periodic transformations within $\tilde{G}$ since the physical states are interpreted as the corresponding $\tilde{G}_0$-orbits, and the center-symmetric states correspond to the center-invariant $\tilde{G}_0$-orbits. We shall now characterize the elements of $\tilde{G}_0$ and then identify the center-invariant $\tilde{G}_0$-orbits using the concept of Weyl chambers. In fact, we know already part of the elements of $\tilde{G}_0$ since the Weyl transformations are trivially periodic. The other generators of $\tilde{G}_0$ are the periodic winding transformations, which we now characterize.

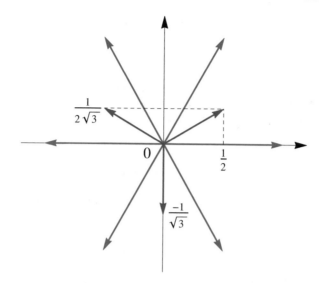

Fig. 5.2 Weight diagram for the defining representation of SU(3), together with the corresponding root diagram (which is nothing but the non-zero weight diagram of the adjoint representation)

5.3.1 Periodic Winding Transformations

To identify the periodic winding transformations, we assume that the H_j can all be diagonalized.[5] Since they commute with each other, they can in fact be diagonalized simultaneously,

$$H_j|\rho\rangle = \rho_j|\rho\rangle,$$ (5.22)

where the labels ρ are called the weights of the defining representation or *defining weights* for short and are just a convenient way to gather all eigenvalues ρ_j of a given eigenvector $|\rho\rangle$. The defining weights are again vectors of $\mathbb{R}^{dc}$ and can, therefore, be represented together with the root diagram. The reason for representing them on the same copy of $\mathbb{R}^{dc}$ appears more clearly once quarks are introduced back in the analysis [98]. In the SU(2) case, the weights of the defining representation are $\pm 1/2$. In the SU(3) case, we find $(1/2, 1/(2\sqrt{3}))$, $(-1/2, 1/(2\sqrt{3}))$, and $(0, -1/\sqrt{3})$ (see Fig. 5.2 for a representation of the corresponding weight diagram and Appendix A for the construction of the defining weights for any value of N).

In terms of the weights, a general winding transformation writes

$$e^{i\frac{\tau}{\beta}s_j H_j} = \sum_{\rho} e^{i\frac{\tau}{\beta}s\cdot\rho}|\rho\rangle\langle\rho|,$$ (5.23)

---

[5] In the case of SU(N), we have chosen the H_j already in an explicitly diagonalized form.

where $s \cdot \rho \equiv s_j \rho_j$. Periodic winding transformations are obtained by requiring that

$$s \cdot \rho \in 2\pi \, \mathbb{Z} \, , \forall \rho. \tag{5.24}$$

The solution to (5.24) can then be found by repeating the previous arguments used to solve (5.10). We first introduce a set of defining weights or their opposites[6] $\{\rho^{(j)}\}$ that generate any other weight through linear combination with integer coefficients. The general solution to (5.24) is then

$$s = n_k \bar{\rho}^{(k)} \, , \tag{5.25}$$

with $n_k \in \mathbb{Z}$ and where $\{\bar{\rho}^{(k)}\}$ is a basis of the lattice dual to the one generated by the weights:

$$\rho^{(j)} \cdot \bar{\rho}^{(k)} = 2\pi \, \delta_{jk} \, . \tag{5.26}$$

It follows that, in the restricted background space, periodic winding transformations are generated by translations along the vectors $\bar{\rho}^{(k)}$.

We mention that, because periodic winding transformations are just particular winding transformations, the lattice generated by $\{\bar{\rho}^{(j)}\}$ needs to be a sub-lattice of the one generated by $\{\bar{\alpha}^{(j)}\}$. This can be understood from the fact that the adjoint representation is generally obtained by decomposing the tensor product of the defining representation and the associated contragredient representation (whose weights are opposite to those of the defining representation). We can thus construct the eigenstates of ad_{H_j} in terms of the eigenstates of H_j, and we find that the roots can always be written as differences of weights of the defining representation (an explicit proof is given in Appendix A for the SU(N) case). The lattice generated by $\{\alpha^{(j)}\}$ is then a sub-lattice of the one generated by $\{\rho^{(j)}\}$, which implies the opposite relation for the corresponding dual lattices.

In the SU(2) case, we can choose $\rho^{(1)} = 1/2$ and therefore $\bar{\rho}^{(1)} = 4\pi$. Thus, the periodic winding transformations are generated by the shift $r \to r + 4\pi$. In the SU(3) case, we can choose $\rho^{(1)} = (1/2, -1/(2\sqrt{3}))$ and $\rho^{(2)} = (1/2, 1/(2\sqrt{3}))$. Then, the periodic winding transformations are generated by translations along the vectors $\bar{\rho}^{(1)} = (2\pi, -2\pi\sqrt{3})$ and $\bar{\rho}^{(2)} = (2\pi, 2\pi\sqrt{3})$. We note that, in both cases, we could choose the various bases such that $\bar{\alpha}^{(k)} = 4\pi \rho^{(k)}$ and $\bar{\rho}^{(k)} = 4\pi \alpha^{(k)}$ (see the previous section for the values of $\alpha^{(k)}$ and $\bar{\alpha}^{(k)}$). This property generalizes to higher values of N (see Appendix A) and boils down to the fact that the roots and the defining weights form dual lattices in the sense that

$$\alpha^{(j)} \cdot \rho^{(k)} = \frac{\delta_{jk}}{2} \, . \tag{5.27}$$

---

[6] The opposite of a defining weight is a weight of the corresponding contragredient representation.

As a consequence, if we define the *reduced background* as $\bar{r} \equiv r/4\pi$, we can identify the reduced and the dual background spaces and represent all the above geometrical transformations together with the root and weight diagrams: Weyl transformations correspond to reflections with respect to hyperplanes orthogonal to roots and containing the origin, periodic winding transformations are generated by translations along the roots, whereas generic winding transformations are generated by translations along the weights of the defining and contragredient representations. In the next section, for simplicity, we define the Weyl chambers in the reduced background space. However, when representing the latter graphically, we do so in the restricted background space by applying a factor 4π.

We mention finally that, according to Eq. (5.23), the center element associated with a winding transformation of parameter s is given by $e^{is\cdot\rho}$.[7] In particular, this fixes the phase in Eq. (5.20) to $\varphi(k) = \bar{\alpha}^{(k)} \cdot \rho = 4\pi\rho^{(k)} \cdot \rho$, and we show in Appendix A that, in the case of SU(N), this is always $\mp 2\pi/N$, with the sign depending on whether $\rho^{(k)}$ is a weight or the opposite of a weight.

5.3.2 Weyl Chambers and Invariant States

We are now ready to characterize the center-invariant $\tilde{\mathcal{G}}_0$-orbits, that is, the center-symmetric states, in the reduced background space $\mathbb{R}^{d_C}$. One possible strategy would be to work directly with the orbits as collections of background configurations and, by trial and error, find which orbits are globally invariant under center transformations. In Fig. 5.3, we have tried to illustrate how painful such a procedure could be, in the SU(3) case. The difficulty here is that, in order to explore the various orbits, one varies a given orbit locally (by changing any of the backgrounds that represent the orbit), with little control on how the orbit changes globally and therefore little chance of finding the center-invariant orbits. We now introduce a simpler strategy based on the notion of Weyl chambers which allows us to treat the orbits as a whole.

The idea is to use the geometrical interpretation of the transformations of $\tilde{\mathcal{G}}_0$ in the reduced background space to subdivide the latter into cells which are connected to each other by the transformations of $\tilde{\mathcal{G}}_0$ and which are, therefore, physically equivalent. These are known as *Weyl chambers*[8] whose points can be seen as representing the various possible $\tilde{\mathcal{G}}_0$-orbits and thus the various possible states of the system.

Now, as we have seen in the previous chapter, the action of a center transformation $\mathcal{U} \in \mathcal{G}/\mathcal{G}_0 \simeq \tilde{\mathcal{G}}/\tilde{\mathcal{G}}_0$ can be defined directly on the $\mathcal{G}_0$-orbits and similarly on the $\tilde{\mathcal{G}}_0$-orbits. This action should then appear as a geometrical transformation

---

[7] This center element does not depend on the choice of ρ since $s = n_k \bar{\alpha}^{(k)}$, and the difference of two weights is always a root, which implies that $s \cdot \rho - s \cdot \rho'$ is a multiple of 2π.

[8] Here the definition of the Weyl chambers differs substantially from the one taken by mathematicians, where it corresponds to subdivisions associated with Weyl transformations only [168].

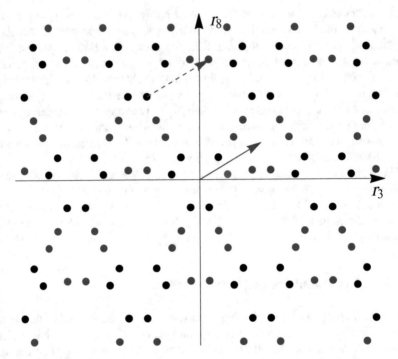

Fig. 5.3 A given $\tilde{\mathcal{G}}_0$-orbit (black dots) and its transformation (blue dots) under a non-periodic winding transformation (translation along the blue vector). To vary the original orbit, one varies locally one of its representing backgrounds (pink dot) with little control on how the orbit changes globally, making it difficult to identify the center-invariant orbits

of a given Weyl chamber into itself. To see how this is achieved in practice, one first chooses any representative $U \in \mathcal{G}$ of the center transformation $\mathcal{U}$ and applies the transformation to the chosen Weyl chamber. This typically moves the Weyl chamber away from its original location. However, without altering the physical interpretation, we can bring the Weyl chamber back to its original location by using transformations of $\tilde{\mathcal{G}}_0$. In doing so, one defines the action of $\tilde{\mathcal{G}}$ directly on the chosen Weyl chamber. Moreover, two transformations of $\tilde{\mathcal{G}}$ that correspond to the same center transformation in $\tilde{\mathcal{G}}/\tilde{\mathcal{G}}_0$ lead to the same geometrical transformation of the Weyl chamber, thereby defining the action of the center symmetry group $\tilde{\mathcal{G}}/\tilde{\mathcal{G}}_0$ on the chosen Weyl chamber.

In summary, center transformations correspond to certain geometrical transformations of the Weyl chambers into themselves, and the symmetric states appear as the fixed points of the Weyl chambers under these geometrical transformations. Similar considerations apply to other physical transformation, in particular to charge conjugation.

5.3.3 Explicit Construction for SU(N)

In the case of SU(N), we have seen that, in the reduced background space, the generators of $\tilde{G}_0$ are the translations by a root and the reflections with respect to hyperplanes orthogonal to a root and containing the origin. In order to identify the Weyl chambers, it is convenient to find a generating set of $\tilde{G}_0$ made only of reflections. Such a set is given by all the reflections with respect to hyperplanes orthogonal to a root and translated by any multiple of half that root. Indeed, it is easily checked that these reflections are elements of $\tilde{G}_0$ and that the combination of such two adjacent reflections generates the translation by the corresponding root. The benefit of this generating set is that it allows to identify the Weyl chambers as the regions delimitated by the corresponding network of hyperplanes. Moreover, the reflections with respect to the facets of the so-obtained Weyl chambers allow us to transform them into one another without altering the physical interpretation.

To further understand the structure of the Weyl chambers, consider first the network of hyperplanes generated by the roots of a generating basis $\{\alpha^{(j)}\}$. It is easily seen (use a bi-dimensional example first) that the intersections of these hyperplanes define a lattice generated by a certain basis $\{x^{(j)}\}$ and whose elements obey the constraints

$$x^{(j)} \cdot \alpha^{(k)} = 0, \text{ for } k \neq j \quad \text{and} \quad (x^{(j)} - \alpha^{(j)}/2) \cdot \alpha^{(j)} = 0. \tag{5.28}$$

Using that the roots are all of norm unity, these constraints read more simply

$$x^{(j)} \cdot \alpha^{(k)} = \frac{\delta_{jk}}{2}, \tag{5.29}$$

and we see that they are solved by the defining weights (see Eq. (5.27)). The SU(N) Weyl chambers are then sub-pavings of the parallelepipeds generated by the weights $\rho^{(j)}$. To finally reach the Weyl chambers, we need to see how the remaining hyperplanes (those associated with roots other than the $\alpha^{(j)}$) further divide these parallelepipeds. We shall do this here in the cases $N = 2$ and $N = 3$ and leave the discussion of $N = 4$ for Chap. 8.

In the case of SU(2), there is only one family of hyperplanes, associated with $\alpha^{(1)} = 1$. In the reduced background space, the Weyl chambers are then directly given by the intervals generated by $\rho^{(1)} = 1/2$, that is, $[k/2, (k + 1)/2]$, with $k \in \mathbb{Z}$. It will be enough to work in the interval $[0, 1/2]$ which we refer to as the *fundamental Weyl chamber*. For SU(3), there are three families of hyperplanes associated with $\alpha^{(1)} = (1/2, -\sqrt{3}/2)$, $\alpha^{(2)} = (1/2, \sqrt{3}/2)$ and $\alpha^{(3)} = -\alpha^{(1)} - \alpha^{(2)}$. The fundamental parallelepiped associated with the first two is given in terms of $\rho^{(1)} = (1/2, -1/(2\sqrt{3}))$ and $\rho^{(2)} = (1/2, 1/(2\sqrt{3}))$. It is further divided by the family of hyperplanes associated with $\alpha^{(3)}$. One finds eventually that the Weyl chambers are equilateral triangles (see Fig. 5.4), the fundamental Weyl chamber being defined as the one with vertices $(0, 0)$, $(1/2, -1/(2\sqrt{3}))$, and $(1/2, 1/(2\sqrt{3}))$.

We can now locate the center-symmetric states. We have seen that translations along the $\rho^{(k)}$ (the edges of the fundamental parallelepiped) correspond to elemen-

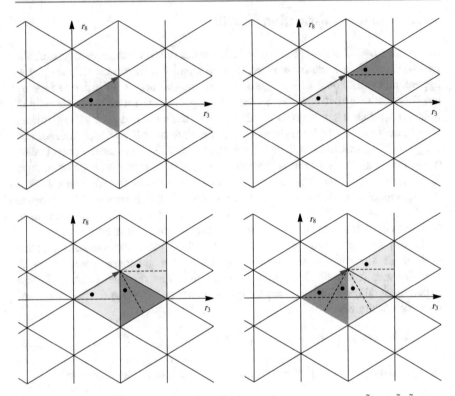

Fig. 5.4 SU(3) Weyl chambers and their transformations under the action of $\tilde{G}$ and $\tilde{G}/\tilde{G}_0$. The first figure shows the fundamental Weyl chamber together with one of its axes and an arbitrary point (representing an arbitrary state) to ease orientation as we transform the Weyl chamber. In each figure, the darker color indicates the actual position of the Weyl chamber, while the light colors indicate previous positions of the Weyl chamber. The blue arrow represents one of the two non-trivial center transformations in the case of SU(3)

tary center transformations with a phase $\varphi(k) = \mp 2\pi/N$ with the sign depending on whether $\rho^{(k)}$ is a weight or the opposite of a weight. Moreover, reflections with respect to the facets of the Weyl chamber are elements of $\tilde{G}_0$, by construction, and allow one to move the Weyl chamber around without altering the physical interpretation. In the SU(2) case, under the translation by $\rho^{(1)}$, the fundamental Weyl chamber $[0, 1/2]$ is transformed into $[1/2, 1]$. The latter can be brought back to its original location by means of a reflection with respect to $1/2$. As a result, we find that center transformations act on the fundamental Weyl chamber as $\bar{r} \to 1/2 - \bar{r}$. There is only one fixed point in the fundamental Weyl chamber, one confining state, $\bar{r} = 1/4$ (or $r = \pi$). In the SU(3) case, the fundamental equilateral triangle is such that the two edges connected to the origin correspond, respectively, to a weight and the opposite of a weight. Therefore, translations along these edges represent the two non-trivial elements of $\mathbb{Z}_3$. After any of these translations (see Fig. 5.4), we need two reflections with respect to the edges to bring the fundamental triangle back to its original location, resulting in a rotation by an angle $\pm 2\pi/3$. Again, there is only one fixed point under any of these transformations, the center of the triangle $\bar{r} = (1/3, 0)$ (or $r = (4\pi/3, 0)$). In order to analyze the deconfinement transition, we need therefore to evaluate the background-field effective potential on the fundamental Weyl chamber and monitor the position of its minimum with respect to the center-symmetric point, as the temperature is varied. We shall do so in various situations starting from the next chapter.

Similarly, we can locate the charge conjugation invariant states. In the SU(2) case, charge conjugation acts on the restricted backgrounds just as a Weyl reflection (see Problem 5.5). There is therefore no constraint from charge conjugation invariance in that case. In the SU(3) case, applying the transformation $r \to -r$ to the fundamental Weyl chamber and bringing it back to its original location by means of an element of $\tilde{\mathcal{G}}_0$, one finds that the physical charge conjugation is realized as a reflection of the fundamental Weyl chamber about the $r_8 = 0$ axis (see Fig. 5.5). In other words, charge conjugation invariance of the YM system implies that one

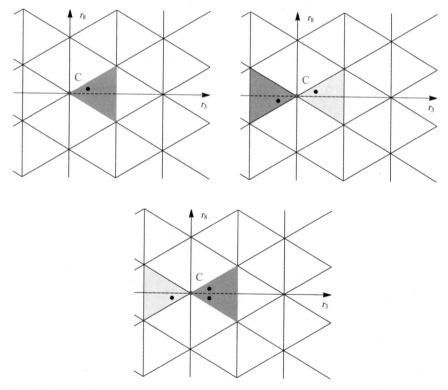

Fig. 5.5 SU(3) Weyl chambers and their transformations under charge conjugation. The first figure shows the fundamental Weyl chamber together with one of its axes and an arbitrary point (representing an arbitrary state) to ease orientation as we transform the Weyl chamber. In each figure, the darker color indicates the actual position of the Weyl chamber, while the light colors indicate previous positions of the Weyl chamber

can restrict the analysis to restricted backgrounds of the form $r = (r_3, 0)$ in the fundamental Weyl chamber. We shall see in Chap. 6 that this is indeed what is found from the minimization of the background-field effective potential. This will remain true in the presence of quarks with a vanishing chemical potential. However, for a non-vanishing chemical potential, due to the explicit breaking of charge conjugation invariance, we expect the restricted background to develop a non-zero r_8 component. We shall see precisely how this happens in Chaps. 9 and 10.

Appendix: Euclidean Spacetime Symmetries

To conclude this chapter, let us critically revisit the original assumption of a constant and temporal background. Recall that the reason behind this choice was that such backgrounds, and in turn, the corresponding G_0-orbits, are invariant both under temporal and spatial translations and under spatial rotations, in line with the homogeneity and isotropy of the YM system at finite temperature.

However, as we also pointed out, there could be other states compatible with homogeneity and isotropy, such that none of the representing backgrounds are invariant under translations or rotations, but only invariant modulo non-trivial true gauge transformations. Using Eq. (4.30), such backgrounds would be such that:

Homogeneity

$$\text{(Hom)} \qquad \forall x, u \in [0, \beta] \times \mathbb{R}^3, \ \exists U_u(x) \in G_0, \ \bar{A}_\mu(x + u) = \bar{A}_\mu^{U_u}(x) \tag{5.30}$$

and

Isotropy

$$\text{(Iso)} \qquad \forall x \in [0, \beta] \times \mathbb{R}^3, \forall R \in SO(3), \ \exists U_R(x) \in G_0, \ \begin{cases} \bar{A}_0(R^{-1}x) = \bar{A}_0^{U_R}(x) \\ R_{ij}\bar{A}_j(R^{-1}x) = \bar{A}_i^{U_R}(x) \end{cases} \tag{5.31}$$

In writing these conditions, we have introduced the notation $R x \equiv (\tau, R\mathbf{x})$. It is trivially verified that any background in the same orbit than an explicitly translation and rotation invariant background obeys these properties. Here, however, we wonder about the existence of backgrounds that obey (Hom) and (Iso) but whose orbit does not contain any constant temporal representative.

Problem 5.6 investigates the existence of such backgrounds, in the SU(2) case. In addition to the orbits containing constant temporal backgrounds, we find orbits that, while obeying (Hom) and (Sym), do not contain any representative that is explicitly translation and rotation invariant. One example is the orbit stemming from the configuration

$$\bar{A}_0(x) = 0, \quad \bar{A}_i^a(x) = \bar{A}\,\delta_i^a, \tag{5.32}$$

with $\bar{A} \neq 0$. This configuration obeys (Hom) in a trivial way. It also obeys condition (Iso) because, due to the locking δ_i^a of space and color indices, any spatial rotation can be absorbed back in the form of a color rotation. Moreover, the orbit stemming from this configuration does not contain any constant and temporal background representative because the associated field-strength tensor $F_{ij}^a = \bar{A}^2 \varepsilon_{aij}$ does not vanish. Aside from this particular example, the other possible solutions that we find are characterized by $\bar{A}_0 = 0$. Thus, although one should in principle evaluate the background-field effective action on these configurations and compare it to the result obtained with constant temporal backgrounds, we expect these exotic configurations (in particular the space-color locked configurations) to play no role in the confining phase since they are not center-invariant.

It would be interesting to extend the analysis to the SU(3) case and see if other types of configurations could be relevant in the confining phase. Similar considerations should be considered in the vacuum with regard to the Euclidean version of Lorentz invariance.

Problems

5.1 Restricted Twisted Gauge Transformations (⋆)
Consider a restricted gauge transformation as defined in Sect. 5.1.2.

(a) Show that it can only depend on the Euclidean time τ, and deduce that it obeys a first-order differential equation.
(b) Integrate formally this differential equation to show that any restricted twisted gauge transformation takes the form

$$U(\tau) = e^{i\frac{\tau}{\beta}s_j H_j}\,W,$$

for some vector of components s_j and some color rotation W that leaves the Cartan sub-algebra (globally) invariant.

5.2 SU(3) Root Diagram (⋆)
Construct the roots of SU(3). *Tip:* start by identifying SU(2) sub-algebras in terms of the Gell-Mann matrices.

5.3 Winding Transformations (⋆)

Show that the winding transformation $e^{i\frac{\tau}{\beta}s_j H_j}$ acts on the gauge field $A_\mu(x)$ as in (5.8)–(5.9). *Tip:* this is a simple and direct application of Eq. (5.7).

5.4 Weyl Rotations (⋆⋆)

Consider the Lie algebra element $w_\alpha = i(\theta_\alpha E_\alpha + \theta_{-\alpha} E_{-\alpha})$ (no summation over α implied), with $\theta_{-\alpha} = \theta_\alpha^*$ and where E_α is defined in Eq. (5.7).

(a) Show that

$$\mathrm{ad}_{w_\alpha}^{2p}(H_j) = i^{2p}\left(2|\theta_\alpha|^2\alpha^2\right)^p \frac{\alpha_k H_k}{\alpha^2}\alpha_j\,,$$

for $p > 0$, and

$$\mathrm{ad}_{w_\alpha}^{2p+1}(H_j) = -i^{2p+1}\left(2|\theta_\alpha|^2\alpha^2\right)^p(\theta_\alpha E_\alpha - \theta_{-\alpha}E_{-\alpha})\alpha_j\,,$$

for $p \geq 0$. Tip: proceed recursively using Eq. (5.7) and $[E_\alpha, -E_\alpha] = \alpha_k H_k$.

(b) Deduce that

$$e^{\mathrm{ad}_{w_\alpha}}H_j = H_j + \left(\cos\left(\sqrt{2\alpha^2}|\theta_\alpha|\right) - 1\right)\frac{\alpha_k H_k}{\alpha^2}\alpha_j$$

$$-\frac{i\alpha_j}{\sqrt{2\alpha^2}|\theta_\alpha|}\sin\left(\sqrt{2\alpha^2}|\theta_\alpha|\right)(\theta_\alpha E_\alpha - \theta_{-\alpha}E_{-\alpha})\,.$$

5.5 SU(2) Charge Conjugation as a Weyl Rotation (⋆)

Show that in the case of SU(2), charge conjugation is nothing but a Weyl rotation. *Tip:* construct a color rotation that changes the sign of the first and third color components.

5.6 Homogeneity and Isotropy Modulo Gauge Transformations (⋆⋆⋆)

We look for SU(2) background configurations obeying the homogeneity and isotropy constraints (Hom) and (Iso) given in the Appendix above.

(a) Show that the conditions (Hom) and (Iso) are gauge-invariant, that is, if $\bar{A}_\mu(x)$ is a configuration obeying these conditions, then for any periodic gauge transformation $U \in \mathcal{G}_0$, the transformed configuration $\bar{A}_\mu^U(x)$ also obeys these properties. Without loss of generality, one can then look for such configurations in an appropriate gauge.

(b) Express the consequences of the condition (Hom) on the field-strength tensor $\bar{F}_{\mu\nu}(x)$. Deduce that $\bar{F}_{\mu\nu}(x) = U(x)\bar{F}_{\mu\nu}(0)U^\dagger(x)$ for some SU(2) matrix field

$U(x)$. The question is now whether $U(x)$ can be gauged away. This is not obvious since $U(x)$ is not necessarily an element of $\mathcal{G}_0$.

(c) Use the β-periodicity of $\bar{F}_{\mu\nu}(x)$ to write a condition on $U(x)$ that also depends on $\bar{F}_{\mu\nu}(0)$. Assuming that two of the Lorentz components of $\bar{F}_{\mu\nu}(0)$ lie in different color directions, show that the constraint on $U(x)$ rewrites $U(\tau + \beta, \mathbf{x}) = U(\tau, \mathbf{x})$ and thus that $U(x)$ can indeed be gauged away, that is, one can assume $\bar{F}_{\mu\nu}(x) = \bar{F}_{\mu\nu}(0)$. We shall concentrate on this case for the moment and leave the situation in which the various components $\bar{F}_{\mu\nu}(0)$ lie at most along one color direction for the next problem.

(d) Proceed similarly using (Iso), and show that all the Lorentz components $\bar{F}_{\mu\nu}(0)$ are invariant under a real, three-dimensional representation $\mathcal{R}(R)$ of the rotation group which can be either the trivial representation $\mathcal{R}(R) = \mathbb{1}$ or the adjoint representation. Show that the first case leads to $\bar{F}_{\mu\nu}(0) = 0$. In the other case, show that $\bar{F}^a_{0i}(x) = \alpha\delta_{ia}$ and $\bar{F}^a_{ij}(x) = \beta\epsilon_{aij}$ for some coefficients α and β which do not vanish simultaneously.

(f) We now investigate whether such a configuration of $\bar{F}_{\mu\nu}(x)$ derives from a background $\bar{A}_\mu(x)$. Show that the Bianchi identities for $\bar{F}_{\mu\nu}$ rewrite

$$0 = \beta\bar{A}^b_0(\delta_{ai}\delta_{bj} - \delta_{aj}\delta_{bi}) + \alpha\bar{A}^b_i\varepsilon_{abj} - \alpha\bar{A}^b_j\varepsilon_{abi},$$

$$0 = \beta\left((\bar{A}^k_i - \bar{A}^i_k)\delta_{aj} + (\bar{A}^i_j - \bar{A}^j_i)\delta_{ak} + (\bar{A}^j_k - \bar{A}^k_j)\delta_{ai}\right).$$

From the first one, deduce that $\beta\bar{A}^a_0 = 0$ and $\alpha\bar{A}^a_b = 0$. From the second one, deduce that $0 = \beta(\bar{A}^a_b - \bar{A}^b_a)$. We note that one cannot have α and β simultaneously non-zero and neither α and β simultaneously 0.

(g) In the case where $\beta = 0$ and $\alpha \neq 0$, show that $\bar{A}^a_\mu(x) = \alpha x^a + \gamma^a(\tau)$ and that such a configuration can in no way obey (Hom). *Tip:* consider an infinitesimal spatial translation.

(h) In the case where $\alpha = 0$ and $\beta \neq 0$, show that $\bar{A}^a_0 = 0$ and $\bar{A}^a_i = \bar{A}^i_a$ with $\bar{A}^a_i$ not depending on τ and obeying the equation

$$\beta\varepsilon_{aij} = \partial_i\bar{A}^a_j - \partial_j\bar{A}^a_i + \varepsilon^{abc}\bar{A}^b_i\bar{A}^c_j.$$

Show that one solution is provided by $\bar{A}^a_i = \sqrt{\beta}\delta^a_i$ and that it satisfies the conditions (Hom) and (Iso).

5.7 Degenerate Case (★★★)

We now investigate the case where the Lorentz components $\bar{F}_{\mu\nu}(0)$ are all aligned along the same color direction, that is, $\bar{F}^a_{\mu\nu}(0) = f_{\mu\nu}n^a$.

(a) Show that $f_{\mu\nu}$ is invariant under color rotations and thus that it necessarily vanishes.

(b) We are thus then left with the pure gauge case $\bar{F}_{\mu\nu}(x) = 0$ or, equivalently, $\bar{A}_\mu(x) = -U(x)\partial_\mu U^\dagger(x)$. Neglecting first the question of boundary conditions, show that these configurations obey the conditions (Hom) and (Iso) for certain transformations U_u and U_R that can be expressed in terms of U.

(c) Express the β-periodicity of U_u and U_R, and deduce that $U^\dagger(\tau + \beta, \mathbf{x}) U(\tau, \mathbf{x})$ is a constant SU(2) matrix V.

(d) Introduce M (constant SU(N) matrix) and r_j such that $M^\dagger V^\dagger M = e^{ir_j H_j}$, and consider

$$W(x) = U(x) M e^{-i\frac{\tau}{\beta} r_j H_j} .$$

Show that W is β-periodic and such that $\beta \bar{A}_\mu^{W^\dagger} = ir_j H_j \delta_{\mu 0}$. In other words, this shows that the most general periodic pure gauge configurations compatible with the properties (Hom) and (Iso) belong to the same $\mathcal{G}_0$-orbits as constant temporal backgrounds.

Yang–Mills Deconfinement Transition from the Curci–Ferrari Model at Leading Order

6

After these various introductory chapters on the use of background-field techniques at finite temperature,[1] we can at last put all the developed machinery into work and start investigating the confinement/deconfinement transition from a perturbative perspective. We shall do so using the Landau–DeWitt gauge (the background extension of the Landau gauge) and its infrared completion in the form of a background-extended Curci–Ferrari model. Indeed, given that the original Curci–Ferrari model opens a perturbative window at low energy in the Landau gauge at vanishing temperature, it is legitimate to expect that the corresponding background extension shares the same properties at finite temperature.

In fact, it is the purpose of the rest of the manuscript to review how many features of the QCD phase diagram, usually obtained from non-perturbative methods, can be captured from a simple perturbative expansion within the background extension of the Curci–Ferrari model. In this chapter, we do so at one-loop order, in the case of the pure Yang–Mills theory.

After introducing some generalities about the Landau–DeWitt gauge, we use the properties of the Cartan–Weyl bases introduced earlier to derive a general expression for the one-loop background-field effective potential, valid for any gauge group, and which obeys all the symmetries identified in Chap. 5. We then use this expression to investigate the confinement/deconfinement transition in the SU(2) and SU(3) cases. In particular, we compare our results for the corresponding transition temperatures to both lattice and non-perturbative continuum results. Finally, we discuss some open questions concerning the thermodynamical properties which, in fact, go beyond the mere framework of the Curci–Ferrari model and affect most, if not all, present continuum approaches.

---

[1] Some additional features will be discussed in Chap. 9 for the case of finite density. Moreover, a recently developed, novel implementation of background-field techniques at finite temperature will be discussed in Chap. 11.

© The Author(s), under exclusive license to Springer Nature Switzerland AG 2022
U. Reinosa, *Perturbative Aspects of the Deconfinement Transition*, Lecture Notes
in Physics 1006, https://doi.org/10.1007/978-3-031-11375-8_6

6.1   Landau–DeWitt Gauge

Since the Landau–DeWitt gauge has only been mentioned formally so far, let us derive the associated gauge-fixed action, together with the relevant CF completion.

6.1.1 Faddeev–Popov Action

The Landau–DeWitt gauge corresponds to the following choice of gauge-fixing functional

$$F^a[A] = \bar{D}_\mu (A^a_\mu - \bar{A}^a_\mu),$$ (6.1)

with $\bar{D}_\mu \varphi^a \equiv \partial_\mu \varphi^a + f^{abc} \bar{A}^b_\mu \varphi^c$. Applying the general formula derived in Chap. 1, we find that the corresponding FP gauge-fixing term reads

$$\delta S_{FP} = \int d^d x \left\{ \bar{D}_\mu \bar{c}^a(x)\, D_\mu c^a(x) + i h^a(x)\, \bar{D}_\mu a^a_\mu(x) \right\},$$ (6.2)

with $a^a_\mu(x) \equiv A^a_\mu(x) - \bar{A}^a_\mu(x)$ and where, for convenience, we have used an integration by parts in the ghost term. Of course, all the general properties derived in Chap. 1 apply here. In particular, the gauge-fixed action is BRST invariant, and both the partition function and the expectation value of any gauge-invariant observable are independent of the choice of the background $\bar{A}^a_\mu(x)$, seen as an infinite collection of gauge-fixing parameters.

The gauge-fixing term (6.2) just looks like the corresponding gauge-fixing term in the Landau gauge upon making the replacement $\partial_\mu \to \bar{D}_\mu$ and $A^a_\mu \to a^a_\mu$. Moreover, it is easily checked that

$$F^a_{\mu\nu} = \bar{F}^a_{\mu\nu} + \bar{D}_\mu a^a_\nu - \bar{D}_\nu a^a_\mu + f^{abc} a^b_\mu a^c_\nu,$$ (6.3)

where $\bar{F}^a_{\mu\nu}$ is the field-strength tensor associated with the background (see Problem 6.1). Therefore, for constant, temporal and diagonal backgrounds such as those considered in this work, the above replacement rule applies also to the original YM contribution. This means in particular that the background components can be interpreted as imaginary chemical potentials associated with the (commuting) color charges [96]. Together with the fact that the gauge-fixed action boils down to the one in the Landau gauge in the limit of vanishing background, this ensures the renormalizability of the Landau–DeWitt gauge-fixed action. Moreover, the background is not renormalized, as any other chemical potential.[2]

---

[2] This can be seen as the consequence of the abelian symmetry associated with the color charges.

6.1.2 Curci–Ferrari Completion

Similar to the Landau gauge, the Landau–DeWitt gauge fixing is hampered by the Gribov ambiguity which means that the FP action (6.2) should be considered as an approximation, valid at best at high energies, and that it should be extended at lower energies. Since the CF model has proven to be a good candidate for a completion of the gauge fixing in the case of the Landau gauge, here we complete the FP action using a background extension of the CF model.

The two candidates to construct such a background extension are

$$\int d^d x \, \frac{1}{2} m^2 A_\mu^a A_\mu^a \quad \text{or} \quad \int d^d x \, \frac{1}{2} m^2 a_\mu^a a_\mu^a . \tag{6.4}$$

However, if we want to maintain the order parameter interpretation of the background, it is crucial that the symmetry (4.5) is preserved, which leaves us only with the second possibility.[3] The background extension of the Curci–Ferrari model that we shall consider in the rest of this manuscript is therefore

$$\delta S_{CF} = \int d^d x \left\{ \bar{D}_\mu \bar{c}^a(x) \, D_\mu c^a(x) + i h^a(x) \, \bar{D}_\mu a_\mu^a(x) + \frac{1}{2} m^2 a_\mu^a(x) \, a_\mu^a(x) \right\}. \tag{6.5}$$

Following the same steps as in Chap. 1, it is easily seen that this action is invariant under a modified BRST symmetry

$$s_m A_\mu^a = D_\mu c^a, \quad s_m c^a = -\frac{1}{2} f^{abc} c^b c^c, \quad s_m \bar{c}^a = i h^a, \quad s_m i h^a = m^2 c^a , \tag{6.6}$$

which ensures in particular the renormalizability of the model. Equivalently one can again use the interpretation of the background as a chemical potential.

6.1.3 Order Parameter Interpretation

As we saw in Chap. 4, the order parameter interpretation of the background relied not only on the symmetry property (4.5) but also on the convexity of the generating functional $W[J; \bar{A}]$ and the background independence of $W[0; \bar{A}]$.

In the case of the FP action (6.2), the partition function is background independent as we have shown in Chap. 1, but the measure is not positive definite, which can seriously jeopardize the convexity of $W[J; \bar{A}]$. The presence of the CF mass in (6.5) reduces the positivity violation, and a not-so-large mass could be enough to

---

[3] We note however that, because A^2 and a^2 differ only by an affine contribution in a, the two models are in fact trivially related.

ensure the convexity of the generating functional $W[J; \bar{A}]$. However, the presence of the CF mass violates the background independence of the free energy since we now find

$$\frac{\delta \ln W[0, \bar{A}]}{\delta \bar{A}^a_\mu(x)} = m^2 \left(A^a_{\min,\mu}[\bar{A}](x) - \bar{A}^a_\mu(x) \right). \qquad (6.7)$$

The right-hand side vanishes if the background is self-consistent. Generalizing the arguments given in Chap. 4, this is enough to show that the self-consistent backgrounds are extrema of the background-field effective action $\tilde{\Gamma}[\bar{A}]$, but it is not clear anymore which extrema they correspond to.

One can try to restore the background independence of the partition function, and thus the identification of the self-consistent backgrounds with the absolute minima of $\tilde{\Gamma}[\bar{A}]$, by introducing a background dependence of the CF mass. After all, the CF mass serves as a model of the terms that are lacking in the FP procedure within a given gauge. Since the FP gauge-fixed action changes with $\bar{A}$, it is reasonable to expect that so would the terms required beyond the FP gauge fixing. It remains to be investigated, however, whether this leads to a consistent system of equations and also whether this is enough to ensure the background independence of other observables. These questions are beyond the scope of the present discussion, and, in a first approximation, we should neglect the possible background dependence of the mass.

Moreover, it should be kept in mind that the actual action beyond the FP prescription may contain other operators that help in restoring the background independence of the partition function. For this reason, in what follows, we will assume that the state of the system is indeed described by the minima of $\tilde{\Gamma}[\bar{A}]$. In fact, the previous difficulties of principle are present in most continuum approaches, and there is always an implicit assumption made that the correct recipe is that of minimizing $\tilde{\Gamma}[\bar{A}]$.[4] As already mentioned, in Chap. 11, we will present a novel implementation of the background-field method that does not rely on the background independence of the free energy and which, therefore, should lead to more robust results.

---

[4] This is not just a formal discussion. In Chap. 10, we will see that the non-positivity of the measure due to the presence of quarks at finite density modifies this recipe substantially.

6.2   Background-Field Effective Potential

The evaluation of the background-field effective potential $\tilde{V}(r)$ at one-loop order requires only the quadratic part of the gauge-fixed action (6.5) with respect to the fields $a_\mu = A_\mu - \bar{A}_\mu$, h, c and $\bar{c}$.[5] It reads

$$S_0 \equiv \int d^d x \left\{ \frac{1}{2} \left(\bar{D}_\mu a_\nu ; \bar{D}_\mu a_\nu - \bar{D}_\nu a_\mu \right) + \frac{m^2}{2} \left(a_\mu ; a_\mu \right) + \left(ih; \bar{D}_\mu a_\mu \right) + \left(\bar{D}_\mu \bar{c}; \bar{D}_\mu c \right) \right\},$$

$$(6.8)$$

where, for later purpose, we have considered the rescaling $A \to gA$, $\bar{A} \to g\bar{A}$, $h \to h/g$, as well as $m^2 \to m^2/g^2$.[6] In particular, the background covariant derivative reads now $\bar{D}_{\mu\,-} \equiv \partial_{\mu\,-} - g\,[\bar{A}_\mu, \_]$. We have also used the intrinsic notation introduced in the first chapter, in order to facilitate the change from Cartesian color bases to Cartan–Weyl color bases that we shall consider below.

The presence of a preferred color direction, as provided by the background $\beta g \bar{A}_\mu(x) = i r_j H_j \delta_{\mu 0}$, renders the color structure a bit more complicated than usual. Indeed, the background covariant derivative in a Cartesian basis writes

$$\bar{D}_\mu^{ab} = \partial_\mu \delta_{ab} + g f^{acb} \bar{A}_\mu^c,$$

$$(6.9)$$

which is not diagonal in color space for $\mu = 0$. This calls for the introduction of bases that diagonalize the action of the background covariant derivative in color space. Since the background is assumed to lie in the Cartan sub-algebra, these bases are nothing but the Cartan–Weyl bases introduced in Chap. 5.

6.2.1   Notational Convention

In order to fully exploit the properties of Cartan–Weyl bases, we introduce the following notation. Given an element X of the algebra, we write its decomposition into a Cartan–Weyl basis $\{i H_j, i E_\alpha\}$ as $X = i(X_j H_j + X_\alpha E_\alpha) \equiv i X_\kappa t_\kappa$, where the label κ can take two types of values, $\kappa = 0^{(j)}$ or $\kappa = \alpha$, with $t_{0^{(j)}} \equiv H_j$ and $t_\alpha \equiv E_\alpha$. The label $0^{(j)}$ is referred to as a *zero* and should be taken literally as corresponding to the nul vector in the same space as the one containing the roots α (the dual background space in the terminology of the previous chapter). There is of course only one such nul vector. However, there are as many color labels associated with zero as there are elements in the Cartan sub-algebra; hence the label j used to denote the various zeros.

---

[5] More generally, the effective action $\Gamma[\tilde{A}, \bar{A}]$ is obtained by expanding the action in the fields $A - \tilde{A}$, h, c and $\bar{c}$. We shall compute it to one-loop order in Chap. 11.

[6] It is also convenient to rescale the parameter of the BRST transformation by a factor g or equivalently to do $s_m \to s_m/g$.

Given these notational conventions, the definition of a Cartan–Weyl basis can be written compactly as

$$[t_j, t_\kappa] = \kappa_j t_\kappa \,, \tag{6.10}$$

where, of course, $0_j^{(k)} = 0$, no matter the values of j and k. The vectors κ appear therefore as the weights of the adjoint representation. Now, since the background is assumed to lie in the Cartan sub-algebra, the background covariant derivative acting on X can be shown to write

$$\bar{D}_\mu X = i \bar{D}_\mu^\kappa X_\kappa t_\kappa \,, \tag{6.11}$$

with $\bar{D}_\mu^\kappa \equiv \partial_\mu - i T r \cdot \kappa \, \delta_{\mu 0}$ and $r \cdot \kappa \equiv r_j \kappa_j$ (see Problem 6.2). As announced, the background covariant derivative is diagonal in color space when expressed in a Cartan–Weyl basis. This simplifies considerably the evaluation of the background field effective potential. Similarly, in order to evaluate the background-dependent Polyakov loop, it will be convenient to diagonalize the defining action of the H_j as $H_j|\rho\rangle = \rho_j|\rho\rangle$, with ρ the weights of the defining representation.

6.2.2 General One-Loop Expression

Coming back to the quadratic action (6.8) and choosing the Cartan–Weyl basis such that $(it_\kappa; it_\lambda) = \delta_{\kappa, -\lambda}$ (see Appendix A), we find

$$S_0 = \int d^d x \left\{ \frac{1}{2} (\bar{D}_\mu^{-\kappa} a_\nu^{-\kappa} \bar{D}_\mu^\kappa a_\nu^\kappa - \bar{D}_\nu^{-\kappa} a_\mu^{-\kappa} \bar{D}_\mu^\kappa a_\nu^\kappa) + \frac{m^2}{2} a_\mu^{-\kappa} a_\mu^\kappa \right.$$

$$\left. + i h^{-\kappa} \bar{D}_\mu^\kappa a_\mu^\kappa + \bar{D}_\mu^{-\kappa} \bar{c}^{-\kappa} \bar{D}_\mu^\kappa c^\kappa \right\}. \tag{6.12}$$

In Fourier space, using that $\varphi^{-\kappa}(x) = \varphi^\kappa(x)^*$ for bosonic fields (see Appendix A), this reads

$$S_0 = \int_Q^T \left\{ \frac{1}{2} a_\mu^\kappa(Q)^* [(Q_\kappa^2 + m^2)\delta_{\mu\nu} - Q_\mu^\kappa Q_\nu^\kappa] a_\nu^\kappa(Q) \right.$$

$$\left. + h^\kappa(Q)^* Q_\mu^\kappa a_\mu^\kappa(Q) + \bar{c}^{-\kappa}(-Q) \, Q_\kappa^2 c^\kappa(Q) \right\}, \tag{6.13}$$

where $Q_\kappa \equiv Q + T r \cdot \kappa n$ is referred to as the *generalized momentum*, combining the standard momentum Q and a shift $T r \cdot \kappa n$ along the frequency direction $n \equiv (1, 0)$. We recall that, at finite temperature, the β-periodicity of the fields along the Euclidean time direction implies that the frequencies are discrete. This means that $Q = (\omega_q, q)$ where $\omega_q = 2\pi q T$ is a bosonic Matsubara frequency and $q \in \mathbb{Z}$.[7] The vacuum d-dimensional integrals are then turned into a discrete sum and a $(d-1)$-dimensional integral

$$\int_Q^T f(Q) \equiv T \sum_{q \in \mathbb{Z}} \int \frac{d^{d-1}q}{(2\pi)^{d-1}} f(\omega_q, q) \tag{6.14}$$

known as Matsubara sum-integrals.

From the quadratic action (6.13), a standard Gaussian integration leads to the one-loop potential

$$\tilde{V}_{1\text{loop}}(r) = \sum_\kappa \left[\frac{d-1}{2} \int_Q^T \ln \left(Q_\kappa^2 + m^2 \right) - \frac{1}{2} \int_Q^T \ln Q_\kappa^2 \right]. \tag{6.15}$$

For later purpose, it is important to emphasize that the massive contribution originates from the transverse gluonic modes, whereas the massless contribution results from an over-cancellation between one massless longitudinal mode and two massless ghost modes. It is also worth noting that the zero-temperature limit of the above expression at fixed $\hat{r} \equiv Tr$ does not depend on $\hat{r}$. Indeed, in this limit, the Matsubara sum-integral becomes a d-dimensional integral, and the frequency shift $\hat{r} \cdot \kappa$ hidden in Q_κ can be absorbed using a simple change of variables in the frequency integral.[8] If we disregard this background-independent, zero-temperature limit, we can set $d = 4$ and use an integration by parts (with respect to the variable q)[9] to find

$$\tilde{V}_{1\text{loop}}(r) = \sum_\kappa \left[-\int_Q^T \frac{q^2}{Q_\kappa^2 + m^2} + \frac{1}{3} \int_Q^T \frac{q^2}{Q_\kappa^2} \right], \tag{6.16}$$

which involves convergent Matsubara sums unlike expression (6.15).

Using standard contour integration techniques (see Appendix B), we arrive at (again, we disregard the zero-temperature limit)

$$\tilde{V}_{1\text{loop}}(r) = \frac{1}{2\pi^2} \sum_\kappa \left[-\int_0^\infty dq \, \frac{q^4}{\varepsilon_q} \, \mathrm{Re} \, n_{\varepsilon_q - i\hat{r}\cdot\kappa} + \frac{1}{3} \int_0^\infty dq \, q^3 \, \mathrm{Re} \, n_{q - i\hat{r}\cdot\kappa} \right], \tag{6.17}$$

---

[7] The variable q in ω_q denotes an integer $q \in \mathbb{Z}$, not to be mistaken with the vector q. There is no source of confusion possible since the former variable always appear as subscript to a Matsubara frequency.

[8] As we show in the next section, this property generalizes in fact to any loop order.

[9] The boundary terms can be neglected in the finite temperature contribution since they are typically exponentially suppressed by Bose factors.

with $n(x) \equiv 1/(e^{\beta x} - 1)$ the Bose-Einstein distribution function and $\varepsilon_q \equiv \sqrt{q^2 + m^2}$. Using the integration by parts backward, this can also be rewritten as

$$\tilde{V}_{1\mathrm{loop}}(r) = \frac{T}{2\pi^2} \sum_{\kappa} \int_0^{\infty} dq \, q^2 \, \mathrm{Re} \, \ln \frac{\left(1 - e^{-\beta \varepsilon_q + i r \cdot \kappa}\right)^3}{1 - e^{-\beta q + i r \cdot \kappa}}. \tag{6.18}$$

Moreover, given that the root diagram is invariant under $\alpha \to -\alpha$, we can remove the explicit real parts in the previous expressions. Alternatively, we can evaluate these real parts to find

$$\tilde{V}_{1\mathrm{loop}}(r) = \frac{1}{2\pi^2} \sum_{\kappa} \Bigg[-\int_0^{\infty} dq \, \frac{q^4}{\varepsilon_q} \frac{e^{\beta \varepsilon_q} \cos(r \cdot \kappa) - 1}{e^{2\beta \varepsilon_q} - 2e^{\beta \varepsilon_q} \cos(r \cdot \kappa) + 1}$$

$$+ \frac{1}{3} \int_0^{\infty} dq \, q^3 \, \frac{e^{\beta q} \cos(r \cdot \kappa) - 1}{e^{2\beta q} - 2e^{\beta q} \cos(r \cdot \kappa) + 1} \Bigg], \tag{6.19}$$

or, equivalently,

$$\tilde{V}_{1\mathrm{loop}}(r) = \frac{T}{4\pi^2} \sum_{\kappa} \int_0^{\infty} dq \, q^2 \, \ln \frac{\left(e^{-2\beta \varepsilon_q} - 2e^{-\beta \varepsilon_q} \cos(r \cdot \kappa) + 1\right)^3}{e^{-2\beta q} - 2e^{-\beta q} \cos(r \cdot \kappa) + 1}. \tag{6.20}$$

6.2.3   Checking the Symmetries

As a cross-check, let us verify that the above formulas fulfil all the symmetries that we identified in Chap. 5.

Winding Transformations First of all, the potential should be invariant under winding transformations which appear in the restricted background space as translations along any of the vectors $\bar{\alpha}^{(k)}$ that generate the lattice dual to the one generated by the roots:

$$\tilde{V}\left(r + \bar{\alpha}^{(k)}\right) = \tilde{V}(r), \quad \forall \bar{\alpha}^{(k)}. \tag{6.21}$$

This identity is trivially fulfilled by Eq. (6.15) because, under a winding transformation, the generalized momentum Q_κ becomes $Q_\kappa + T\bar{\alpha}^{(k)} \cdot \kappa$ and the shift $T\bar{\alpha}^{(k)} \cdot \kappa \in 2\pi T \mathbb{Z}$ can be absorbed using a change of variables in the Matsubara sum. Similarly, $e^{i r \cdot \kappa}$ and $\cos(r \cdot \kappa)$, which appear, for instance, in Eqs. (6.18) and (6.20), respectively, remain invariant under winding transformations.

As a consequence of the identity (6.21), we can show that the background independence of the zero-temperature limit of the potential, which we observed at one loop, generalizes in fact to any loop order (see Problem 6.3). Another consequence of the symmetry identity (6.21) is that the analysis of the background

field effective potential can be restricted to the fundamental parallelepiped defined
by the vectors $\bar{\alpha}^{(j)}$. In fact, it is convenient to decompose the restricted background
as $r = x_j \bar{\alpha}^{(j)}$, with $2\pi x_j = r \cdot \alpha^{(j)}$, and to express the potential as a function of
the $x_i \in [0, 1]$. We shall use this decomposition in the various plots to be shown in
what follows.

Weyl Transformations The potential should also be invariant under Weyl transformations which appear as reflections with respect to hyperplanes orthogonal to the
roots:

$$\tilde{V}\left(r - 2\frac{r \cdot \alpha}{\alpha^2}\alpha\right) = \tilde{V}(r), \ \forall \alpha . \tag{6.22}$$

This identity is indeed fulfilled by Eq. (6.15) because, under a Weyl transformation,
the momentum Q_κ becomes $Q + T\left(r \cdot \kappa - 2\frac{r \cdot \alpha\alpha \cdot \kappa}{\alpha^2}\right) = Q_{\kappa - 2\frac{\kappa \cdot \alpha}{\alpha^2}\alpha}$ and this change
of κ can be absorbed using a redefinition of the sum over κ for it is generally true
that the root diagram is invariant under Weyl transformations. The same remark
applies to $e^{ir \cdot \kappa}$ and $\cos(r \cdot \kappa)$.

Charge Conjugation Finally, the potential should be invariant under charge conjugation which appears as the reflection about the origin of the restricted background
space:

$$\tilde{V}(-r) = \tilde{V}(r) . \tag{6.23}$$

This identity is fulfilled by Eq. (6.15) because, under charge conjugation, the
momentum Q_κ is changed to $Q_{-\kappa}$ and this change of κ can again be absorbed using
a redefinition of the sum over κ since the root diagram is invariant under $\alpha \to -\alpha$.
The same remark applies to $e^{ir \cdot \kappa}$ and $\cos(r \cdot \kappa)$.

6.3 SU(2) and SU(3) Gauge Groups

We now use the formulas derived in the previous section to investigate the
deconfinement transition in the SU(2) and SU(3) cases.

6.3.1 Deconfinement Transition

SU(2) In the SU(2) case, we can work over the fundamental Weyl chamber
$r \in [0, 2\pi]$ whose only center-symmetric point, or confining state, is at $r = \pi$.
Using that $\kappa \in \{-1, 0, +1\}$ and writing only the background-dependent contribu-

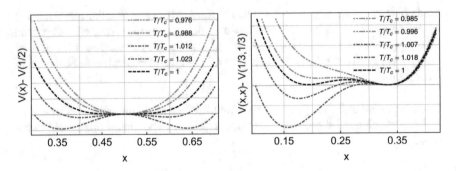

Fig. 6.1 Left: SU(2) background-field effective potential in the fundamental parallelepiped $r = x\bar{\alpha}^{(1)}$, with $\bar{\alpha}_1 = 2\pi$ and $0 < x < 1$, which is also the fundamental Weyl chamber. The center-symmetric or confining point is located at $x = 1/2$. Right: SU(3) background-field effective potential along the line $x_1 = x_2 \equiv x$ of the fundamental parallelepiped. This fundamental parallelepiped contains two Weyl chambers, and we show the potential in the fundamental Weyl chamber only. The center-symmetric or confining point is located at $x = 1/3$. In both cases, we have subtracted the value of the potential at the confining point, for more readability

tions to the potential, we find

$$\tilde{V}^{SU(2)}_{1loop}(r) = \frac{T}{2\pi^2} \int_0^\infty dq\, q^2 \ln \frac{\left(e^{-2\beta\varepsilon_q} - 2e^{-\beta\varepsilon_q}\cos(r) + 1\right)^3}{e^{-2\beta q} - 2e^{-\beta q}\cos(r) + 1}. \qquad (6.24)$$

This potential is plotted in Fig. 6.1 (left) for various temperatures. As the temperature is increased, the minimum of the potential moves continuously from the confining, center-symmetric state at $r = \pi$ to a pair of degenerate states $(r, 2\pi - r)$ connected by center symmetry. The transition is then of the second-order type, in agreement with the results of lattice simulations [159–163] or continuum non-perturbative approaches [48, 49, 51, 73, 74, 76, 169–174].

The transition temperature T_c can be determined in terms of m by requiring that the curvature of the potential vanishes for $r = \pi$. This condition reads

$$0 = \frac{T}{\pi^2} \int_0^\infty dq\, q^2 \left[\frac{e^{\beta q}}{(e^{\beta q} + 1)^2} - \frac{3e^{\beta\varepsilon_q}}{(e^{\beta\varepsilon_q} + 1)^2} \right], \qquad (6.25)$$

from which we find $T^{SU(2)}_{c,1loop} \simeq 0.336m$.

SU(3) The corresponding one-loop expression for the potential in the SU(3) case can be obtained by observing that each pair of roots is associated with a SU(2) subgroup. It follows that (again we keep only the background-dependent contributions)

$$\tilde{V}^{SU(3)}_{1loop}(r) = \tilde{V}^{SU(2)}_{1loop}(r_3) + \tilde{V}^{SU(2)}_{1loop}\left(\frac{r_3 - r_8\sqrt{3}}{2}\right) + \tilde{V}^{SU(2)}_{1loop}\left(\frac{r_3 + r_8\sqrt{3}}{2}\right). \qquad (6.26)$$

In Fig. 6.2, we show contour graphs of the potential in the fundamental Weyl chamber for various temperatures. We observe that, as the temperature is increased, the minimum of the potential jumps eventually from the confining, center-symmetric state at $r = (4\pi/3, 0)$ to a triplet of degenerate states connected by center symmetry. We find that one of the absolute minima lies always along the $r_8 = 0$ axis. As we have explained in the previous chapter, this is a consequence of charge conjugation invariance.[10] For us, this means that we can restrict the analysis to the $r_8 = 0$ axis, along which the potential reads more simply

$$\tilde{V}_{1\mathrm{loop}}^{SU(3)}(r) = \tilde{V}_{1\mathrm{loop}}^{SU(2)}(r_3) + 2\tilde{V}_{1\mathrm{loop}}^{SU(2)}\left(\frac{r_3}{2}\right). \tag{6.27}$$

In Fig. 6.1, we show the potential along this axis, in the fundamental Weyl chamber, for various temperatures, with a clear signal of a first-order-type transition, in agreement with other approaches.

From center symmetry, it follows that the confining point is always an extremum, so it is possible to locate the highest spinodal by requiring the curvature of the potential to vanish at this point. To this purpose, we notice first that

$$\left.\frac{\partial^2}{\partial r^2}\tilde{V}_{1\mathrm{loop}}^{SU(2)}(r)\right|_{r=4\pi/3} = \left.\frac{\partial^2}{\partial r^2}\tilde{V}_{1\mathrm{loop}}^{SU(2)}(r)\right|_{r=2\pi/3}. \tag{6.28}$$

Using this identity, the equation for a vanishing curvature at the confining point is found to be

$$0 = \frac{T}{2\pi^2}\int_0^\infty dq\, q^2 \left[\frac{e^{-3\beta q} + 4e^{-2\beta q} + e^{-\beta q}}{(e^{-2\beta q} + e^{-\beta q} + 1)^2} - 3\frac{e^{-3\beta\varepsilon_q} + 4e^{-2\beta\varepsilon_q} + e^{-\beta\varepsilon_q}}{(e^{-2\beta\varepsilon_q} + e^{-\beta\varepsilon_q} + 1)^2}\right], \tag{6.29}$$

from which we deduce that $T_{\mathrm{s,1loop}}^{SU(3)} \simeq 0.382m$, while the transition temperature is found to be $T_{\mathrm{c,1loop}}^{SU(3)} \simeq 0.364m$.

Transition Temperatures Given that the mass scale m can be fixed by fitting lattice Landau gauge propagators with the corresponding one-loop expressions obtained within the CF model,[11] we can estimate the values for the transition temperatures. Table 6.1 summarizes our results and compares them to other approaches. The obtained values for T_c are already quite reasonable given the simplicity of the one-loop approximation that we have considered. In the next chapter, we shall see

---

[10] The presence of two other directions along which the minima move is in line with center symmetry and just means that there are three equivalent ways to define charge conjugation symmetry.

[11] We fix the parameters of the model at zero temperature, in which case the background $\hat{r} = rT$ vanishes and the Landau–DeWitt gauge becomes the Landau gauge.

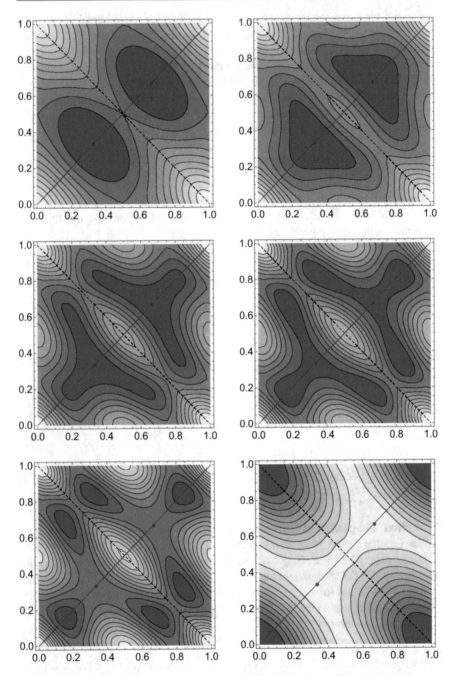

Fig. 6.2 SU(3) background-field effective potential in the fundamental parallelepiped $r = x_1 \bar{\alpha}^{(1)} + x_2 \bar{\alpha}^{(2)}$, with $\bar{\alpha}_1 = 2\pi \, (1, -1/\sqrt{3})$, $\bar{\alpha}_2 = 2\pi \, (1, 1/\sqrt{3})$, and $0 < x_i < 1$. The fundamental parallelepiped is made of two Weyl chambers (separated by the dashed line) which contain each one center-symmetric or confining point, located $(1/3, 1/3)$ and $(2/3, 2/3)$, respectively (red dots). Finally, the charge conjugation invariant states are indicated by a red line

Table 6.1 Transition temperatures for the SU(2) and SU(3) deconfinement transitions as computed from various approaches

T_c (MeV)	Lattice [175]	fRG [49]	Variational [174]	CF model [92]
SU(2)	295	230	239	238
SU(3)	270	275	245	185

that the two-loop corrections further improve these values. In Chap. 11, we will find that an even better improvement can be achieved by rethinking the way the background-field method is applied at finite temperature.

6.3.2 Inversion of the Weiss Potential

The previous results can be understood in a simple way by comparing the high- and low-temperature forms of the background-field effective potential.

For high temperatures ($T \gg m$), the mass can be neglected, and the gluonic contributions combine with the massless contributions to yield

$$\tilde{V}_{1\text{loop}}(r) \simeq T^4 v_{\text{Weiss}}(r), \quad \text{for } T \gg m, \tag{6.30}$$

where

$$
v_{\text{Weiss}}(r) \equiv \frac{1}{2\pi^2} \sum_\kappa \int_0^\infty dq\, q^2 \ln\left(e^{-2q} - 2e^{-q}\cos(r\cdot\kappa) + 1\right)
$$

$$
= -\frac{1}{3\pi^2} \sum_\kappa \int_0^\infty dq\, q^3 \frac{e^q \cos(r\cdot\kappa) - 1}{e^{2q} - 2e^q \cos(r\cdot\kappa) + 1} \tag{6.31}
$$

is known as the *Weiss potential* [176, 177]. This potential rewrites

$$
v_{\text{Weiss}}(r) = -\frac{1}{3\pi^2} \sum_\kappa P_4(r\cdot\kappa), \tag{6.32}
$$

where

$$
P_4(r) \equiv -\frac{(\pi - r)^4}{8} + \frac{\pi^2(\pi - r)^2}{4} - \frac{7\pi^4}{120}, \tag{6.33}
$$

is related to the Bernouilli polynomial of degree 4 and the notation $r\cdot\kappa$ stands for the real number between 0 and 2π that differs from $r\cdot\kappa$ by a multiple of 2π.

In contrast, for low temperatures ($T \ll m$), the gluonic contributions can be neglected, and one obtains again the Weiss potential but inverted with respect to the high-temperature regime:

$$\tilde{V}_{1\text{loop}}(r) \simeq -\frac{T^4}{2} v_{\text{Weiss}}(r), \quad \text{for } T \ll m. \tag{6.34}$$

This inversion of the Weiss potential (due to the presence of the scale m) triggers the deconfinement transition if the center-symmetric state of the system corresponds to the absolute maximum of the Weiss potential. It is a simple exercise to check that this is so in the SU(2) and SU(3) cases.

The inversion mechanism is present in other continuum approaches and was in fact originally proposed in Ref. [73]. Let us emphasize that it relies on the dominance of ghost degrees of freedom at low temperatures, which obviously poses some conceptual questions. One such question is to know how this temperature regime can be *a priori* dominated by unphysical degrees of freedom while maintaining basic principles of thermodynamical consistency. We shall analyze this question below and try to answer it, at least partially.

6.3.3 Polyakov Loops

We can also evaluate the background-dependent Polyakov loop $\ell(r)$ which becomes the physical order parameter when evaluated at the minimum of $\tilde{V}(r)$.

Let us recall that $\ell(r)$ is nothing but the expectation value of $\Phi_A(\mathbf{x})$ in the presence of a source $J[\bar{A}]$ such that $\langle A \rangle_{\bar{A}} = \bar{A}$. Therefore, to evaluate $\ell(r)$, we expand $\Phi_A(\mathbf{x})$ in powers of $a_\mu(x) = A_\mu(x) - \bar{A}_\mu(x)$ and substitute any one-point function $\langle a_\mu(x) \rangle_{\bar{A}}$ by 0 in the evaluation of the expectation value. At leading order, we write $\Phi_A(\mathbf{x}) \simeq \Phi_{\bar{A}}(\mathbf{x}) = \text{tr}\, e^{i\beta g \bar{A}}/N$, where we have used that the background is constant to get rid of the time-ordering. We find

$$\ell(r) = \frac{1}{N}\text{tr}\, e^{ir_j H_j} = \frac{1}{N}\sum_\rho e^{ir\cdot\rho}, \tag{6.35}$$

$$\bar{\ell}(r) = \frac{1}{N}\text{tr}\, e^{-ir_j H_j} = \frac{1}{N}\sum_\rho e^{-ir\cdot\rho}, \tag{6.36}$$

where ρ are the weights of the defining representation and $-\rho$ those of the corresponding contragredient representation.

These formulas obey the expected symmetries. Under a winding transformation $r \to r + \bar{\alpha}^{(k)}$ (representing a center transformation, with center element $e^{i\bar{\alpha}^{(k)}\cdot\rho}$), the phase $e^{i\rho\cdot r}$ becomes $e^{i\rho\cdot r}e^{i\rho\cdot\bar{\alpha}^{(k)}}$ with $e^{i\rho\cdot\bar{\alpha}^{(k)}}$ not depending on the chosen weight ρ, as we have seen in the Chap. 5. Therefore

$$\ell\left(r + \bar{\alpha}^{(k)}\right) = e^{i\bar{\alpha}^{(k)}\cdot\rho}\ell(r), \quad \forall\bar{\alpha}^{(k)}, \tag{6.37}$$

as it should for a winding transformation. Under a Weyl transformation, $e^{ir\cdot\rho}$ is transformed into

$$e^{i\left(r-\frac{\alpha\cdot\rho}{\alpha^2}\alpha\right)\cdot\rho} = e^{ir\cdot\rho}e^{-2i\frac{r\cdot\alpha\alpha\cdot\rho}{\alpha^2}} = e^{ir\cdot\left(\rho-\frac{\alpha\cdot\rho}{\alpha^2}\alpha\right)}, \tag{6.38}$$

and the change of ρ can be absorbed in a redefinition of the sum over ρ for it is generally true that the weight diagram is globally invariant under Weyl transformations. It follows that

$$\ell\left(r - 2\frac{r\cdot\alpha}{\alpha^2}\alpha\right) = \ell(r), \tag{6.39}$$

as it should for a Weyl transformation. Finally, we observe that $\ell(-r) = \bar{\ell}(r)$, in line with charge conjugation invariance (see Chap. 4). As we also saw there, we have $\ell(-r) = \ell^*(r)$ and, therefore, $\bar{\ell}(r) = \ell^*(r)$.

For illustration, let us discuss some specific cases. In the SU(2) case, we find

$$\ell(r) = \bar{\ell}(r) = \cos(r/2), \tag{6.40}$$

whereas in the SU(3) case (see Problem 6.4),

$$\ell(r) = \frac{e^{-i\frac{r_8}{\sqrt{3}}} + 2e^{i\frac{r_8}{2\sqrt{3}}}\cos(r_3/2)}{3}, \tag{6.41}$$

$$\bar{\ell}(r) = \frac{e^{i\frac{r_8}{\sqrt{3}}} + 2e^{-i\frac{r_8}{2\sqrt{3}}}\cos(r_3/2)}{3}. \tag{6.42}$$

We mention that for the SU(2) gauge group, the constraint from charge conjugation invariance can also be seen as arising from Weyl symmetry, and therefore, as we mentioned already, there is no real constraint from charge conjugation invariance in this case. The reason is that there exists a Weyl transformation $W = i\sigma_2$ such that $(i\sigma_2)^\dagger \bar{A}_0(i\sigma_2) = \bar{A}_0^C$. From this, it is easily deduced that $\ell(-r) = \bar{\ell}(r)$ and this without invoking charge conjugation invariance in this case. Since Weyl transformations also imply $\ell(-r) = \ell(r)$, it follows that $\ell(r) = \bar{\ell}(r)$ for any value of r. This last property is in general not true in the SU(3) case, except for a charge conjugation invariant state $r = (r_3, 0)$ for which

$$\ell(r) = \bar{\ell}(r) = \frac{1 + 2\cos(r_3/2)}{3}. \tag{6.43}$$

That ℓ and $\bar{\ell}$ become equal and real in this case follows from hermiticity and charge conjugation invariance, as we discuss more generally in Chap. 9.

The physical Polyakov loops, as obtained from the one-loop potential derived above, are shown in Fig. 6.3. We mention that both Polyakov loops saturate to 1 at a temperature $T_\star \simeq 1.45m$, corresponding to the background reaching 0 at this

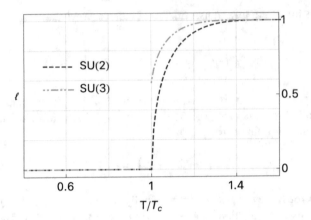

Fig. 6.3 Polyakov loops for the SU(2) and SU(3) defining representations computed at one-loop order within the CF model using self-consistent backgrounds, as functions of T/T_c

temperature. This occurs with a discontinuity in the second derivative of the order parameter which leads to artificial singularities in the thermodynamical observables. A similar feature is observed in other approaches (see, for instance, [174]). We will see in the next chapter that these artefacts disappear at two-loop order. We also mention that the Polyakov loops obtained here feature a too fast growth at the transition which prevents a faithful comparison to the lattice data. This question will be revisited in Chap. 11.

6.4   Thermodynamics

In principle, the free-energy density of the system is obtained as the background field effective potential is evaluated at its minimum $r(T)$:

$$f(T) = \tilde{V}(r(T); T), \tag{6.44}$$

where we have made the T-dependence of the potential explicit. From this, all other thermodynamical observables can be derived. In particular the pressure and the entropy density are given, respectively, by

$$p(T) = -f(T) = -\tilde{V}(r(T); T), \tag{6.45}$$

$$s(T) = p'(T) = -\left.\frac{\partial \tilde{V}(r; T)}{\partial T}\right|_{r=r(T)}. \tag{6.46}$$

The energy density can then be computed as

$$e(T) = f(T) + Ts(T) = T^2(p/T)', \tag{6.47}$$

and the interaction measure is

$$i(T) = \frac{e(T) - 3p(T)}{T^4} = T(p/T^4)'. \tag{6.48}$$

6.4.1 High- and Low-Temperature Behavior

In the high-temperature limit, all modes contributing to the potential (6.15) become massless, and the resulting potential counts effectively two degrees of freedom per color mode. This is the result of the usual cancellation between the two unphysical components of the gauge field A_μ and the two ghost degrees of freedom which, due to their Grassmannian nature, contribute to the potential with an extra minus sign.

In the low-temperature phase, the situation is in general more delicate and still open to debate [93, 94, 178]. In the present approach, since three of the four components of the gauge field become massive while the ghosts remain massless, the potential is essentially dominated by ghost degrees of freedom. While this *ghost domination* is not specific to the present approach and is in fact a key ingredient in the Weiss potential inversion mechanism described above, one may fear the presence of inconsistencies in the behavior of the thermodynamical observables, such as a negative (thermal) pressure or a negative entropy density. As we now explain, this unphysical behavior does not seem to be present.

The point is that, although the low-temperature phase is indeed dominated by the one uncanceled ghost degree of freedom, the latter is surrounded by a background $\bar{A}_0$. Instead of cancelling each other, these two unphysical degrees of freedom are subtlety combined in such a way that the thermodynamics remains consistent at low temperatures. More precisely, the confining value of the background $\bar{A}_0$ operates a transmutation of the thermal distribution function of the ghost field from a negative distribution to a positive distribution.

Let us first illustrate this transmutation mechanism in the SU(2) case where it is particularly compelling. From Eqs. (6.45) and (6.17), the contribution to the pressure at low temperatures writes

$$p(T) = \frac{1}{6\pi^2} \int_0^\infty dq\, q^3 \left[-n_q - n_{q-i\hat{r}(T)} - n_{q+i\hat{r}(T)} \right]. \tag{6.49}$$

The bracket contains the three-color-mode contributions, corresponding to $\kappa \in \{-1, 0, +1\}$. The neutral mode $\kappa = 0$ is blind to the background and contributes negatively to the thermal pressure, as we would expect from a ghost degree of freedom. However, in the presence of a confining background $\hat{r}(T) = \pi T$, the charge modes $\kappa = \pm 1$ contribute with the transmuted distribution functions

$$- n_{q-i\hat{r}(T)} = -\frac{1}{e^{\beta q - i\pi} - 1} = \frac{1}{e^{\beta q} + 1} \equiv f_q > 0, \tag{6.50}$$

and thus behave effectively as true fermions! Finally, using $f_q = n_q - 2n_{2q}$ and a simple change of variables, one obtains the total contribution to the thermal pressure as

$$p(T) = \frac{3}{24\pi^2} \int_0^\infty dq \, q^3 \, n_q, \tag{6.51}$$

which is positive. Since $s(T)/T^3 \sim 4p(T)/T^4$, the same conclusion is reached for the entropy density. We can proceed identically in the SU(3) case. The thermal pressure reads

$$p(T) = \frac{1}{6\pi^2} \int_0^\infty dq \, q^3 \left[-2n_q - 2\mathrm{Re}\, n_{q-i\hat{r}(T)} - 4\mathrm{Re}\, n_{q-i\hat{r}(T)/2} \right]. \tag{6.52}$$

In the confining phase $r(T) = 4\pi/3$ and

$$- \mathrm{Re}\, n_{q-i\hat{r}(T)} = -\mathrm{Re}\, n_{q-i\hat{r}(T)/2} = \frac{e^{\beta q}/2 + 1}{e^{2\beta q} + e^{\beta q} + 1} > 0. \tag{6.53}$$

To check that the total contribution to the pressure is positive, we can use the identity (see Problem 6.5)

$$\sum_{k=0}^{N-1} n_{q-i\frac{2\pi}{N}k} = Nn_{Nq}. \tag{6.54}$$

After a simple change of variables, we find

$$p(T) = \frac{8}{54\pi^2} \int_0^\infty dq \, q^3 \, n_q. \tag{6.55}$$

We conclude therefore that, despite the ghost dominance at low temperatures, the thermodynamical properties seem consistent.

6.4.2 Vicinity of the Transition

The transmutation mechanism identified in the previous section could lead to inconsistencies in the vicinity of the transition. Indeed, as one approaches the transition temperature from below, the gluonic degrees of freedom start playing a role, while the background remains confining. Then, because the gluonic distribution functions are positive in the absence of background, those orthogonal to the background transmute into negative distribution functions in the confining phase and can potentially turn the entropy density or the thermal pressure negative. A calculation reveals that a very tiny violation occurs at leading order over a narrow range of temperatures below T_c (see [93, 94]). In the next chapter, we investigate whether higher loop corrections can cure this behavior.

We mention finally that, in the SU(3) case, the first-order transition is characterized by a discontinuity of the entropy density, associated with a latent heat $\Delta\epsilon/T_c^4 = \Delta s/T_c^3$. At one-loop order in the CF approach, we find 0.41 that is only one third of the lattice value [179]. It remains to be investigated whether this result improves when including higher-order corrections.

Problems

6.1 Field-Strength Tensor and Background Covariant Derivatives (⋆)
Show formula (6.3).

6.2 Covariant Derivative in a Cartan–Weyl Basis (⋆)
Show formula (6.11). *Tip:* decompose both X and the background along a Cartan–Weyl basis, and use Eq. (6.10).

6.3 Background Independence at Zero Temperature (⋆⋆)
Consider the background-field effective potential $\tilde{V}(r)$ as a function of $\hat{r} \equiv Tr$ rather than r, and show that its $T \to 0$ limit at fixed $\hat{r}$ does not depend on $\hat{r}$. *Tip:* apply identity (6.21) n times along the direction $\bar{\alpha}^{(k)}$, and take the zero-temperature limit in an appropriate way to show that the background-field effective potential does not depend on the background.

6.4 SU(3) Polyakov Loops (⋆)
Show that the SU(3) Polyakov loops are given by

$$\ell(r) = \frac{e^{-i\frac{r_8}{\sqrt{3}}} + 2e^{i\frac{r_8}{2\sqrt{3}}}\cos(r_3/2)}{3},$$

$$\bar{\ell}(r) = \frac{e^{i\frac{r_8}{\sqrt{3}}} + 2e^{-i\frac{r_8}{2\sqrt{3}}}\cos(r_3/2)}{3}.$$

Tip: use the general expressions (6.35) and (6.36) together with the defining weights of SU(3).

6.5 Summation Formula (⋆⋆)

(a) Show formula (6.54). *Tip:* write the Bose factor as a geometric series, and invert the order of summations.

(b) This formula can also be seen as a consequence of the following more general result. Given a Matsubara sum $S(T) \equiv T \sum_{n \in \mathbb{Z}} f(i\omega_n)$, show that the related double sum $\sum_{k=0}^{N-1} T \sum_{n \in \mathbb{Z}} f\left(i\omega_n + iT\frac{2\pi}{N}k\right)$ is nothing but $NS(T/N)$. *Tip:* interpret the double sum as a single Matsubara sum.

Yang-Mills Deconfinement Transition from the Curci-Ferrari Model at Next-to-Leading Order

We have seen in the previous chapter that the background extension of the Curci-Ferrari model captures the physics of the confinement/deconfinement transition in Yang-Mills theories already at leading order. In particular, we have obtained rather good estimates for the SU(2) and SU(3) transition temperatures, given the simplicity of the considered approximation.

In this chapter, we investigate the next-to-leading-order corrections to the background-field effective potential and to the background-dependent Polyakov loop. Not only are these corrections necessary to assess the convergence properties of the approach, but they are also needed to try to cure some of the problematic features identified at leading order.

We shall see that the two loop corrections help in improving the values for the transition temperatures and cure some of the problematic features. There will remain some open questions, however. As we argue, these questions are not restricted to the mere Curci-Ferrari approach but pose in fact challenges to most, if not all, present continuum approaches.

7.1 Feynman Rules and Color Conservation

In order to evaluate the two-loop corrections to the background-field effective potential, we first need to determine the Feynman rules in the Landau-DeWitt gauge. Due to our previous experience with the one-loop calculation, we suspect these rules to become simpler when expressed in a Cartan-Weyl basis. We will check below that this is indeed the case. But let us first see how this result can be anticipated using the symmetries of the gauge-fixed action.

© The Author(s), under exclusive license to Springer Nature Switzerland AG 2022
U. Reinosa, *Perturbative Aspects of the Deconfinement Transition*, Lecture Notes in Physics 1006, https://doi.org/10.1007/978-3-031-11375-8_7

7.1.1 Color Conservation

In fact, there is still one invariance that we have not yet exploited, namely, the invariance under background gauge transformations ($\varphi^U \equiv U\varphi U^\dagger$ for $\varphi = h, c$, or $\bar{c}$)

$$S_{\mathrm{CF}}[A^U, h^U, c^U, \bar{c}^U; \bar{A}^U] = S_{\mathrm{CF}}[A, h, c, \bar{c}; \bar{A}], \tag{7.1}$$

in the case where the background is left invariant, $\bar{A}^U = \bar{A}$. Indeed, since only the dynamical fields are transformed in this case, this invariance implies certain conservation rules that should appear explicitly at the level of the action provided the fields are decomposed into the eigenmodes of the corresponding Noether charges.

In the case where the background is constant, temporal, and diagonal, a class of transformations that leave the background invariant are color rotations of the form $e^{i\theta_j H_j}$, with θ_j constant. The corresponding conserved charges on the Lie algebra are therefore the operators ad_{H_j}, whose eigenstates are nothing but the elements t_κ of a Cartan-Weyl basis, with associated charges given by the components κ_j of κ.

Using Eqs. (6.3) and (6.5) and the Killing form introduced in Chap. 1, the background extension of the CF model writes

$$S_{\mathrm{CF}} = \int d^d x \left\{ \frac{1}{2}(\bar{D}_\mu a_\nu; \bar{D}_\mu a_\nu - \bar{D}_\nu a_\mu) + \frac{m^2}{2}(a_\mu; a_\mu) + (ih; \bar{D}_\mu a_\mu) + (\bar{D}_\mu \bar{c}; \bar{D}_\mu c) \right.$$

$$\left. - (\bar{D}_\mu \bar{c}; [a_\mu, c]) - (\bar{D}_\mu a_\nu; [a_\mu, a_\nu]) + \frac{1}{4}([a_\mu, a_\nu]; [a_\mu, a_\nu]) \right\}. \tag{7.2}$$

Decomposing each field along a Cartan-Weyl basis, we find

$$S_{\mathrm{CF}} = \int d^d x \left\{ \frac{1}{2} \bar{D}_\mu^{-\kappa} a_\nu^{-\kappa} (\bar{D}_\mu^\kappa a_\nu^\kappa - \bar{D}_\nu^\kappa a_\mu^\kappa) + \frac{m^2}{2} a_\mu^{-\kappa} a_\mu^\kappa + ih^{-\kappa} \bar{D}_\mu^\kappa a_\mu^\kappa + \bar{D}_\mu^{-\kappa} \bar{c}^{-\kappa} \bar{D}_\mu^\kappa c^\kappa \right.$$

$$\left. - i f_{\kappa\lambda\tau} (\bar{D}_\mu^\kappa \bar{c}^\kappa) a_\mu^\lambda c^\tau - i f_{\kappa\lambda\tau} (\bar{D}_\mu^\kappa a_\nu^\kappa) a_\mu^\lambda a_\nu^\tau - \frac{1}{4} f_{\kappa\lambda\xi} f_{(-\xi)\tau\sigma} a_\mu^\kappa a_\nu^\lambda a_\mu^\tau a_\nu^\sigma \right\}. \tag{7.3}$$

Here, we have used $(t_\kappa; t_\lambda) = -\delta_{\kappa,-\lambda}$, and we have introduced the notation $f_{\kappa\lambda\tau} \equiv -(t_\kappa; [t_\lambda, t_\tau])$. This tensor is totally antisymmetric, owing to the antisymmetry of the bracket and the cyclicality of the Killing form (see Chap. 1). In fact, it plays the role of the structure constant tensor in the Cartan-Weyl basis since $[t_\lambda, t_\tau] = f_{\kappa\lambda\tau} t_{-\kappa}$ as it is easily shown. In particular, this has been used to write the four-gluon interaction.

We may now ask how is the conservation of color manifest at the level of the action (7.3). To answer this question, let us first notice that $f_{\kappa\lambda\tau}$ vanishes if two or three labels are zeros. This is because $[t_{0(j)}, t_{0(k)}] = 0$. Moreover, since $[t_{0j}, t_\alpha] = \alpha_j t_\alpha$, we have $f_{0(j)\alpha\beta} = \alpha_j \delta_{\beta(-\alpha)}$. Finally, given two roots α and β, it can be shown that

$$\mathrm{ad}_{t_{0(j)}} \mathrm{ad}_{t_\alpha} t_\beta = (\alpha + \beta)_j \mathrm{ad}_{t_\alpha} t_\beta, \tag{7.4}$$

(see Problem 7.1). So, either $\mathrm{ad}_{t_\alpha} t_\beta = [t_\alpha, t_\beta] = 0$, or it has to be collinear to $t_{\alpha+\beta}$ (in which case $\alpha + \beta$ has to be a root). In all cases, we notice that $f_{\kappa\lambda\tau} = 0$ if $\kappa + \lambda + \tau \neq 0$, which ensures color conservation at the level of the action (7.3).

Finally, let us mention that upon complex conjugation in the complexified algebra (see Appendix A) and using that it is possible to choose the t_κ such that $\bar{t}_\kappa = -t_{-\kappa}$ (where the bar denotes the conjugation in the complexified algebra), we find $[t_\lambda, t_\tau] = -f^*_{(-\kappa)(-\lambda)(-\tau)} t_{-\kappa}$ which implies

$$f_{\kappa\lambda\tau} = -f^*_{(-\kappa)(-\lambda)(-\tau)}. \tag{7.5}$$

7.1.2 Feynman Rules

The action (7.3) is pretty similar to the action in the Landau gauge, up to the missing conventional factor of i in the structure constants and the presence of background covariant derivatives that depend on the color charge of the field they act upon. In Fourier space, these covariant derivatives become generalized momenta, the color charge of the field entering as a shift $Tr \cdot \kappa$ of the frequencies, similar to an imaginary chemical potential for the color charge. This means that the Feynman rules for the Landau-DeWitt gauge in a Cartan-Weyl basis are pretty similar to those for the Landau gauge in a conventional Cartesian basis. For the gluon and ghost propagators, we find, respectively,

$$P^\perp_{\mu\nu}(Q_\kappa) G_m(Q_\kappa) \tag{7.6}$$

and

$$G_0(Q_\kappa), \tag{7.7}$$

where we have introduced

$$G_m(Q) \equiv \frac{1}{Q^2 + m^2}, \tag{7.8}$$

as well as the transverse projector $P^\perp_{\mu\nu}(Q) \equiv \delta_{\mu\nu} - Q_\mu Q_\nu / Q^2$. These propagators are represented diagrammatically in Fig. 7.1. Each line carries both a momentum Q and a color charge κ. We note the property $Q_\kappa = -(-Q)_{-\kappa}$ which implies

Fig. 7.1 Diagrammatic representation of the propagators in the Landau-DeWitt gauge

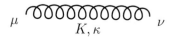

K, κ

μ ν

K, κ

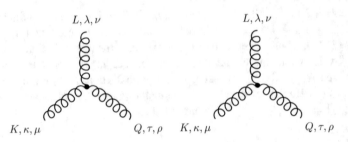

L, λ, ν L, λ, ν

K, κ, μ Q, τ, ρ K, κ, μ Q, τ, ρ

Fig. 7.2 Diagrammatic representation of the derivative vertices

$G_m(Q_\kappa) = G_m((-Q)_{-\kappa})$ and $P^{\perp}_{\mu\nu}(Q_\kappa) = P^{\perp}_{\mu\nu}((-Q)_{-\kappa})$ and thus that the common orientation of momentum and charge can be chosen arbitrarily.

As for the vertices, the ghost-antighost-gluon vertex is given by

$$g f_{\kappa\lambda\tau} K^{\kappa}_{\nu}, \tag{7.9}$$

where K is the momentum of the outgoing antighost, κ is the corresponding outgoing color charge, and similarly λ and τ are the charges of the outgoing gluon and ghost. Similarly, the three-gluon vertex is given by

$$\frac{g}{6} f_{\kappa\lambda\tau} \left\{ \delta_{\mu\rho}(K^{\kappa}_{\nu} - Q^{\tau}_{\nu}) + \delta_{\nu\mu}(L^{\lambda}_{\rho} - K^{\kappa}_{\rho}) + \delta_{\rho\nu}(Q^{\tau}_{\mu} - L^{\lambda}_{\mu}) \right\}, \tag{7.10}$$

where K_κ, L_λ and Q_τ and L_τ denote the shifted outgoing momenta associated with the indices μ, ν, and ρ. These derivative vertices are represented in Fig. 7.2. Finally, the four-gluon vertex is given by

$$\frac{g^2}{24} \sum_{\eta} \left\{ f_{\kappa\lambda\eta} f_{\tau\xi(-\eta)} (\delta_{\mu\rho}\delta_{\nu\sigma} - \delta_{\mu\sigma}\delta_{\nu\rho}) \right.$$

$$+ f_{\kappa\tau\eta} f_{\lambda\xi(-\eta)} (\delta_{\mu\nu}\delta_{\rho\sigma} - \delta_{\mu\sigma}\delta_{\nu\rho})$$

$$\left. + f_{\kappa\xi\eta} f_{\tau\lambda(-\eta)} (\delta_{\mu\rho}\delta_{\nu\sigma} - \delta_{\mu\nu}\delta_{\sigma\rho}) \right\}, \tag{7.11}$$

where κ, λ, τ, and ξ represent the outgoing charges. This vertex is represented in Fig. 7.3. Using Eq. (7.5), the momenta/charges in the vertices of Figs. 7.2 and 7.3 can also all be considered incoming, provided one replaces Eqs. (7.9), (7.10), and (7.11) by the complex-conjugated expressions.

Fig. 7.3 Diagrammatic representation of the four-gluon vertex

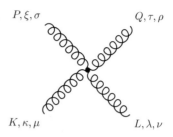

P, ξ, σ Q, τ, ρ

K, κ, μ L, λ, ν

7.2 Two-Loop Effective Potential

With the Feynman rules at our disposal, we can now evaluate corrections to the one-loop background-field effective potential. There are three diagrams contributing at two-loop order. We can label them according to the number of gluon propagators, $\delta \tilde{V}_{2\text{loop}}^{1\text{gl}}$, $\delta \tilde{V}_{2\text{loop}}^{2\text{gl}}$, and $\delta \tilde{V}_{2\text{loop}}^{3\text{gl}}$, respectively. Applying the Feynman rules listed above, we find

$$\delta \tilde{V}_{2\text{loop}}^{1\text{gl}} = \sum_{\kappa\lambda\tau} C_{\kappa\lambda\tau} \left[-\frac{g^2}{2} \int_Q^T \int_K^T Q_\kappa \cdot P^\perp(L_\tau) \cdot K_\lambda \, G_0(Q_\kappa) \, G_0(K_\lambda) \, G_m(L_\tau) \right],$$

$$\delta \tilde{V}_{2\text{loop}}^{2\text{gl}} = \frac{g^2}{4} \sum_{\kappa\lambda\tau} C_{\kappa\lambda\tau} \int_Q^T \int_K^T \left[\operatorname{tr} P^\perp(Q_\kappa) \operatorname{tr} P^\perp(K_\lambda) \right.$$

$$\left. - \operatorname{tr} P^\perp(Q_\kappa) \, P^\perp(K_\lambda) \right] G_m(Q_\kappa) \, G_m(K_\lambda),$$

$$\delta \tilde{V}_{2\text{loop}}^{3\text{gl}} = -g^2 \sum_{\kappa\lambda\tau} C_{\kappa\lambda\tau} \int_Q^T \int_K^T \mathcal{B}(Q_\kappa, K_\lambda, L_\tau) \, G_m(Q_\kappa) \, G_m(K_\lambda) \, G_m(L_\tau),$$

$$(7.12)$$

where we have defined the tensor $C_{\kappa\lambda\tau} \equiv |f_{\kappa\lambda\tau}|^2$. Since $f_{\kappa\lambda\tau}$ vanishes if $\kappa + \lambda + \tau \neq 0$, the same is true for $C_{\kappa\lambda\tau}$, and we can assume that the color sums in Eq. (7.12) are constrained by color conservation, meaning that $\kappa + \lambda + \tau = 0$. Since we have chosen the momenta such that $Q + K + L = 0$, the generalized momenta are also conserved in the sense that

$$Q_\kappa + K_\lambda + L_\tau = 0. \qquad (7.13)$$

We shall repeatedly make use of this property in what follows. In particular, under this constraint, one can show the following very useful result, left as an exercise in Problem 7.2: the combination $\Delta(Q_\kappa, K_\lambda, L_\tau) \equiv Q_\kappa^2 K_\lambda^2 - (Q_\kappa \cdot K_\lambda)^2$ that shall appear frequently below is totally symmetric under permutations of $(Q_\kappa, K_\lambda, L_\tau)$.

Let us also mention that, for the last diagram, we have introduced the function

$$\mathcal{B}(Q_\kappa, K_\lambda, L_\tau) \equiv - \left[Q_\kappa \cdot P^\perp(L_\tau) \cdot K_\lambda \right] \text{tr} \left[P^\perp(K_\lambda) P^\perp(Q_\kappa) \right]$$
$$- 2 L_\tau \cdot P^\perp(Q_\kappa) \cdot P^\perp(L_\tau) \cdot P^\perp(K_\lambda) \cdot L_\tau . \quad (7.14)$$

Owing to the symmetry of the corresponding summand/integrand in Eq. (7.12) with respect to permutations of the triplet $(Q_\kappa, K_\lambda, L_\tau)$, we can replace this function by its symmetrized version $\mathcal{B}_{\text{sym}}(Q_\kappa, K_\lambda, L_\tau)$. A lengthy but straightforward calculation shows that

$$\mathcal{B}_{\text{sym}}(Q_\kappa, K_\lambda, L_\tau) = \frac{1}{3} \left\{ \left(d - \frac{3}{2} \right) \left(\frac{1}{Q_\kappa^2} + \frac{1}{K_\lambda^2} + \frac{1}{L_\tau^2} \right) \right.$$
$$\left. + \frac{Q_\kappa^4 + K_\lambda^4 + L_\tau^4}{4 Q_\kappa^2 K_\lambda^2 L_\tau^2} \right\} \Delta(Q_\kappa, K_\lambda, L_\tau), \quad (7.15)$$

(see Problem 7.3).

7.2.1 Reduction to Scalar Sum-Integrals

It is convenient to reduce the sum-integrals in Eq. (7.12) to a set of simpler sum-integrals. To this purpose, we proceed in two steps. We first reduce everything to the set

$$J_\alpha^\kappa \equiv \int_Q^T G_\alpha(Q_\kappa), \qquad J_{\mu\nu}^\kappa \equiv \int_Q^T \frac{Q_\mu^\kappa Q_\nu^\kappa}{Q_\kappa^2} G_m(Q_\kappa). \quad (7.16)$$

and

$$I_{\alpha\beta\gamma}^{\kappa\lambda\tau} \equiv \int_Q^T \int_K^T \Delta(Q_\kappa, K_\lambda, L_\tau) G_\alpha(Q_\kappa) \, G_\beta(K_\lambda) \, G_\gamma(L_\tau). \quad (7.17)$$

For instance, using

$$\text{tr } P^\perp(Q_\kappa) = d - 1 \quad \text{and} \quad \text{tr } P^\perp(Q_\kappa) P^\perp(K_\lambda) = d - 2 + \frac{(Q_\kappa \cdot K_\lambda)^2}{Q_\kappa^2 K_\lambda^2}, \quad (7.18)$$

we find

$$\delta \tilde{V}_{2\text{loop}}^{2\text{gl}} = \frac{g^2}{4} \sum_{\kappa,\lambda,\tau} C_{\kappa\lambda\tau} \left[(d^2 - 3d + 3) J_m^\kappa J_m^\lambda - J_{\mu\nu}^\kappa J_{\mu\nu}^\lambda \right], \tag{7.19}$$

for the diagram with two-gluon propagators. The diagram with one-gluon propagator can be treated using Eq. (7.13) together with the identity

$$G_0(L_\tau) G_m(L_\tau) = \frac{1}{m^2} \left[G_0(L_\tau) - G_m(L_\tau) \right]. \tag{7.20}$$

We find

$$\delta \tilde{V}_{2\text{loop}}^{1\text{gl}} = \frac{g^2}{2m^2} \sum_{\kappa,\lambda,\tau} C_{\kappa\lambda\tau} \left[I_{000}^{\kappa\lambda\tau} - I_{00m}^{\kappa\lambda\tau} \right]. \tag{7.21}$$

As for the diagram with three-gluon propagators, similar considerations using the symmetrized function (7.15) lead to

$$\delta \tilde{V}_{2\text{loop}}^{3\text{gl}} = -\frac{g^2}{4} \sum_{\kappa,\lambda,\tau} C_{\kappa\lambda\tau} \left\{ J_m^\kappa J_m^\lambda - J_{\mu\nu}^\kappa J_{\mu\nu}^\lambda \right. $$
$$\left. - \frac{4}{m^2} \left[\left(d - \frac{5}{4} \right) I_{mmm}^{\kappa\lambda\tau} - (d-1) I_{mm0}^{\kappa\lambda\tau} + \frac{1}{4} I_{m00}^{\kappa\lambda\tau} \right] \right\}. \tag{7.22}$$

Adding up the three diagrams, we arrive at

$$\delta \tilde{V}_{2\text{loop}} = g^2 \sum_{\kappa,\lambda,\tau} C_{\kappa\lambda\tau} \left\{ \frac{(d-1)(d-2)}{4} J_m^\kappa J_m^\lambda + \frac{1}{2m^2} I_{000}^{\kappa\lambda\tau} \right.$$
$$\left. + \frac{1}{m^2} \left[\left(d - \frac{5}{4} \right) I_{mmm}^{\kappa\lambda\tau} - (d-1) I_{mm0}^{\kappa\lambda\tau} - \frac{1}{4} I_{m00}^{\kappa\lambda\tau} \right] \right\}, \tag{7.23}$$

where we note that the contributions involving $J_{\mu\nu}^\kappa$ have cancelled out.

In the second step, we reduce all the sum-integrals to the set J_α^κ,

$$\tilde{J}_\alpha^\kappa \equiv \int_Q^T Q_\kappa^0 G_\alpha(Q_\kappa) \quad \text{and} \quad S_{\alpha\beta\gamma}^{\kappa\lambda\tau} \equiv \int_Q^T \int_K^T G_\alpha(Q_\kappa) G_\beta(K_\lambda) G_\gamma(L_\tau). \tag{7.24}$$

To this purpose, we use

$$Q_\kappa \cdot K_\lambda = \frac{1}{2} \left[\alpha + \beta - \gamma + L_\tau^2 + \gamma - (Q_\kappa^2 + \alpha) - (K_\lambda^2 + \beta) \right] \tag{7.25}$$

to obtain, after a lengthy but straightforward calculation

$$
I_{\alpha\beta\gamma}^{\kappa\lambda\tau} = \frac{1}{4}\left[(\gamma - \alpha - \beta)\, J_\alpha^\kappa J_\beta^\lambda + (\alpha - \beta - \gamma)\, J_\beta^\lambda J_\gamma^\tau + (\beta - \gamma - \alpha -)\, J_\gamma^\tau J_\alpha^\kappa\right]
$$
$$
- \frac{1}{4}\left(\alpha^2 + \beta^2 + \gamma^2 - 2\alpha\beta - 2\beta\gamma - 2\gamma\alpha\right) S_{\alpha\beta\gamma}^{\kappa\lambda\tau}
$$
$$
- \frac{1}{2}\left[\tilde{J}_\alpha^\kappa \tilde{J}_\beta^\lambda + \tilde{J}_\beta^\lambda \tilde{J}_\gamma^\tau + \tilde{J}_\gamma^\tau \tilde{J}_\alpha^\kappa\right], \tag{7.26}
$$

(see Problem 7.4). Plugging this identity back into Eq. (7.23), we arrive finally at

$$
\delta\tilde{V}_{2\text{loop}} = g^2 \sum_{\kappa,\lambda,\tau} C_{\kappa\lambda\tau}\Bigg\{ \frac{1}{4}\left(d^2 - 4d + \frac{15}{4}\right) J_m^\kappa J_m^\lambda + \frac{1}{8} J_0^\kappa J_m^\lambda - \frac{1}{16} J_0^\kappa J_0^\lambda
$$
$$
- \frac{1}{m^2}\left(d - \frac{11}{8}\right) \tilde{J}_m^\kappa \tilde{J}_m^\lambda + \frac{1}{m^2}\left(d - \frac{3}{4}\right) \tilde{J}_0^\kappa \tilde{J}_m^\lambda - \frac{5}{8m^2} \tilde{J}_0^\kappa \tilde{J}_0^\lambda
$$
$$
+ \frac{3m^2}{4}\left(d - \frac{5}{4}\right) S_{mmm}^{\kappa\lambda\tau} + \frac{m^2}{16} S_{m00}^{\kappa\lambda\tau} \Bigg\}. \tag{7.27}
$$

This expression is convenient because the Matsubara sum-integrals J_α^κ, $\tilde{J}_\alpha^\kappa$, and $S_{\alpha\beta\gamma}^{\kappa\lambda\tau}$ are essentially those found in a scalar theory and can be easily evaluated using standard techniques (see Ref. [93] and Appendix B).

7.2.2   Thermal Decomposition

After evaluating the elementary sum-integrals identified above, one can typically decompose them as

$$
J_\alpha^\kappa = J_\alpha(0n) + J_\alpha^\kappa(1n), \tag{7.28}
$$
$$
\tilde{J}_\alpha^\kappa = \tilde{J}_\alpha(0n) + \tilde{J}_\alpha^\kappa(1n), \tag{7.29}
$$
$$
S_{\alpha\beta\gamma}^{\kappa\lambda\tau} = S_{\alpha\beta\gamma}(0n) + S_{\alpha\beta\gamma}^{\kappa\lambda\tau}(1n) + S_{\alpha\beta\gamma}^{\kappa\lambda\tau}(2n), \tag{7.30}
$$

where (in) refers to the number of loops that are cut off by the presence of thermal factors n^1 (see [93] and Appendix B). Plugging these decompositions into Eq. (7.27), we arrive at a similar decomposition for the two-loop correction to the background-field effective potential:

$$
\delta\tilde{V}_{2\text{loop}} = \delta\tilde{V}_{2\text{loop}}(0n) + \delta\tilde{V}_{2\text{loop}}(1n) + \delta\tilde{V}_{2\text{loop}}(2n). \tag{7.31}
$$

---

[1] More precisely, these should be factors of n with an argument whose real part is positive, so that the loop momentum is indeed cut off.

As we have shown in the previous chapter, the zero-temperature piece $\delta \tilde{V}_{2\text{loop}}(0n)$ does not depend on the background and can be ignored in the following analysis.

The contribution $\delta \tilde{V}_{2\text{loop}}(1n)$ involves products of the form $J_\alpha(0n) J_\beta^\lambda(1n)$ or $\tilde{J}_\alpha(0n) \tilde{J}_\beta^\lambda(1n)$, as well as $S_{\alpha\beta\gamma}^{\kappa\lambda\tau}(1n)$. This latter quantity can be expressed in terms of $J_\alpha^\kappa(1n)$ as

$$S_{\alpha\beta\gamma}^{\kappa\lambda\tau}(1n) = J_\alpha^\kappa(1n) \, \text{Re} \, \tilde{I}_{\beta\gamma}(\varepsilon_{\alpha,q} + i0^+; q)$$

$$J_\beta^\lambda(1n) \, \text{Re} \, \tilde{I}_{\gamma\alpha}(\varepsilon_{\beta,k} + i0^+; k)$$

$$J_\gamma^\tau(1n) \, \text{Re} \, \tilde{I}_{\alpha\beta}(\varepsilon_{\gamma,l} + i0^+; l), \qquad (7.32)$$

where $\text{Re} \, \tilde{I}_{\beta\gamma}(\varepsilon_{\alpha,q} + i0^+; q)$ is defined as the real part of the vacuum bubble integral

$$I_{\beta\gamma}(Q) \equiv \int_K^{T=0} G_\beta(K) \, G_\gamma(L), \qquad (7.33)$$

analytically continued to real frequencies $Q = (\omega_q, q) \to (-iq_0 + \epsilon, q)$ and evaluated on the mass shell $q_0^2 - q^2 = \alpha^2$. For the cases of interest here, we find

$$\text{Re} \, \tilde{I}_{00}(\varepsilon_q + i0^+; q) = \frac{1}{16\pi^2} \left[\frac{1}{\epsilon} + \ln \frac{\bar{\mu}^2}{m^2} + 2 \right], \qquad (7.34)$$

$$\text{Re} \, \tilde{I}_{m0}(q + i0^+; q) = \frac{1}{16\pi^2} \left[\frac{1}{\epsilon} + \ln \frac{\bar{\mu}^2}{m^2} + 1 \right], \qquad (7.35)$$

$$\text{Re} \, \tilde{I}_{mm}(\varepsilon_q + i0^+; q) = \frac{1}{16\pi^2} \left[\frac{1}{\epsilon} + \ln \frac{\bar{\mu}^2}{m^2} + 2 - \frac{\pi}{\sqrt{3}} \right], \qquad (7.36)$$

where we note that the right-hand sides do not depend on q.[2] From these relations, together with the standard results

$$J_\alpha(0n) = -\frac{\alpha^2}{16\pi^2} \left[\frac{1}{\epsilon} + \ln \frac{\bar{\mu}^2}{\alpha^2} + 1 \right], \qquad (7.37)$$

and $\tilde{J}_\alpha(0n) = 0$, one arrives at

$$\delta \tilde{V}_{2\text{loop}}(1n) = \frac{g^2 C_{\text{ad}}}{128\pi^2} m^2 \left(\frac{35}{\epsilon} + 35 \ln \frac{\bar{\mu}^2}{m^2} + \frac{313}{3} - \frac{99\pi}{2\sqrt{3}} \right) \sum_\kappa J_m^\kappa(1n), \qquad (7.38)$$

---

[2] This follows from the definition of $\text{Re} \, \tilde{I}_{\beta\gamma}(\varepsilon_{\alpha,q} + i0^+; q)$ and Lorentz invariance.

where we have made use of the symmetry of $C_{\kappa\lambda\tau}$ together with the Casimir identity $\sum_{\tau} C_{\kappa\lambda\tau} = C_{\text{ad}}\,\delta_{\kappa,-\lambda}$ [94].

Finally, the contribution $\delta\tilde{V}_{\text{2loop}}(2n)$ with two thermal factors is UV finite by construction, and one can evaluate it directly in $d = 4$ dimensions. One finds

$$
\delta\tilde{V}_{\text{2loop}}(2n) = \frac{3g^2}{8} \sum_{\kappa,\lambda,\tau} C_{\kappa\lambda\tau} \left[\frac{5}{2} U^{\kappa} V^{\lambda} - \frac{7}{m^2} \tilde{U}^{\kappa} \tilde{V}^{\lambda} \right]
$$

$$
+ \frac{g^2 m^2}{16} \sum_{\kappa,\lambda,\tau} C_{\kappa\lambda\tau} \left[33 S^{\kappa\lambda\tau}_{mmm}(2n) + S^{\kappa\lambda\tau}_{m00}(2n) \right], \tag{7.39}
$$

with $\quad U^{\kappa} \equiv J^{\kappa}_m(1n) + J^{\kappa}_0(1n)/3, \quad V^{\kappa} \equiv J^{\kappa}_m(1n) - J^{\kappa}_0(1n)/5, \quad \tilde{U}^{\kappa} \equiv \tilde{J}^{\kappa}_m(1n)$
$- \tilde{J}^{\kappa}_0(1n)$ and $\tilde{V}^{\kappa} \equiv \tilde{J}^{\kappa}_m(1n) - 5\tilde{J}^{\kappa}_0(1n)/21$.

7.2.3 Counterterm Contribution and Renormalization

One could be worried by the presence of a temperature-dependent divergence in $\delta V_{\text{2loop}}(1n)$. However, at two-loop order, one should also consider one-loop diagrams involving counterterms.[3] These are obtained by first rewriting the action in terms of renormalized fields and renormalized parameters as

$$
a_{\mu} \to Z_a^{1/2} a_{\mu}, \quad c \to Z_c^{1/2} c, \quad m^2 \to Z_{m^2} m^2, \quad g \to Z_g g, \tag{7.40}
$$

writing $Z = 1 + \delta Z$ and treating the various δZ as new interaction vertices. At the considered order, we find the counterterm contribution

$$
\delta\tilde{V}^{\text{ct}}_{\text{2loop}} = -\delta Z_c \sum_{\kappa} \int_{Q}^{T} Q_{\kappa}^2 G_0(Q_{\kappa})
$$

$$
+ \frac{1}{2} \sum_{\kappa} \int_{Q}^{T} \left(\delta Z_a(Q_{\kappa}^2 + m^2) + m^2 \delta Z_{m^2} \right) \text{tr}\, P^{\perp}(Q_{\kappa})\, G(Q_{\kappa}). \tag{7.41}
$$

Using $\text{tr}\, P^{\perp}(Q_{\kappa}) = d - 1$ together with the fact that $\int_{Q}^{T} 1 = 0$ in dimensional regularization, this rewrites

$$
\delta\tilde{V}^{\text{ct}}_{\text{2loop}} = \frac{d - 1}{2} m^2 \delta Z_{m^2} \sum_{\kappa} J^{\kappa}_m. \tag{7.42}
$$

---

[3] The counterterms should be evaluated at one-loop order, so that the counterterm diagram counts effectively as a two-loop order diagram.

We note that this is precisely of the same form as the divergence that needs to be cancelled out in Eq. (7.38). In fact, the value of δZ_{m^2} is fixed from the renormalization of the gluon two-point function at one-loop order [83]. In the zero-momentum scheme considered in this reference, one finds

$$\delta Z_{m^2} = \frac{g^2 N}{192\pi^2}\left[-\frac{35}{\epsilon} + z_f\right],\tag{7.43}$$

and

$$\begin{aligned}
z_f = &-\frac{1}{s^2} + \frac{111}{2s} - \frac{287}{6} - 35\ln(\bar{s}) - \frac{1}{2}\left(s^2 - 2\right)\ln(s)\\
&+ \left(s^2 - 10s + 1\right)\left(\frac{1}{s} + 1\right)^3\ln(s + 1)\\
&+ \frac{1}{2}(s^2 - 20s + 12)\left(\frac{4}{s} + 1\right)^{3/2}\ln\left(\frac{\sqrt{4/s + 1} - 1}{\sqrt{4/s + 1} + 1}\right),
\end{aligned}\tag{7.44}$$

with $s \equiv \mu^2/m^2$ and $\bar{s} \equiv \bar{\mu}^2/m^2$. The pole in Eq. (7.43) is universal and exactly cancels the one in Eq. (7.38), as expected since the CF model is renormalizable. Consequently, in what follows, we redefine $\delta\tilde{V}_{2\text{loop}}(1n)$ to be the sum of (7.38) and (7.42):

$$\delta\tilde{V}_{2\text{loop}}(1n) = \frac{g^2 C_{\text{ad}}}{128\pi^2}m^2\left(z_f + 35\ln\bar{s}^2 + \frac{313}{3} - \frac{99\pi}{2\sqrt{3}}\right)\sum_\kappa J_m^\kappa(1n).\tag{7.45}$$

7.2.4 UV-Finite Contributions

To complete our evaluation of the two-loop corrections, we just need to recall the expressions for the UV-finite contributions $J_\alpha^\kappa(1n)$, $\tilde{J}_\alpha^\kappa(1n)$, and $S_{\alpha\beta\gamma}^{\kappa\lambda\tau}(2n)$, (see Ref. [94] and Appendix B). Setting $\hat{r} \equiv r\,T$, one has

$$J_\alpha^\kappa(1n) = \frac{1}{2\pi^2}\int_0^\infty dq\,\frac{q^2}{\varepsilon_{\alpha,q}}\text{Re}\,n_{\varepsilon_{\alpha,q} - i\hat{r}\cdot\kappa},\tag{7.46}$$

$$\tilde{J}_\alpha^\kappa(1n) = \frac{1}{2\pi^2}\int_0^\infty dq\,q^2\text{Im}\,n_{\varepsilon_{\alpha,q} - i\hat{r}\cdot\kappa},\tag{7.47}$$

and

$$S_{\alpha\beta\gamma}^{\kappa\lambda\tau}(2n) = \frac{1}{32\pi^4} \int_0^\infty dq\, q \int_0^\infty dk\, k\, \mathrm{Re}\, \frac{n_{\varepsilon_{\alpha,q}} - i\hat{r}\cdot\kappa\, n_{\varepsilon_{\beta,k}} - i\hat{r}\cdot\lambda}{\varepsilon_{\alpha,q}\, \varepsilon_{\beta,k}}$$

$$\times \mathrm{Re}\, \ln \frac{(\varepsilon_{\alpha,q} + \varepsilon_{\beta,k} + i0^+)^2 - (\varepsilon_{\gamma,k+q})^2}{(\varepsilon_{\alpha,q} + \varepsilon_{\beta,k} + i0^+)^2 - (\varepsilon_{\gamma,k-q})^2}$$

$$+ \frac{1}{32\pi^4} \int_0^\infty dq\, q \int_0^\infty dk\, k\, \mathrm{Re}\, \frac{n_{\varepsilon_{\alpha,q}} - i\hat{r}\cdot\kappa\, n_{\varepsilon_{\beta,k}} + i\hat{r}\cdot\lambda}{\varepsilon_{\alpha,q}\, \varepsilon_{\beta,k}}$$

$$\times \mathrm{Re}\, \ln \frac{(\varepsilon_{\alpha,q} - \varepsilon_{\beta,k} + i0^+)^2 - (\varepsilon_{\gamma,k+q})^2}{(\varepsilon_{\alpha,q} - \varepsilon_{\beta,k} + i0^+)^2 - (\varepsilon_{\gamma,k-q})^2}$$

$$+ \text{cyclic permutations of } (\alpha, \kappa), (\beta, \lambda), \text{ and } (\gamma, \tau). \tag{7.48}$$

These formulas, together with Eqs. (7.45) and (7.39), form the basis for the evaluation of the two-loop corrections to the background-field effective potential.

7.3   Next-to-Leading-Order Polyakov Loop

Corresponding to the two-loop background-field effective potential, we can evaluate the background-dependent Polyakov loop at next-to-leading order. When evaluated at the minimum of the two-loop background-field effective potential, this gives access to the physical Polyakov loop at next-to-leading order. As already emphasized, we do not really need to evaluate the physical Polyakov loop in order to assess the phase transition since the background that minimizes the background field effective potential is an equally relevant order parameter for center symmetry, simpler to compute in practice. Computing the Polyakov loop is nevertheless interesting for it is a gauge-independent quantity and can therefore be compared, in principle, with non-gauge-fixed lattice simulations. Another motivation for evaluating the next-to-leading-order corrections to the Polyakov loop is to assess the role of higher-order corrections in curing some of the artificial behaviors observed at leading order.

7.3.1   Setting Up the Expansion

As emphasized in Chap. 4, the background-dependent Polyakov loop is defined in the presence of a source that forces the background to be self-consistent, that is, $\bar{A} = \langle A \rangle_{\bar{A}}$ or $\langle a \rangle_{\bar{A}} = 0$. We thus need to expand the action and the Wilson line in powers of $a \equiv A - \bar{A}$ and take an average where we set $\langle a \rangle_{\bar{A}} = 0$.

Due to the presence of time-ordering, the expansion of the Wilson line $L_A(\mathbf{x})$ in $a = A - \bar{A}$ (and not A) is cumbersome. We can however use the following trick.[4] Consider the gauge transformation

$$U(\tau) = e^{-g\bar{A}\tau}. \tag{7.49}$$

It is such that $\bar{A}^U = 0$ and therefore

$$L_{\bar{A}+a}(\mathbf{x}) = U^\dagger(\beta) L_{a^U}(\mathbf{x}), \tag{7.50}$$

with $a^U \equiv U a U^\dagger$ (see Problem 7.5). The expansion in a^U is easily done, despite the presence of the time-ordering. To order g^2, we find

$$L_{\bar{A}+a}(\mathbf{x}) = U^\dagger(\beta) \left[\mathbb{1} + g \int_0^\beta d\tau\, a_0^U(\tau, \mathbf{x}) + \frac{g^2}{2} \int_0^\beta d\tau \int_0^\beta d\sigma\, \mathcal{P} a_0^U(\tau, \mathbf{x}) a_0^U(\sigma, \mathbf{x}) \right]. \tag{7.51}$$

Upon taking the trace and averaging with $\langle a \rangle_{\bar{A}} = 0$, we find

$$\ell(\bar{A}) = \ell_0(\bar{A}) - \frac{g^2}{2N} \int_0^\beta d\tau \int_0^\tau d\sigma\, G_{00}^{\kappa\lambda}(\tau - \sigma, \mathbf{0})\, \mathrm{tr}\left[U^\dagger(\beta - \tau + \sigma) t^\kappa U^\dagger(\tau - \sigma) t^\lambda \right]$$
$$- \frac{g^2}{2N} \int_0^\beta d\tau \int_\tau^\beta d\sigma\, G_{00}^{\kappa\lambda}(\sigma - \tau, \mathbf{0})\, \mathrm{tr}\left[U^\dagger(\beta - \sigma + \tau) t^\lambda U^\dagger(\sigma - \tau) t^\kappa \right], \tag{7.52}$$

where we have made use of the cyclicality of the trace and where

$$G_{00}^{\kappa\lambda}(\tau - \sigma, \mathbf{0}) \equiv \langle \mathcal{P} a_0^\kappa(\tau, \mathbf{x}) a_0^\lambda(\sigma, \mathbf{x}) \rangle_{\bar{A}} \tag{7.53}$$

denotes the Euclidean time component of the gluon propagator, computed at tree level. We can make the changes of variables $\sigma \to \beta - \sigma$ and $\sigma \to \beta + \tau - \sigma$ in each of the σ-integrals, respectively. We arrive at

$$\ell(\bar{A}) = \ell_0(\bar{A}) - \frac{g^2}{2N} \int_0^\beta d\tau \int_0^\tau d\sigma\, G_{00}^{\kappa\lambda}(\sigma, \mathbf{0})\, \mathrm{tr}\left[U^\dagger(\beta - \sigma) t^\kappa U^\dagger(\sigma) t^\lambda \right]$$
$$- \frac{g^2}{2N} \int_0^\beta d\tau \int_\tau^\beta d\sigma\, G_{00}^{\kappa\lambda}(\sigma - \beta, \mathbf{0})\, \mathrm{tr}\left[U^\dagger(\beta - \sigma) t^\kappa U^\dagger(\sigma) t^\lambda \right], \tag{7.54}$$

---

[4] We have checked that the brute force calculation leads to the same result.

where we have again used the cyclicality of the trace. Using the periodicity of the propagator, this rewrites eventually as

$$\ell(\bar{A}) = \ell_0(\bar{A}) - \frac{g^2 \beta}{2N} \int_0^\beta d\sigma \, G_{00}^{\kappa\lambda}(\sigma, \mathbf{0}) \, \text{tr} \left[U^\dagger(\beta - \sigma) \, t^\kappa \, U^\dagger(\sigma) \, t^\lambda \right], \qquad (7.55)$$

where the τ-integral has become trivial, producing a factor β.

7.3.2 Using the Weights

To proceed with the calculation, we need to evaluate the trace. Before doing so, let us notice that, with little modifications, the previous formula applies to the Polyakov loop in any representation. We have indeed

$$\ell_R(\bar{A}) = \ell_{R,0}(\bar{A}) - \frac{g^2 \beta}{2d_R} \int_0^\beta d\sigma \, G_{00}^{\kappa\lambda}(\sigma, \mathbf{0}) \, \text{tr} \left[U^\dagger(\beta - \sigma) \, t_R^\kappa \, U^\dagger(\sigma) \, t_R^\lambda \right].$$
$$(7.56)$$

We shall continue with this more generic formula from now on. Now, the color trace is easily evaluated using the weights ρ of the representation, defined from $t_R^{0^{(j)}} | \rho \rangle = \rho_j | \rho \rangle$. Since the background is taken along the Cartan sub-algebra, we have $U^\dagger(\sigma)|\rho\rangle = e^{i\frac{\sigma}{\beta} r \cdot \rho} |\rho\rangle$ and then

$$\text{tr} \left[U^\dagger(\beta - \sigma) \, t^\kappa \, U^\dagger(\sigma) \, t^\lambda \right] = \sum_{\rho\rho'} e^{i\frac{\beta-\sigma}{\beta} r \cdot \rho + i\frac{\sigma}{\beta} r \cdot \rho'} \langle \rho | t_R^\kappa | \rho' \rangle \langle \rho' | t^\lambda | \rho \rangle. \qquad (7.57)$$

Since the free gluon propagator has non-zero components only for $\kappa = \lambda = 0^{(j)}$ and $\kappa = -\lambda = -\alpha$, the color trace has to be considered in those cases only. In the first case, from the definition of the weights, the color trace rewrites

$$\text{tr} \left[U^\dagger(\beta - \sigma) \, t^{0^{(j)}} \, U^\dagger(\sigma) \, t^{0^{(j)}} \right] = \sum_\rho \rho_j^2 \, e^{i r \cdot \rho}. \qquad (7.58)$$

In the second case, similar to what we did for the adjoint representation in Eq. (7.4), it is easily argued that $t^\alpha |\rho\rangle$ is either zero or has a definite weight equal to $\rho + \alpha$ (a necessary condition is that $\rho + \alpha$ belongs to the weight diagram). Therefore, in the second case, the color trace rewrites

$$\text{tr} \left[U^\dagger(\beta - \sigma) \, t^{-\alpha} \, U^\dagger(\sigma) \, t^\alpha \right] = \sum_\rho e^{i r \cdot \rho + i\frac{\sigma}{\beta} r \cdot \alpha} |\langle \rho + \alpha | t_R^\alpha | \rho \rangle|^2, \qquad (7.59)$$

where it is understood that $|\rho - \alpha\rangle = 0$ if $\rho - \alpha$ is not a weight. In fact, Eqs. (7.58) and (7.59) can be written as a single result. One has just to replace α by

κ in Eq. (7.59). The next-to-leading-order correction to the background-dependent Polyakov loop reads, therefore,

$$\Delta \ell_R(\bar{A}) = -\frac{g^2 \beta}{2 d_R} \sum_\rho e^{ir \cdot \rho} \sum_\kappa |\langle \rho + \kappa | t_R^\kappa | \rho \rangle|^2 \int_0^\beta d\sigma \, e^{i \frac{\sigma}{\beta} r \cdot \kappa} \, G_{00}^{(-\kappa)\kappa}(\sigma, \mathbf{0}) .$$

(7.60)

7.3.3 Completing the Calculation

We finally note that

$$G_{00}^{(-\kappa)\kappa}(\sigma, \mathbf{0}) = \int_Q^T \frac{q^2}{(\omega_n + Tr \cdot \kappa)^2 + q^2} \frac{e^{i\omega_n \sigma}}{(\omega_n + Tr \cdot \kappa)^2 + q^2 + m^2}$$

$$= \frac{1}{m^2} \left[\int_Q^T \frac{q^2 \, e^{i\omega_n \sigma}}{(\omega_n + Tr \cdot \kappa)^2 + q^2} - \int_Q^T \frac{q^2 \, e^{i\omega_n \sigma}}{(\omega_n + Tr \cdot \kappa)^2 + q^2 + m^2} \right].$$

(7.61)

Standard contour techniques for the evaluation of Matsubara sums (see Appendix B) lead to (recall that $0 < \sigma < \beta$)

$$\int_Q^T \frac{q^2 \, e^{i\omega_n \sigma}}{(\omega_n + Tr \cdot \kappa)^2 + q^2 + m^2}$$

$$= \int \frac{d^3 q}{(2\pi)^3} \frac{q^2}{2\varepsilon_q} \left[e^{(\varepsilon_q - iTr \cdot \kappa)\sigma} n_{\varepsilon_q - iTr \cdot \kappa} - e^{(-\varepsilon_q - iTr \cdot \kappa)\sigma} n_{-\varepsilon_q - iTr \cdot \kappa} \right],$$

(7.62)

which generalizes the well-known result for $\kappa = 0$. It follows that

$$\int_0^\beta d\sigma \, e^{i \frac{\sigma}{\beta} r \cdot \kappa} \, G_{00}^{(-\kappa)\kappa}(\sigma, \mathbf{0})$$

$$= \frac{1}{2m^2} \int \frac{d^3 q}{(2\pi)^3} \left\{ \left[(e^{\beta q} - 1) n_{q - iTr \cdot \kappa} + (e^{-\beta q} - 1) n_{-q - iTr \cdot \kappa} \right] \right.$$

$$\left. - \frac{q^2}{q^2 + m^2} \left[(e^{\beta \varepsilon_q} - 1) n_{\varepsilon_q - iTr \cdot \kappa} + (e^{-\beta \varepsilon_q} - 1) n_{-\varepsilon_q - iTr \cdot \kappa} \right] \right\}.$$

(7.63)

Finally, we write

$$(e^{\beta \varepsilon_q} - 1)n_{\varepsilon_q - iT r \cdot \kappa} = e^{ir \cdot \kappa} + (e^{ir \cdot \kappa} - 1)n_{\varepsilon_q - iT r \cdot \kappa} \tag{7.64}$$

from which it follows that

$$
\begin{aligned}
&(e^{\beta \varepsilon_q} - 1)n_{\varepsilon_q - iT r \cdot \kappa} + (e^{-\beta \varepsilon_q} - 1)n_{-\varepsilon_q - iT r \cdot \kappa} \\
&= 2e^{ir \cdot \kappa} + (e^{ir \cdot \kappa} - 1)(n_{\varepsilon_q - iT r \cdot \kappa} + n_{-\varepsilon_q - iT r \cdot \kappa}) \\
&= e^{ir \cdot \kappa} + 1 + (e^{ir \cdot \kappa} - 1)(n_{\varepsilon_q - iT r \cdot \kappa} - n_{\varepsilon_q + iT r \cdot \kappa}),
\end{aligned} \tag{7.65}
$$

and thus that

$$
\begin{aligned}
&\int_0^\beta d\sigma \, e^{i\frac{\sigma}{\beta} r \cdot \kappa} \, G_{00}^{(-\kappa)\kappa}(\sigma, \mathbf{0}) \\
&= \frac{e^{ir \cdot \kappa} + 1}{2} \int \frac{d^3 q}{(2\pi)^3} \frac{1}{q^2 + m^2} \\
&+ i \frac{e^{ir \cdot \kappa} - 1}{m^2} \int \frac{d^3 q}{(2\pi)^3} \left\{ \operatorname{Im} n_{q - iT r \cdot \kappa} - \frac{q^2}{q^2 + m^2} \operatorname{Im} n_{\varepsilon_q - iT r \cdot \kappa} \right\},
\end{aligned} \tag{7.66}
$$

which we conveniently rewrite as

$$
\begin{aligned}
&\int_0^\beta d\sigma \, e^{i\frac{\sigma}{\beta} r \cdot \kappa} \, G_{00}^{(-\kappa)\kappa}(\sigma, \mathbf{0}) \\
&= \frac{e^{ir \cdot \kappa} + 1}{2} \left[\int \frac{d^3 q}{(2\pi)^3} \frac{1}{q^2 + m^2} + \frac{2}{m^2} \sin^2 \left(\frac{r \cdot \kappa}{2} \right) \right. \\
&\left. \times \int \frac{d^3 q}{(2\pi)^3} \left\{ \frac{q^2}{\varepsilon_q^2} \frac{1}{\cosh(\varepsilon_q / T) - \cos(r \cdot \kappa)} - (m \to 0) \right\} \right].
\end{aligned} \tag{7.67}
$$

We can now plug this expression back into Eq. (7.60) by noticing that

$$\sum_\rho e^{ir \cdot \rho} \sum_\kappa |\langle \rho + \kappa | t_R^\kappa | \rho \rangle|^2 e^{ir \cdot \kappa} e_\kappa = \sum_\rho e^{ir \cdot \rho} \sum_\kappa |\langle \rho + \kappa | t_R^\kappa | \rho \rangle|^2 e_\kappa, \tag{7.68}$$

for any e_κ such that $e_{-\kappa} = e_\kappa$. We arrive at

$$
\begin{aligned}
\Delta \ell_R(\bar{A}) &= \frac{g^2}{8\pi} \frac{C_R}{d_R} \frac{m}{T} \sum_\rho e^{ir \cdot \rho} \\
&+ \frac{g^2}{2\pi^2} \frac{1}{d_R} \frac{m}{T} \sum_\rho e^{ir \cdot \rho} \sum_\kappa |\langle \rho + \kappa | t_R^\kappa | \rho \rangle| a(T, r \cdot \kappa) \sin^2 \left(\frac{r \cdot \kappa}{2} \right)
\end{aligned} \tag{7.69}
$$

where we have used that $\sum_\kappa |\langle \rho + \kappa | t_R^\kappa | \rho \rangle|^2 = C_R$ is the Casimir of the representation,

$$\int \frac{d^3 q}{(2\pi)^3} \frac{1}{q^2 + m^2} = -\frac{m}{4\pi} \tag{7.70}$$

in dimensional regularization, and we have introduced the function

$$a(T, r \cdot \kappa) \equiv \int_0^\infty \frac{dq \, q^2}{m^3} \left(\frac{1}{\cosh(q/T) - \cos(r \cdot \kappa)} - \frac{q^2}{\varepsilon_q^2} \frac{1}{\cosh(\varepsilon_q/T) - \cos(r \cdot \kappa)} \right). \tag{7.71}$$

7.4 Results

We use a renormalization scheme at zero temperature, more precisely the zero-momentum scheme alluded to above. In this limit, the Landau-DeWitt gauge boils down to the Landau gauge. We can therefore adjust the values of m and g to those for which the CF propagators (computed within the same renormalization scheme) best fit the Landau gauge lattice propagators at zero temperature, namely, $g = 4.9$ and $m = 540\,\mathrm{MeV}$ (see Ref. [83]).

7.4.1 SU(2) and SU(3) Transitions

We find once more that the transition is second order in the SU(2) case and first order in the SU(3) case. Our updated values for the transition temperatures are shown in Table 7.1 and have neatly improved as compared to the leading-order estimates.

7.4.2 Polyakov Loops

The Polyakov loop is shown in Fig. 7.4 for both the SU(2) and SU(3) groups. Let us first mention that the singularity that we found at leading order (in addition to the one at T_c) seems to have disappeared. Interestingly, the Polyakov loop

Table 7.1 Transition temperatures for the SU(2) and SU(3) deconfinement transitions as computed from various approaches, compared to the values obtained within the Curci-Ferrari model at two-loop order

T_c (MeV)	Lattice [175]	fRG [49]	Variational [174]	2-loop CF [94]
SU(2)	295	230	239	284
SU(3)	270	275	245	254

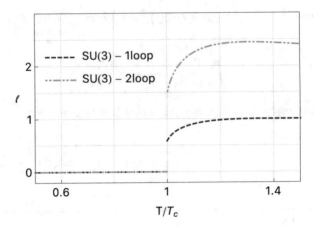

Fig. 7.4 Leading (dashed) and next-to-leading (dash-dotted)-order Polyakov loops for the SU(3) defining representation, as functions of T/T_c

displays a similar non-monotonous behavior as the Polyakov loop evaluated on the lattice, approaching its large-temperature limit from above. However, the rise of the Polyakov loop above the transition is faster than on the lattice, which seems to be a common feature of many calculations in the continuum, with the noticeable exception of [180]. In Chap. 11, we put forward a different implementation of background-field techniques at finite temperature where the rise of the Polyakov loop is slower and a faithful comparison to the lattice data can be performed within the Curci-Ferrari model, up to temperatures of the order of twice the transition temperature.

7.4.3 Vicinity of the Transition

Now that we have the two-loop corrections to the potential, we can revisit the vicinity of the transition. We find that the slight violation of positivity at the level of the entropy density that was found at one loop has disappeared (see Fig. 7.5). However, we find only a slight change of the latent which remains roughly 1/3 of the lattice value.

7.4.4 Low-Temperature Behavior

A simple analysis [93] shows that, in the absence of RG improvement at least, the two-loop corrections are sub-leading with respect to the one-loop ones in the low-temperature regime. This seems a good news since, as we saw, the latter are thermodynamically consistent despite being dominated by the ghosts. However, one important problem remains since these contributions are massless and contribute

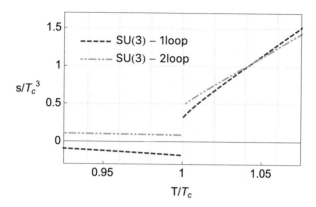

Fig. 7.5 SU(3) entropy density at one-loop (dashed) and two-loop (dash-dotted) order, in the vicinity of T_c

to the thermodynamical observables polynomially in T, a behavior which is not observed on the lattice.

This problem is not specific to the CF model and occurs in fact in most continuum approaches in the Landau gauge (at least those that do not dispose of the ghost degrees of freedom by hand) (see, for instance, Ref. [178]) and illustrates their inability, so far, to properly exclude the massless degrees of freedom from the thermodynamical quantities.

Problems

7.1 Color Conservation ($\star$)
Show that

$$\mathrm{ad}_{t_{0(j)}} \mathrm{ad}_{t_\alpha} t_\beta = (\alpha + \beta)_j \mathrm{ad}_{t_\alpha} t_\beta .$$

Tip: rewrite the LHS in terms of $[t_{0(j)}, t_\alpha]$ and $[t_{0(j)}, t_\beta]$. Why does this identity encode color conservation?

7.2 Symmetry of $\Delta(Q_\kappa, K_\lambda, L_\tau)$ ($\star$)
Show that, under the constraint, $Q_\kappa + K_\lambda + L_\tau = 0$, the function $\Delta(Q_\kappa, K_\lambda, L_\tau) \equiv Q_\kappa^2 K_\lambda^2 - (Q_\kappa \cdot K_\lambda)^2$ is totally symmetric under permutations of $(Q_\kappa, K_\lambda, L_\tau)$. *Tip:* rewrite this combination in terms of $P^\perp(Q_\kappa)$, and use the momentum conservation constraint.

7.3 Contractions (⋆⋆⋆)

(a) Show that the first term in the RHS of (7.14) rewrites as

$$
\frac{\Delta(Q_\kappa, K_\lambda, L_\tau)}{L_\tau^2} \left[d - 2 + \frac{(K_\lambda \cdot Q_\kappa)^2}{K_\lambda^2 Q_\kappa^2} \right],
$$

whereas the second term rewrites as

$$
2 \frac{\Delta(Q_\kappa, K_\lambda, L_\tau)}{Q_\kappa^2 K_\lambda^2 L_\tau^2} (Q_\kappa \cdot L_\tau)(L_\tau \cdot K_\lambda).
$$

(b) Upon symmetrization, obtain Eq. (7.15). *Tip:* use the constraint (7.13) to simplify $(K_\lambda \cdot Q_\kappa)^2 + 2(Q_\kappa \cdot L_\tau)(L_\tau \cdot K_\lambda) +$ permutations of $(Q_\kappa, K_\lambda, L_\tau)$.

7.4 Reduction Formula (⋆⋆⋆)

Show Eq. (7.26). *Tip:* use Eq. (7.25).

7.5 Gauging Away the Background in the Polyakov Loop Operator (⋆)

With U given in Eq. (7.49), show that $\bar{A}^U = 0$ and that Eq. (7.50) holds true.

Relation Between the Center Symmetry Group and the Deconfinement Transition

The confinement/deconfinement transition in pure Yang-Mills theories has been studied for a variety of gauge groups using lattice simulations [181–183] or non-perturbative continuum methods [76]. Not only is this useful as a benchmark for the various theoretical approaches to non-abelian gauge theories, but it also helps refining the connection between the confinement/deconfinement transition, the Polyakov loop, and center symmetry breaking. Here, we provide a general discussion of these questions in the case of the SU(N) gauge group and investigate them in more detail in the case of the SU(4) gauge group using the background-extended Curci-Ferrari model.

In Chap. 3, we saw that, if the center symmetry group $\simeq Z_N \equiv \{e^{i\frac{2\pi}{N}k}\mathbb{1}|k = 0, \ldots, N-1\}$ is realized in the Wigner-Weyl sense, the Polyakov loop associated with the defining representation needs to vanish. We also assumed the reciprocal property to hold true. This is far from obvious, however, because one could imagine scenarios where only a subgroup of the center symmetry group is broken, in which case the defining Polyakov loop would still be equal to zero. Although this is not a possibility for any prime value of N since Z_N does not have any non-trivial subgroup in this case,[1] for other values of N, the question needs to be posed how to characterize the Wigner-Weyl realization of center symmetry in terms of order parameters. To this purpose, it is convenient to consider Polyakov loops in representations other than the defining representation and introduce the concept of N-ality. We shall then start with a brief review of these notions (see Ref. [7, 184] for a more complete discussion) before summarizing the results obtained within the Curci-Ferrari model for the SU(4) gauge group.

We also investigate the order parameter in various representations and test the Casimir scaling hypothesis.

---

[1] This includes the cases $N = 2$ and $N = 3$ treated so far.

© The Author(s), under exclusive license to Springer Nature Switzerland AG 2022
U. Reinosa, *Perturbative Aspects of the Deconfinement Transition*, Lecture Notes
in Physics 1006, https://doi.org/10.1007/978-3-031-11375-8_8

8.1   Polyakov Loops in Other Representations

To a given representation $R : t^a \mapsto t_R^a$ of dimension d_R, one associates a temporal Wilson line as

$$L_{A_R}(\mathbf{x}) \equiv \mathcal{P} \exp\left\{ i \int_0^\beta d\tau\, A_0^a(\tau, \mathbf{x})\, t_R^a \right\}. \tag{8.1}$$

The corresponding Polyakov loop is defined as

$$\ell_R \equiv \langle \Phi_{A_R}(\mathbf{x}) \rangle \equiv \frac{\int \mathcal{D}A\, \Phi_{A_R}(\mathbf{x})\, e^{-S_{YM}[A]}}{\int \mathcal{D}A\, e^{-S_{YM}[A]}}, \tag{8.2}$$

with $\Phi_{A_R}(\mathbf{x}) \equiv \operatorname{tr} L_{A_R}(\mathbf{x}) / d_R$.

8.1.1 N-Ality of a Representation

To any representation, one can associate its N-ality that characterizes the way the corresponding Polyakov loop transforms under center transformations. In general, a given representation R can be obtained by decomposing tensor products of the form $N \otimes N \otimes \ldots \bar{N} \otimes \ldots$ involving a certain number n of defining representations and a certain number $\bar{n}$ of contragredient representations. Using the two properties

$$L_{A_{R_1 \oplus R_2}} = L_{A_{R_1}} \oplus L_{A_{R_2}}, \tag{8.3}$$

$$L_{A_{R_1 \otimes R_2}} = L_{A_{R_1}} \otimes L_{A_{R_2}}, \tag{8.4}$$

it is then easily seen that the Polyakov loops in all representations extracted from the same tensor product transform in the same way under center transformations. In place of the transformation rule (3.10), we now have

$$\ell_{\alpha - \frac{2\pi}{N} k}^R = e^{i\frac{2\pi}{N} \nu_R k}\, \ell_\alpha^R, \tag{8.5}$$

where $\nu_R \equiv n - \bar{n}$ is known as the N-ality of the representation. In fact, the N-ality is defined only modulo multiples of N, and, for convenience, we shall make the choice $1 \leq \nu_R \leq N$ in what follows.

Of particular interest are those representations whose N-ality divides N. Indeed, it is easily seen in this case that the values of k that make the phase equal to 1 in Eq. (8.5) are multiples of N/ν_R, that is, $k = (N/\nu_R)k'$. The corresponding elements of Z_N are

$$e^{i\frac{2\pi}{N} k} = e^{i\frac{2\pi}{\nu_R} k'}. \tag{8.6}$$

They form the group Z_{ν_R} which appears then as a subgroup of Z_N.[2] For later purpose, it will be useful to recall that the subgroups of Z_N are all of this form, that is, Z_ν with $1 \leq \nu \leq N$ a divisor of N, and that, for each ν dividing N, there is only one such subgroup in Z_N, with $Z_1 = \{1\}$ and $Z_{\nu=N} = Z_N$.

Now, since ℓ_R is not transformed under Z_{ν_R}, the vanishing of ℓ_R does not tell us anything about the way center symmetry is realized in this subgroup. However, and interestingly enough, it tells us that center symmetry is necessarily realized in the Wigner-Weyl sense by some elements of Z_N not included in Z_{ν_R}.[3]

8.1.2 Center Symmetry Characterization

From the previous considerations, it is easily deduced that, in order to characterize the Wigner-Weyl realization of center symmetry, it is enough to consider a collection of representations that exhaust all N-alities $1 \leq \nu < N$ that divide N. Indeed the vanishing of the Polyakov loops in all these representations implies that center symmetry needs to be realized for elements that do not belong to any of the strict subgroups of Z_N. Since these elements necessarily generate Z_N upon repeated iteration (since otherwise they would belong to one of the strict subgroups Z_ν), any element of Z_N will then realize the symmetry in the Wigner-Weyl sense. The reciprocal of this result is obvious. We have thus arrived at the following characterization of center symmetry:

> **Center Symmetry Characterization (v1)**
> Center symmetry is manifest in the Wigner-Weyl sense iff all Polyakov loops vanish within a collection of representations covering all possible N-alities dividing N but not equal to N (i.e., $\nu \mid N$ and $\nu \neq N$).

We stress that this characterization does not include representations with $\nu_R = N$, also known as representations with *vanishing N-ality* (since N equals 0 modulo N). The Polyakov loops in such representations are not transformed under any center

---

[2] For a generic N-ality ν_R, the phase in Eq. (8.5) is equal to 1 for any k multiple of $\mathrm{lcm}(N, \nu_R) / \nu_R$, where $\mathrm{lcm}(p, q)$ stands for the least common multiple of p and q. This corresponds to elements of Z_N in the subgroup $Z_{N\nu_R / \mathrm{lcm}(N, \nu_R)} = Z_{\gcd(N, \nu_R)}$, where $\gcd(p, q)$ stands for the greatest common divisor of p and q.

[3] We are here implicitly excluding the presence of symmetries other than the center that would make the Polyakov loops vanish. We mention also that Polyakov loops could vanish accidentally, without the need of a symmetry. However, if this happens, it most probably does for isolated values of the external parameters (temperature, ...) and we should not consider these exceptional cases in the characterization of center symmetry to be given below.

transformation and, therefore, are not constrained to vanish.[4] It is then no surprise that they do not appear in the characterization of center symmetry. Let us also mention that, once the criterion applies, the Polyakov loops in any representation with non-vanishing N-ality need to vanish.

In fact, the characterization of center symmetry can be further simplified as follows. The set of divisors of N that are strictly smaller than N is naturally equipped with a partial ordering corresponding to the relation *being a divisor of*. We call *maximal N-alities,* the maximal elements for this partial ordering over this set.[5] We can now state that:

> **Center Symmetry Characterization (v2)**
> Center symmetry is manifest in the Wigner-Weyl sense iff all Polyakov loops vanish within a collection of representations covering all possible maximal N-alities.

Indeed, for two N-alities v_1 and v_2 (dividing N) that are ordered in the sense that v_1 divides v_2, we have $Z_{v_1} \subset Z_{v_2}$. It follows that the union of the subgroups associated with the maximal N-alities is equal to the union of all the strict subgroups of Z_N and, then, the vanishing of all the Polyakov loops with maximal N-alities leads to the same consequences than the vanishing of all the Polyakov loops with non-vanishing N-alities dividing N.

We mention finally that the characterization of center symmetry can also be formulated in terms of the confinement of color charges: the Wigner-Weyl realization of center symmetry is equivalent to the confinement of color charges within a collection of representations covering all possible maximal N-alities. Again, once the criterion is obeyed, all types of color charges with non-vanishing N-ality are confined.

8.1.3 Fundamental Representations

The prototypes for representations covering all possible N-alities are the *fundamental representations* obtained by antisymmetrizing successive tensor products of the defining representation.[6] These correspond to the one-column Young tableaux with

---

[4] In particular, the present formalism does not allow to address the confinement of colored objects in representations with vanishing N-ality, such as gluons.

[5] For instance, if $N = 24 = 2^3 \times 3$, the divisors that are strictly smaller than N are 1, 2, 3, 4, 6, 8, and 12. Since 1, 2, 3, 4, and 6 divide either 8 or 12, the maximal N-alities are 8 and 12. If $N = 3$, the only maximal N-ality is 1.

[6] Another, equivalent possibility, followed by [185], is to consider expectation values of traced powers of the defining Wilson line, $\langle \text{tr}\, L_A^k \rangle / N^k$.

a number of boxes equal to the N-ality. From those, we need only to consider those with maximal N-alities.[7]

Consider the example of SU(4). There are three fundamental representations 4, 6, and $\bar{4}$, of respective N-alities 1, 2, and 3. According to the previous criterion, in order to fully characterize the center, we need only to consider the representation 6 whose N-ality 2 is the only maximal divisor of 4 (within the set of divisors of 4 not equal to 4). The subgroup which is not probed by ℓ_6 is Z_2. The vanishing of ℓ_6 implies that the symmetry is realized in the Wigner-Weyl sense for center transformations outside of Z_2. However, since this is the only non-trivial subgroup of Z_4, these center transformations generate the other center transformations of Z_4 which are then also realized in the Wigner-Weyl sense. In contrast, the vanishing of ℓ_4 is not enough to characterize the centre since this could also be compatible with a partial breaking of the symmetry from Z_4 to the Z_2 subgroup.

In practice, the breaking pattern depends on the dynamics and, therefore, on the considered model and approximation. Its determination requires an explicit calculation. In Sect. 8.3, we shall investigate the breaking pattern in the SU(4) case using the CF model in the presence of a background. Before doing so, in Sect. 8.2, we review the properties of the Weyl chambers in this case, in particular the location of the center-invariant states. We recall that, in this framework, center symmetry is characterized by the fact that the background is found at particular points in the Weyl chambers, those points that are left invariant under the action of the center symmetry group on the Weyl chambers. In fact, it is enough to look for points in the Weyl chamber that are invariant under a center element that generates, upon repeated iteration, the complete center symmetry group, in other words, a center element that does not belong to any of the strict subgroups of the center symmetry group. Such transformations are precisely those that act non-trivially on all the Polyakov loops with maximal N-ality, and, therefore, if the background is located at such invariant points, the corresponding background-dependent Polyakov loops vanish, which in turn is enough to ensure that the center symmetry is fully manifest.

8.2   SU(4) Weyl Chambers

We saw in Chap. 5 that, in the reduced background space ($\bar{r} \equiv r/4\pi$), the SU(N) Weyl chambers are sub-pavings of the parallelepipeds generated by $N-1$ weights $\rho^{(j)}$ of the defining representation, with $1 \leq j \leq N-1$, dual to a generating set of roots $\{\alpha^{(j)}\}$ in the sense of Eq. (5.27). In practice, it is convenient to choose the

---

[7] We mention that the fundamental representations are naturally organized in pairs of N-alities $(v, N-v)$. The representations in each pair are contragredient of each other which, in the case of pure YM theory, implies that the corresponding Polyakov loops are equal to each other (see Chap. 4). This is in line with the fact that, in the above characterization of center symmetry, at most one of the representations in each pair is considered, since when v divides N, $N-v$ does not in general divide N (the only exception is when N is even and $v = N/2 = N - v$, but, in this case, there is anyway only one representation in the pair).

generating set of roots as $\alpha^{(j)} = \rho^{(j)} - \rho^{(N)}$, with $1 \leq j \leq N - 1$ in which case the dual set of weights is given by $\rho^{(j)}$ as one can verify using the properties of weights given in Appendix A. To construct the Weyl chambers, one needs to study how these parallelepipeds are subdivided by each network of hyperplanes orthogonal to a given of the remaining roots, $\rho^{(j)} - \rho^{(j')}$ with $1 \leq j < j' \leq N - 1$, and translated by multiples of half that root. The corresponding hyperplane passing through the origin contains $\rho^{(k)}$ for $k \neq j$ and $k \neq j'$, as well as $\rho^{(j)} + \rho^{(j')}$.

The defining weights of SU(N) can be found in Appendix A. For SU(4), we may choose

$$\rho^{(1)} = \frac{1}{2} \begin{pmatrix} -1 \\ \sqrt{1/3} \\ \sqrt{1/6} \end{pmatrix}, \quad \rho^{(2)} = \frac{1}{2} \begin{pmatrix} 0 \\ -\sqrt{4/3} \\ \sqrt{1/6} \end{pmatrix}, \quad \rho^{(3)} = \frac{1}{2} \begin{pmatrix} 0 \\ 0 \\ -\sqrt{3/2} \end{pmatrix},$$

(8.7)

the remaining weight being $\rho^{(4)} = -(\rho^{(1)} + \rho^{(2)} + \rho^{(3)})$. Consider the fundamental parallelepiped generated by the $\rho^{(j)}$ with $1 \leq j \leq 3$. It is further divided in Weyl chambers by the planes containing $\rho^{(3)}$ and $\rho^{(1)} + \rho^{(2)}$, $\rho^{(1)}$ and $\rho^{(2)} + \rho^{(3)}$, and finally $\rho^{(2)}$ and $\rho^{(3)} + \rho^{(1)}$, as shown in Fig. 8.1. Altogether they divide the

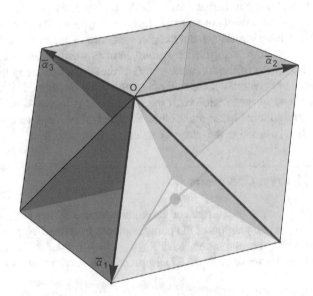

Fig. 8.1 The fundamental parallelepiped and its division into SU(4) Weyl chambers in the restricted background space. The vectors $\bar{\alpha}^{(1)}$, $\bar{\alpha}^{(2)}$, and $\bar{\alpha}^{(3)}$ (blue) define the fundamental parallelepiped. The latter is composed of six tetrahedral Weyl chambers: $\{O, \bar{\alpha}^{(1)}, \bar{\alpha}^{(1)} + \bar{\alpha}^{(2)}, \bar{\alpha}^{(1)} + \bar{\alpha}^{(2)} + \bar{\alpha}^{(3)}\}$, $\{O, \bar{\alpha}^{(2)}, \bar{\alpha}^{(1)} + \bar{\alpha}^{(2)}, \bar{\alpha}^{(1)} + \bar{\alpha}^{(2)} + \bar{\alpha}^{(3)}\}$ (blue), $\{O, \bar{\alpha}^{(1)}, \bar{\alpha}^{(1)} + \bar{\alpha}^{(3)}, \bar{\alpha}^{(1)} + \bar{\alpha}^{(2)} + \bar{\alpha}^{(3)}\}$, $\{O, \bar{\alpha}^{(3)}, \bar{\alpha}^{(1)} + \bar{\alpha}^{(3)}, \bar{\alpha}^{(1)} + \bar{\alpha}^{(2)} + \bar{\alpha}^{(3)}\}$ (red), $\{O, \bar{\alpha}^{(2)}, \bar{\alpha}^{(2)} + \bar{\alpha}^{(3)}, \bar{\alpha}^{(1)} + \bar{\alpha}^{(2)} + \bar{\alpha}^{(3)}\}$, $\{O, \bar{\alpha}^{(3)}, \bar{\alpha}^{(2)} + \bar{\alpha}^{(3)}, \bar{\alpha}^{(1)} + \bar{\alpha}^{(2)} + \bar{\alpha}^{(3)}\}$ (yellow). Each chamber has two edges longer than the other four, namely, the one connecting the origin to the sum of two $\bar{\alpha}$'s and the one connecting one $\bar{\alpha}$ to the sum of the three $\bar{\alpha}$'s, as illustrated in the figure (red lines). Reprinted from Ref. [94]

fundamental parallelepiped into six tetrahedra (see Fig. 8.1). A simple calculation shows that these tetrahedra have two nonadjacent edges that are longer than the other four, by a factor of $2/\sqrt{3}$.

8.2.1 Symmetries

Such an irregular tetrahedron, made of four identical isosceles triangles, is called a tetragonal disphenoid, whose symmetry group is the dihedral group D_{2d}, with eight elements. These are the identity, the three rotations by an angle π around any of the axes which relate the midpoints of nonadjacent edges, the reflections about the two planes perpendicular to one of the long edges and containing the other, and two other elements that can be obtained by combining the former. Using the method described in Chap. 5, it is easily seen that these transformations correspond to the center transformations, charge conjugation, or any combination of these, as we now discuss in more details.

Let us first consider center transformations. As we have seen in Chap. 5, any translation along the defining weights $\rho^{(j)}$ corresponds to a center transformation with center element $e^{-i2\pi/N} = -i$. In order to generate the other non-trivial transformations, we consider sums of two or three weights. For a given Weyl chamber connected to the origin, these correspond to translations along the three edges connected to the origin. It is easily seen that among those, the translation along the longer edge corresponds to a center element -1 and is associated with a rotation by an angle π around the axis that connects the midpoints of the two long edges of the Weyl chamber. The other two center transformations, corresponding to $\pm i$, are obtained by combining the two other rotations with any of the reflection planes described above. Following the same method, one can see that charge conjugation corresponds to one of the reflection planes—the other one being a combination of charge conjugation and the center transformation associated with -1.

Let us mention finally that one very convenient way to guess the various geometrical transformations of the Weyl chamber associated with physical transformations is to notice that the values taken by the leading-order background-dependent Polyakov loop in the defining representation at the nodes of the Weyl chamber span the center of the group (this is because it equals 1 at the origin and translations along the edges of the Weyl chamber span all possible center transformations). Upon a given physical transformation, these values are permuted in a certain way, which in turn allows one to infer the corresponding geometrical transformation of the Weyl chamber, together with its invariant points.

8.2.2   Invariant States

As mentioned above, center symmetry is characterized once the invariant states for a center transformation that generates the whole center symmetry group have been identified. Any such center transformation can be used, and here these correspond to the center elements i or $-i$. These transformations have only one fixed point, the barycenter of the tetrahedron,[8] which is then the only point compatible with center symmetry.

In contrast, the center transformation with center element -1 has a whole line of fixed points, the one connecting the midpoints of the two long edges, which contains in particular the barycenter of the tetrahedron. For any point on this line, except for the barycenter, the Polyakov loop ℓ_4 vanishes but not ℓ_6, corresponding to a scenario where center symmetry is partially broken down to the subgroup Z_2. The complete breaking of the symmetry corresponds to points away from this line.

8.3   One-Loop Results

It is convenient to locate the states in the fundamental parallelepiped in terms of their coordinates in the basis $\rho^{(j)}$. For a restricted background r, we write $r = x_j \bar{\alpha}^{(j)}$, where the $0 \leq x_j \leq 1$ can be obtained as $2\pi x_j = r \cdot \alpha^{(j)}$. In this basis, the center-symmetric point represented in Fig. 8.1 (dot in the figure) is located at the coordinates $(3/4, 1/2, 1/4)$. The Z_2 invariant line in the same Weyl chamber is defined by the equations $x_1 = x_2 + x_3$ and $x_2 = 1/2$. To simplify the discussion, we note that we can restrict to charge conjugation invariant states. In the considered Weyl chamber, these are the points in the plane $x_1 = x_2 + x_3$. In fact, it is even more convenient to work in a basis of this plane $r = y_2(\bar{\alpha}^{(1)} + \bar{\alpha}^{(2)}) + y_3(\bar{\alpha}^{(1)} + \bar{\alpha}^{(3)})$. The confining point is located at $(1/2, 1/4)$, and the Z_2 line corresponds to $y_2 = 1/2$.

8.3.1   Deconfinement Transition

Figure 8.2 shows contour plots of the potential in this plane. At low temperatures, the minimum of the potential is at the center-symmetric point (upper panels), whereas we find a Z_4 quadruplet of degenerate minima at high temperatures (lower panels), located pairwise in the two reflection-symmetry planes of the tetrahedron but away from the line of Z_2 symmetry. The breaking of center symmetry is thus

---

[8] In general, for SU(N), it is pretty obvious that the barycenter of any Weyl chamber is invariant under any center transformation [185] and in fact under any physical transformation (commuting with the periodic gauge transformations in the sense of Eq. (4.28)). A less trivial question is to identify all possible invariant states associated with a given physical transformation. Here we shall limit our analysis to the case of SU(4).

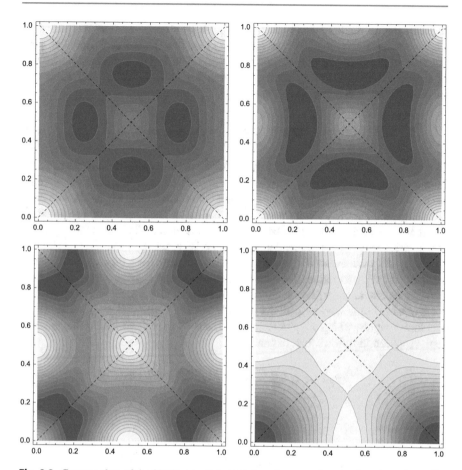

Fig. 8.2 Contour plots of the SU(4) potential at LO in a charge conjugation invariant plane. The latter intersects four Weyl chambers separated by dashed lines. The Z_2 invariant lines are the vertical and horizontal lines (not shown) passing through the origin. The upper (respectively, lower) plots correspond to temperatures below (respectively, above) the transition temperature

complete. We have checked that the transition is first order with $T_c^{LO}/m \approx 0.367$, close to the value obtained for SU(3) at LO.

8.3.2 Background-Dependent Polyakov Loops

It is interesting to evaluate the (background-dependent) fundamental Polyakov loops. The one in the defining representation rewrites

$$\ell_4(r) = \frac{e^{-i\frac{\pi}{2}(x_1+x_2+x_3)}}{4}\left[1 + e^{2\pi i x_1 \cdot r} + e^{2\pi i x_2 \cdot r} + e^{2\pi i x_3 \cdot r}\right], \tag{8.8}$$

in terms of the coordinates in the fundamental parallelepiped (see Problem 8.1). Then, we can interpret the vanishing of $\ell(r)$ as the closure of a rhombus with external angles $2\pi x_1$, $2\pi(x_2 - x_1)$, $2\pi(x_3 - x_2)$, and $2\pi(x_1 - x_3)$ modulo 2π and up to permutations of x_1, x_2, and x_3. This implies $2\pi x_1 + 2\pi(x_2 - x_1) = \pi$ and $2\pi x_1 = 2\pi(x_3 - x_2)$ modulo 2π, that is,

$$x_2 = \frac{1}{2} \bmod 1 \quad \text{and} \quad x_3 = x_1 - \frac{1}{2} \bmod 1. \tag{8.9}$$

In a given Weyl chamber, this corresponds to the segment joining the centers of the long edges of the tetrahedron, to which belong, in particular, the center-symmetric point. The corresponding segments in the other chambers in the fundamental parallelepiped are obtained by permutations of x_1, x_2, and x_3. These are precisely the states where the defining Polyakov loop needs to vanish if the subgroup Z_2 of center transformations is realized in the Wigner Weyl sense.

Let us now evaluate the Polyakov loop associated with the representation 6. Since fundamental representations are constructed by antisymmetrizing tensor products of defining representations, the weights of the fundamental representation with N-ality ν are obtained by considering all the possible sums of ν different weights. There are C_N^ν such weights, which are precisely the dimension of the considered fundamental representation. It follows that, in general

$$\ell_{C_N^\nu}(r) = \frac{1}{C_N^\nu} \sum_{j_1 < \cdots < j_\nu} e^{i\left(\rho^{(j_1)} + \cdots + \rho^{(j_\nu)}\right) \cdot r}. \tag{8.10}$$

Here we have

$$\ell_6(r) = \frac{1}{6}\left[e^{i(\rho^{(1)} + \rho^{(2)}) \cdot r} + e^{i(\rho^{(1)} + \rho^{(3)}) \cdot r} + e^{i(\rho^{(1)} + \rho^{(4)}) \cdot r} \right.$$
$$\left. + e^{i(\rho^{(2)} + \rho^{(3)}) \cdot r} + e^{i(\rho^{(2)} + \rho^{(4)}) \cdot r} + e^{i(\rho^{(3)} + \rho^{(4)}) \cdot r} \right]. \tag{8.11}$$

After some calculation, one finds (see Problem 8.2),

$$\ell_6(r) = \frac{1}{3} \mathrm{Re}\, e^{-i\pi(x_1 + x_2 + x_3)} \left[e^{2\pi i x_1} + e^{2\pi i x_2} + e^{2\pi i x_3} \right]$$
$$= \frac{1}{3}\left[\cos(\pi(x_1 - x_2 - x_3)) \right.$$
$$+ \cos(\pi(x_2 - x_3 - x_1))$$
$$\left. + \cos(\pi(x_3 - x_1 - x_2)) \right] \tag{8.12}$$

or, equivalently

$$
\begin{aligned}
\ell_6(r) &= \frac{1}{6}\text{Re}\left[(e^{\pi i x_1} + e^{-\pi i x_1})(e^{\pi i x_2} + e^{-\pi i x_2})(e^{\pi i x_3} + e^{-\pi i x_3})\right.\\
&\qquad \left. - e^{\pi i (x_1 + x_2 + x_3)} - e^{-\pi i (x_1 + x_2 + x_3)}\right]\\
&= \frac{1}{3}\left[4\cos(\pi x_1)\cos(\pi x_2)\cos(\pi x_3) - \cos(\pi(x_1 + x_2 + x_3))\right], \quad (8.13)
\end{aligned}
$$

whose vanishing defines a surface in the Weyl chamber, containing in particular the center-symmetric point.

We mention that the fact that the Polyakov loop ℓ_6 can vanish away from the center-symmetric point seems in contradiction with the discussion given above. However, this discussion concerned the physical Polyakov loop, whereas here we are discussing the background-dependent Polyakov loop which corresponds to a situation where one artificially imposes the state of the system to correspond to a given background. In practice, the system does not necessarily explore these states, and, if it does, there is no symmetry principle that would enforce it to remain in such a state for a certain range of temperatures. We can disregard these accidental vanishings for the present discussion. We also expect the location of these accidental vanishing points in the Weyl chamber to change, or even disappear, at next-to-leading order.

8.4 Casimir Scaling

Another application of Polyakov loops in higher representations is Casimir scaling. It has been observed that the Polyakov loops in different representations obey the following scaling law [186]

$$
\ell_R^{1/C_R} = \ell_{R'}^{1/C_{R'}}, \tag{8.14}
$$

where C_R denotes the Casimir of the representation R. This in turn implies that the free-energy cost for bringing a colored object into the medium scales like the square of the corresponding color charge.

To investigate Casimir scaling at leading order in the SU(3) case, we use Eq. (8.3). Taking the traces and normalizing by the dimensions of the representations, we arrive at[9]

$$
(d_1 + d_2)\, \Phi_{A_{R_1 \oplus R_2}} = d_1 \Phi_{A_{R_1}} + d_2 L_{A_{R_2}}, \tag{8.15}
$$

$$
\Phi_{A_{R_1 \otimes R_2}} = \Phi_{A_{R_1}}\, \Phi_{A_{R_2}}. \tag{8.16}
$$

---

[9] These are similar to the identities obeyed by the characters of the corresponding representations.

At leading order, these properties are transferred to the corresponding Polyakov loops. Moreover, if $\bar{R}$ denotes the contragredient of a given representation R, we can once again show that the Polyakov loops are equal and real (on a charge conjugation invariant state). These properties allow one to infer the Polyakov loops in any representations from those in lower representations.

For instance, we know that $3 \otimes 3 = \bar{3} \oplus 6$. From this it follows that $9\ell_3^2 = 3\ell_3 + 6\ell_6$ and then $6\ell_6 = 9\ell_3^2 - 3\ell_3$. Similarly, one finds

$$8\ell_8 = 9\ell_3^2 - 1, \tag{8.17}$$

$$10\ell_{10} = 18\ell_3\ell_6 - 8\ell_8, \tag{8.18}$$

$$15\ell_{15} = 18\ell_3\ell_6 - 3\ell_3, \tag{8.19}$$

$$15\ell_{15'} = 30\ell_3\ell_{10} - 15\ell_{15}, \tag{8.20}$$

$$24\ell_{24} = 30\ell_3\ell_{10} - 6\ell_6, \tag{8.21}$$

$$27\ell_{27} = 36\ell_6^2 - 8\ell_8 - 1, \tag{8.22}$$

(see Ref. [186] for more details).

In Fig. 8.3, we show how Casimir scaling is satisfied in our approach. The scaling is well satisfied down to $\simeq 1.1 T_c$. For temperatures below, we observe partial Casimir scaling, with certain groups of representations obeying scaling (6 and 8; 10 and 15; 15', 24, and 27; ...).

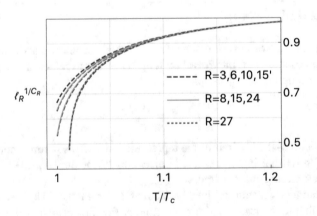

Fig. 8.3 Testing the Casimir scaling at leading order

Problems

8.1 Polyakov Loop in the First Fundamental Representation of SU(4) (★★)
Show that

$$\ell_4(r) = \frac{e^{-i\frac{\pi}{2}(x_1+x_2+x_3)}}{4}\left[1 + e^{2\pi i x_1 \cdot r} + e^{2\pi i x_2 \cdot r} + e^{2\pi i x_3 \cdot r}\right],$$

in terms of the coordinates $2\pi x_j \equiv r \cdot \alpha^{(j)}$ in the fundamental parallelepiped. *Tip:* start from the definition of $\ell_4(r)$ in terms of the defining weights, and use their relation to the roots as well as the constraint on the sum of all weights.

8.2 Polyakov Loop in the Second Fundamental Representation of SU(4) (★★)
Show that

$$\ell_6(r) = \frac{1}{3}\mathrm{Re}\, e^{-i\pi(x_1+x_2+x_3)}\left[e^{2\pi i x_1} + e^{2\pi i x_2} + e^{2\pi i x_3}\right].$$

Tip: same as in Problem 8.1.

Background-Field Gauges: Adding Quarks and Density

9

In the next two chapters, we pursue our investigation of the deconfinement transition by complementing the background-extended Curci-Ferrari model with quark degrees of freedom. Including quarks is in principle straightforward since one has simply to add the usual matter contribution of the QCD action. In practice, however, the presence of quarks leads to new difficulties that require one to revisit the very foundations of the background-field method.

These difficulties concern in fact any continuum approach that relies, in one way or another, on the use of effective actions. They appear at finite density, as the functional integral measure becomes complex, and are of two different types: those that relate to the integration measure not being real-valued and those that relate to its non-positivity. We show how the first type can be cured by exploiting the symmetries of the system at hand. In contrast, the second type is more subtle as it connects directly to the sign problem of lattice QCD [36,37]. A simple recipe will be postulated to handle this second type of difficulty, but its justification will remain an open question beyond the scope of the present manuscript.

9.1 General Considerations

In this section, we derive some general properties that will be used throughout the chapter. We do so by using the QCD action in its non-gauge-fixed version. We shall later extend the discussion to the Landau-DeWitt gauge and to the corresponding Curci-Ferrari model.

The QCD action is obtained by adding the usual matter contribution to the Yang-Mills action, $S_{YM} \to S_{QCD} = S_{YM} + \delta S$, with

$$\delta S[A, \psi, \bar{\psi}; \mu] \equiv \int d^d x \sum_{f=1}^{N_f} \bar{\psi}_f(x) \left(\slashed{D} + M_f - \mu \gamma_0 \right) \psi_f(x). \tag{9.1}$$

© The Author(s), under exclusive license to Springer Nature Switzerland AG 2022
U. Reinosa, *Perturbative Aspects of the Deconfinement Transition*, Lecture Notes
in Physics 1006, https://doi.org/10.1007/978-3-031-11375-8_9

Here, the quarks are taken in the defining representation, and $\mathcal{D}_\mu = \partial_\mu - igA_\mu^a t^a$ is the corresponding covariant derivative. We have also introduced the usual notation $\mathcal{D} \equiv \gamma_\mu \mathcal{D}_\mu$ with γ_μ the Euclidean Dirac matrices [102]. These are related to the standard Minkowski matrices γ_M^μ as $\gamma_0 = \gamma_M^0$ and $\gamma_i = -i\gamma_M^i$. They are thus Hermitian and satisfy the anti-commutation relations $\{\gamma_\mu, \gamma_\nu\} = 2\delta_{\mu\nu}$. In the following, we work in the Weyl basis, where $\gamma_{0,2}^* = \gamma_{0,2}^t = \gamma_{0,2}$ and $\gamma_{1,3}^* = \gamma_{1,3}^t = -\gamma_{1,3}$. Within this basis, one easily derives the two identities (see Problem 9.1)

$$\gamma_2\gamma_0\gamma_\mu\gamma_0\gamma_2 = -\gamma_\mu^t \quad \text{and} \quad \gamma_3\gamma_1\gamma_\mu\gamma_1\gamma_3 = \gamma_\mu^*, \tag{9.2}$$

which will play a major role in the following discussion.

The conjugated field $\bar{\psi}$ is defined as usual as $\bar{\psi} \equiv \psi^\dagger\gamma_0$. It is convenient to extend such conjugation to complex numbers and matrices as $\bar{z} \equiv z^*$ and $\overline{M} \equiv \gamma_0 M^\dagger \gamma_0$. In particular, it is readily checked that $\overline{\gamma_0} = \gamma_0$ and $\overline{\gamma_i} = -\gamma_i$. Moreover, given two matrices M and N, one finds $\overline{MN} = \overline{N}\,\overline{M}$. We also define $\overline{\psi^t} \equiv \bar{\psi}^t = \gamma_0\psi^*$. These properties allow for an easy determination of the conjugate of any expression, similar to the determination of the bra associated with a ket in quantum mechanics. We also introduce $\gamma_5 \equiv \gamma_0\gamma_1\gamma_2\gamma_3$. It obeys the properties $\{\gamma_5, \gamma_\mu\} = 0$, $\gamma_5^2 = 1$, $\gamma_5^\dagger = \gamma_5$, and $\overline{\gamma}_5 = -\gamma_5$.

Finally, let us recall that the last term in Eq. (9.1) corresponds to the baryonic charge of the system. It is proportional to the associated chemical potential $\mu_B \equiv 3\mu$.[1] For reasons that shall become clear below, we allow for both real and imaginary chemical potentials. As usual, the temperature $T \equiv 1/\beta$ enters via the boundary conditions of the fields, and, unlike gluons, quarks obey anti-periodic boundary conditions: $\psi(\tau + \beta, \mathbf{x}) = -\psi(\tau, \mathbf{x})$.

9.1.1   Polyakov Loops

Our investigation of the deconfinement transition will again be achieved by means of the Polyakov loops associated with the defining and contragredient representations,[2]

$$\ell \equiv \left\langle \frac{1}{N} \operatorname{tr}\mathcal{P}\exp\left\{ i \int_0^\beta d\tau\, A_0^a(\tau, \mathbf{x})\, t^a \right\} \right\rangle \tag{9.3}$$

and

$$\bar{\ell} \equiv \left\langle \frac{1}{N} \operatorname{tr}\mathcal{P}\exp\left\{ -i \int_0^\beta d\tau\, A_0^a(\tau, \mathbf{x})\, t_a^t \right\} \right\rangle. \tag{9.4}$$

---

[1] Note that we use the opposite convention for the sign of μ than in Ref. [95].

[2] This is because we are considering the SU(3) case. As we have discussed in Chap. 8, the case of SU(N) requires one to introduce a priori all the fundamental loops.

Those are directly related to the free-energy cost for bringing a quark or an anti-quark into an equilibrated bath of quarks and gluons [37, 159]:

$$\ell \propto e^{-\beta \Delta F} \quad \text{and} \quad \bar{\ell} \propto e^{-\beta \Delta \bar{F}}. \tag{9.5}$$

We mention that this interpretation has been questioned in the presence of a chemical potential due to the non-monotonic behavior of the Polyakov loops as the chemical potential is varied [187]. We will argue instead in the next chapter that the non-monotonicity as a function of μ can be seen precisely as a consequence of the free-energy interpretation.

9.1.2 Symmetries

The other crucial ingredients in the analysis to follow are of course the symmetries of the problem. As we now recall, those are notably modified by the presence of quarks in the defining representation.

First of all, center symmetry is explicitly broken. This is because the defining quarks transform as $\psi^U(x) = U(x)\psi(x)$. So, even though the action remains invariant, $S[A^U, \psi^U, \bar{\psi}^U; \mu] = S[A, \psi, \bar{\psi}; \mu]$, the anti-periodic boundary conditions of the quarks are changed into anti-periodic boundary conditions modulo a phase, $\psi^U(\tau + \beta, \mathbf{x}) = -e^{i\frac{2\pi}{N}k}\psi^U(\tau, \mathbf{x})$, and the symmetry is not manifest at the level of the quantum/thermal expectation values. As a consequence, the Polyakov loops have no reason to vanish anymore. We will see nonetheless that the Polyakov loops possess a singular behavior in the heavy quark regime which qualifies them still as good order parameters.

We expect quarks not to break charge conjugation symmetry as long as the chemical potential is equal to zero. This is easily checked by evaluating the action with charge conjugated fields[3]

$$A^C \equiv -A^t, \quad \psi^C \equiv \gamma_0\gamma_2\bar{\psi}^t, \quad \bar{\psi}^C \equiv -\psi^t\gamma_2\gamma_0, \tag{9.6}$$

and by using the property $\gamma_2\gamma_0\gamma_\mu\gamma_0\gamma_2 = -\gamma_\mu^t$. More generally, it is easily verified that charge conjugation changes the sign of the baryonic charge term in the action, and thus that

$$S[A^C, \psi^C, \bar{\psi}^C; \mu] = S[A, \psi, \bar{\psi}; -\mu], \tag{9.7}$$

(see Problem 9.2). Since charge conjugation does not alter the boundary conditions of the fields, the symmetry applies also to expectation values. In particular, it is

---

[3] The transformation of the field $\bar{\psi}$ is easily obtained from the conjugation rule described above, $\psi^C = \gamma_0\gamma_2\bar{\psi}^t \to \bar{\psi}^C = \overline{\bar{\psi}^t}\bar{\gamma}_2\bar{\gamma}_0 = -\psi^t\gamma_2\gamma_0$, where we have used that $\bar{\gamma}_0 = \gamma_0$, $\bar{\gamma}_2 = -\gamma_2$, and $\overline{\bar{\psi}^t} = \psi^t$.

easily shown that, as a consequence of charge conjugation, $\ell(\mu) = \bar{\ell}(-\mu)$. For $\mu = 0$, we recover the result $\bar{\ell} = \ell$ obtained in Chap. 4.[4]

Another useful transformation is

$$A^{\mathcal{K}} \equiv A^*, \quad \psi^{\mathcal{K}} \equiv \gamma_1\gamma_3\psi, \quad \bar{\psi}^{\mathcal{K}} \equiv \bar{\psi}\gamma_3\gamma_1. \tag{9.8}$$

Indeed, owing to the property $\gamma_3\gamma_1\gamma_\mu\gamma_1\gamma_3 = \gamma_\mu^*$ and assuming that the fermionic fields are real,[5] it is easily checked that

$$S[A^{\mathcal{K}}, \psi^{\mathcal{K}}, \bar{\psi}^{\mathcal{K}}; \mu] = S[A, \psi, \bar{\psi}; \mu^*]^*. \tag{9.9}$$

Although this is strictly speaking not a symmetry for it requires the complex conjugation of the action, it plays a major role in what follows. It implies, for instance, that $\ell(\mu) = \ell(\mu^*)^*$ and $\bar{\ell}(\mu) = \bar{\ell}(\mu^*)^*$ and, thus, that ℓ and $\bar{\ell}$ are both real if the chemical potential is real.

For completeness, let us finally combine the previous two transformations into the transformation

$$A^{\mathcal{KC}} = A, \quad \psi^{\mathcal{KC}} = -\gamma_5\bar{\psi}^{\mathrm{t}}, \quad \bar{\psi}^{\mathcal{KC}} = \psi^{\mathrm{t}}\gamma_5. \tag{9.10}$$

Using $\{\gamma_5, \gamma_\mu\} = 0$ together with $\gamma_\mu^\dagger = \gamma_\mu$, one finds

$$S[A, \psi^{\mathcal{KC}}, \bar{\psi}^{\mathcal{KC}}; \mu] = S[A, \psi, \bar{\psi}; -\mu^*]^*. \tag{9.11}$$

In particular, it follows that $\ell(\mu) = \bar{\ell}(-\mu^*)^*$ and, thus, ℓ and $\bar{\ell}$ become complex conjugate of each other if the chemical potential is imaginary. The transformation (9.10) should not be mistaken with the chiral transformation $\psi \to \gamma_5\psi$, $\bar{\psi} \to -\bar{\psi}\gamma_5$ which leaves the chemical potential unchanged while flipping the sign of all quark masses. We shall also make use of this property below.

9.1.3 Fermion Determinant

Let us end this section by recalling that the fermionic part of the action being quadratic, the fermionic fields can be integrated out exactly. This means that any observable can be written formally as the corresponding integral in the pure Yang-Mills theory but with a functional measure modified by a fermion determinant. For

---

[4] Here, we assume that there is no spontaneously broken symmetry (as we have seen above, center symmetry is explicitly broken in the presence of fundamental quarks). A counterexample is provided by Yang-Mills theory where center symmetry can be spontaneously broken. In that case, $\ell = \bar{\ell}$ does not apply to all possible states of the system.

[5] The Grassmannian fields ψ and $\bar{\psi}$ are a formal device to write the determinant of an operator, $\det M = \int \mathcal{D}[\psi, \bar{\psi}] \exp\{\int d^d x \int d^d y\, \bar{\psi}(x)M(x, y)\psi(y)\}$. Since $(\det M)^* = \det M^*$, it is perfectly consistent to assume that $\psi^* = \psi$ and $\bar{\psi}^* = \bar{\psi}$ (see also Ref. [95]).

instance, the partition function reads

$$Z = \int \mathcal{D}A \, \Delta[A; \mu] \, e^{-S_{YM}[A]}, \tag{9.12}$$

with

$$\Delta[A; \mu] \equiv \int \mathcal{D}[\psi, \bar{\psi}] \, e^{-\delta S[\psi, \bar{\psi}, A; \mu]} = \det \mathcal{M}[A; \mu] \tag{9.13}$$

the determinant of the Dirac operator in the presence of the gauge field A and a chemical potential μ:

$$\mathcal{M}[A; \mu] = \bigotimes_{f=1}^{N_f} (\slashed{\partial} - g\slashed{A} + M_f - \mu\gamma_0). \tag{9.14}$$

A similar rewriting applies of course to the Polyakov loops.

Now, using similar arguments as in the previous section, one finds (see Problem 9.4)

$$\gamma_2\gamma_0 \, \mathcal{M}[A; \mu] \, \gamma_0\gamma_2 = \mathcal{M}[A^C; -\mu]^t, \tag{9.15}$$

$$\gamma_3\gamma_1 \, \mathcal{M}[A; \mu] \, \gamma_1\gamma_3 = \mathcal{M}[A^K; \mu^*]^*, \tag{9.16}$$

$$\gamma_5 \, \mathcal{M}[A; \mu] \, \gamma_5 = \mathcal{M}[A; -\mu^*]^\dagger. \tag{9.17}$$

Again, the last identity was obtained by combining the previous two. If we use instead the chiral transformation $\psi \to \gamma_5\psi$, $\bar{\psi} \to -\bar{\psi}\gamma_5$, we obtain that $-\gamma_5 \, \mathcal{M}[A; \mu] \, \gamma_5$ is the original Dirac operator but with a sign flip of all fermion masses:

$$-\gamma_5 \, \mathcal{M}[A; \mu] \, \gamma_5 = \bigotimes_{f=1}^{N_f} (\slashed{\partial} - g\slashed{A} - M_f - \mu\gamma_0). \tag{9.18}$$

For the fermion determinant, we obtain, correspondingly,

$$\Delta[A; \mu] = \Delta[A^C; -\mu], \tag{9.19}$$

$$\Delta[A; \mu] = \left(\Delta[A^K; \mu^*]\right)^*, \tag{9.20}$$

$$\Delta[A; \mu] = \left(\Delta[A; -\mu^*]\right)^*. \tag{9.21}$$

9.2   Continuum Sign Problem(s)

As it can be seen from Eq. (9.20), the fermion determinant becomes a priori complex when the chemical potential is taken real. This is the source of various difficulties, usually collected under the generic name of *continuum sign problems* [23, 188–193] and which we now discuss in some detail.

We put special care into discriminating between true sign problems that emanate from the non-positivity of the fermion determinant and related difficulties that emanate from its non-real-valuedness. To make the distinction clear, we label the first type as (Pn) and the second type as (Rn). As we argue, this latter type of difficulties is easily handled after one acknowledges the fact that, in the presence of a complex integration measure, a given observable (with real spectrum) does not necessarily lead to a real expectation value. In contrast, the former type is more difficult to handle since they relate directly to the lattice sign problem.

In order to better appreciate the various possible difficulties, we first consider the case of an imaginary chemical potential where these difficulties are absent.[6] Also, we first discuss the various problems in a non-gauge-fixed setting, in a way similar to Ref. [194], and then extend the discussion to background-field gauges.

9.2.1   Imaginary Chemical Potential

In the case of an imaginary chemical potential, the fermion determinant is positive. To see this we follow Ref. [188] and note first that the massless Dirac operator is anti-Hermitian. It is thus diagonalizable with a purely imaginary spectrum. Moreover, the mass operator being proportional to the identity, the massive Dirac operator remains diagonalizable. Finally, it follows from Eq. (9.18) and $\gamma_5^2 = 1$ that, for each eigenstate of eigenvalue $\mu + i\lambda$, there is another eigenstate of eigenvalue $\mu - i\lambda$. The fermion determinant is then the product of positive factors of the form $\mu^2 + \lambda^2$.[7]

The positivity of the fermion determinant is a welcome feature at various levels. On a fundamental level, it ensures the positivity of the partition function (9.12) and thus the real-valuedness of any thermodynamical observable derived from it (since then $\ln Z$ is real).[8] On a practical level, it makes it possible to evaluate the partition function with, discrete, importance sampling Monte Carlo techniques. As we now discuss, it has also its importance within any continuum approach based on the use of effective actions.

---

[6] As we discuss in the next chapter, the case of imaginary chemical potential possesses further interesting features that make it a case worth of study.

[7] For completeness, we mention that the real-valuedness of the fermion determinant (but not its positivity) follows more directly from Eq. (9.21).

[8] We shall not try, for the moment, to give a physical interpretation to the case of imaginary chemical potential. We shall later show that, for certain values of the (imaginary) chemical potential, there is a simple physical interpretation.

Suppose indeed that we introduce sources $\bar{\eta}$ and η coupled, respectively, to the operators Φ_A and $\bar{\Phi}_A$, thus defining the generating functional $W[\eta, \bar{\eta}] \equiv \ln Z[\eta, \bar{\eta}]$, with

$$Z[\eta, \bar{\eta}] \equiv \int \mathcal{D}A \, \Delta[A; \mu] \, e^{-S_{\mathrm{YM}}[A]} \exp\left\{ \int dx \left(\bar{\eta}(\mathbf{x}) \, \Phi_A(\mathbf{x}) + \bar{\Phi}_A(\mathbf{x}) \, \eta(\mathbf{x}) \right) \right\}. \tag{9.22}$$

The sources are a practical way to generate the expectation values of the Polyakov loops or more generally correlations between various Polyakov loops. In the limit of zero sources, one can extract their values from the extremization of the Polyakov loop effective action, defined as the Legendre transform of $W[\eta, \bar{\eta}]$:

$$\Gamma[\ell, \bar{\ell}] \equiv -W[\eta, \bar{\eta}] + \int dx \left(\bar{\eta}(\mathbf{x}) \, \ell(\mathbf{x}) + \bar{\ell}(\mathbf{x}) \, \eta(\mathbf{x}) \right), \tag{9.23}$$

where the conjugated variables ℓ and $\bar{\ell}$ are defined as

$$\ell(\mathbf{x}) \equiv \frac{\delta W}{\delta \bar{\eta}(\mathbf{x})}, \quad \bar{\ell}(\mathbf{x}) \equiv \frac{\delta W}{\delta \eta(\mathbf{x})}, \tag{9.24}$$

and are nothing but the Polyakov and anti-Polyakov loops in the presence of the sources η and $\bar{\eta}$.

For the purpose of generating correlation functions, one could consider the sources to be two independent complex numbers, that is, $(\eta, \bar{\eta}) \in \mathbb{C} \times \mathbb{C}$. Correspondingly, the variables ℓ and $\bar{\ell}$ that enter the effective action are complex and independent (possibly within some region of $\mathbb{C} \times \mathbb{C}$). This is not the most convenient choice, however, because the effective action becomes generically complex, obscuring the fact that it should be real in the limit of zero sources where it corresponds to $-W[0, 0] = -\ln Z$. Moreover, a complex effective action makes it difficult to devise a definite criterion to identify the physical extremum.[9]

As we now explain, we can avoid these difficulties by restricting $(\eta, \bar{\eta})$ to a subspace of $\mathbb{C} \times \mathbb{C}$ such that $\Gamma[\ell, \bar{\ell}]$ is real. This restriction is of course allowed as long as the considered subspace contains the limit of zero sources.[10] However, one problem with restricting the space of sources is that it is not always easy to identify the target space where the conjugated variables ℓ and $\bar{\ell}$ should vary. We now show that it is possible to find a subspace of $\mathbb{C} \times \mathbb{C}$ which is both stabilized by the Legendre transformation and over which the effective action is real.

---

[9] In principle, one expects that the limit of zero sources corresponds to one extremum only or to a collection of degenerated extrema. However, due to the approximations inherent to any approach, additional extrema can appear, and one needs a criterion to identify the correct one.

[10] We assume again that the limit of zero sources does not depend on the way it is taken, which is legitimate here since there is no spontaneously broken symmetry.

Suppose that we restrict the sources such that $\bar{\eta} = \eta^*$, that is, the pair $(\eta, \bar{\eta})$ is taken in the subspace $\Sigma \equiv \{(\eta, \bar{\eta}) \in \mathbb{C} \times \mathbb{C} \mid \bar{\eta} = \eta^*\}$ of complex-conjugated sources. This choice seems natural here since the quantities the sources are coupled to are also complex conjugate of each other, $\bar{\Phi}_A = \Phi_A^*$. We emphasize, however, that this very choice will not be the natural one in the case of a real chemical potential to be treated in the next section. In the present case, this choice of sources, together with the real-valuedness of the fermion determinant, implies that $Z[\eta, \bar{\eta}]$ is real:

Reality 1

$$(\text{R1}) \quad \forall(\eta, \bar{\eta}) \in \Sigma\,, \; Z[\eta, \bar{\eta}] \in \mathbb{R}\,. \tag{9.25}$$

A stronger result is obtained from the positivity of the fermion determinant: $Z[\eta, \bar{\eta}]$ is in fact positive over Σ or, equivalently, that the generating functional $W[\eta, \bar{\eta}]$ is real over Σ:

Positivity 1

$$(\text{P1}) \quad \forall(\eta, \bar{\eta}) \in \Sigma\,, \; W[\eta, \bar{\eta}] \in \mathbb{R}\,. \tag{9.26}$$

Next, we consider the effective action $\Gamma[\ell, \bar{\ell}]$. It is easily seen that, if the sources are taken in the subspace Σ, the conjugated variables belong to the same space:

Reality 2

$$(\text{R2}) \quad (\eta, \bar{\eta}) \in \Sigma \Rightarrow (\ell, \bar{\ell}) \in \Sigma\,. \tag{9.27}$$

This result relies crucially on the real-valuedness of the fermion determinant. Indeed, in the presence of a real-valued integration measure, the expectation values of two complex-conjugated quantities, such as Φ_A and $\bar{\Phi}_A$, are themselves complex conjugate of each other. One can also see (R2) as a consequence of (R1) and the identities $\ell = (\delta Z / \delta \bar{\eta}) / Z$ and $\bar{\ell} = (\delta Z / \delta \eta) / Z$. Moreover, by combining the properties (P1) and (R2), we deduce that $\Gamma[\eta, \bar{\eta}]$ is real over the space Σ:

Positivity 2

$$(P2) \quad \forall (\ell, \bar{\ell}) \in \Sigma, \Gamma[\ell, \bar{\ell}] \in \mathbb{R}. \tag{9.28}$$

Finally, from the positivity of the integration measure, we can use similar arguments as those given in Chap. 4 to show that $W[\eta, \bar{\eta}]$ is convex over Σ and that the the limit of zero sources corresponds to the minimization of $\Gamma[\ell, \bar{\ell}]$ over this space:

Positivity 3

$$(P3) \quad (\eta, \bar{\eta}) \in \Sigma \to (0, 0) \Leftrightarrow \min_{\Sigma} \Gamma[\ell, \bar{\ell}]. \tag{9.29}$$

In summary, we have just shown that, in the case of an imaginary chemical potential, the real-valuedness of the fermion determinant ensures the existence of a subspace over which the functional $Z[\eta, \bar{\eta}]$ is real (R1) and, thus, such it is stabilized by Legendre transformation (R2). On the other hand, the positivity of the fermion determinant ensures that the functionals $W[\eta, \bar{\eta}]$ and $\Gamma[\ell, \bar{\ell}]$ are real over this subspace (P1/P2) and that there exists an unambiguous characterization of the limit of zero sources that allows one to select the correct extremum of the effective action over this subspace (P3). We mention that varying the effective action over Σ is also convenient because it makes sure that the property of the physical Polyakov loops being complex conjugate of each other in the case of an imaginary chemical potential (see above) is automatically fulfilled.

9.2.2 Real Chemical Potential

In the case of a real chemical potential, both the reality and the positivity of the fermion determinant are lost, and we expect the properties (R1), (R2), (P1), (P2), and (P3) not to hold true anymore, at least not in such a simple way as above. As we now recall, the properties (R1) and (R2) can be shown to still hold true with however a new invariant subspace. In contrast, to the best of our knowledge, the properties (P1), (P2), and (P3), for they crucially rely on the positivity of the fermion determinant, have not been extended so far to the case of a real chemical potential.

R1/R2

For a real chemical potential, $Z[\eta, \bar{\eta}]$ is not real in general, even when restricted to Σ. Correspondingly, the Legendre transformation does not stabilize Σ. This problem, however, should not be qualified as a sign problem for it does not originate in the integration measure not being positive, but only in the integration measure not being real. Moreover, it has a simple solution, as originally suggested in the matrix model analysis of Ref. [194]. In fact, using the identity (9.20) as well as $\Phi_{A^{\kappa}} = \Phi_{A}^{*}$, it is easily checked that $Z[\eta, \bar{\eta}]$ is real over the subspace $\mathbb{R} \times \mathbb{R}$ and thus that this subspace is stable under Legendre transformation. Moreover, this is the natural subspace where to consider the effective action, since as we have seen above, the physical Polyakov loops should be real in this case.

The above considerations give a formal basis to the standard rule on how to analyze the Polyakov loop effective action/potential (see, for instance, Refs. [23, 189, 192]), when changing from the case of an imaginary chemical potential to that of a real chemical potential: in contrast to the former case where ℓ and $\bar{\ell}$ are taken complex conjugate of each other, in the latter case, one chooses instead ℓ and $\bar{\ell}$ real and independent. This change of subspace is illustrated in Fig. 9.1.

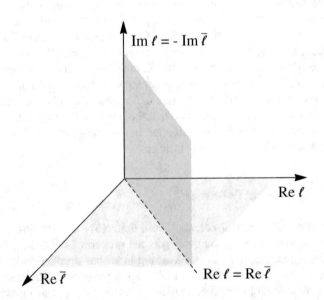

Fig. 9.1 The two different planes where the Polyakov loop effective action should be considered depending on whether $\mu \in i\mathbb{R}$ or $\mu \in \mathbb{R}$. These two planes are sub-manifolds of $\mathbb{C} \times \mathbb{C}$, but for the sake of representing them in a three dimensional figure, we have glued together the axes $\mathrm{Im}\, \ell$ and $-\mathrm{Im}\, \bar{\ell}$

P1/P2

As we have already emphasized, the question of the reality of $Z[\eta, \bar{\eta}]$, and in particular of the partition function $Z \equiv Z[\eta = 0, \bar{\eta} = 0]$, has not much to do with the sign problem and can be addressed using the identity (9.20). The true sign problem appears when one tries to argue that the partition function Z is positive, as it should be the case if $-T \ln Z$ is to represent the free energy of the system. The fact that the fermion determinant is not positive definite makes the proof of the positivity of Z difficult a priori, and, to our knowledge, this question has not been clarified yet. From the reality of $Z[\eta, \bar{\eta}]$, the (assumed) positivity of Z and the expectation that $W[\eta, \bar{\eta}]$ should not diverge at any value of the sources; we also expect $Z[\eta, \bar{\eta}]$ to be always positive. Again, the non-positivity of the fermion determinant makes this statement difficult to prove. These issues are minor, however, in the sense that, within any continuum approach that gives access to an explicit expression for $\Gamma[\ell, \bar{\ell}]$, one can always check whether $\Gamma[\ell, \bar{\ell}]$ is real over $\mathbb{R} \times \mathbb{R}$. For instance, in the next chapter, we check that this is the case for the one-loop Polyakov loop potential computed within the Curci-Ferrari model.

P3

A slightly more serious sign problem is that, even after checking that $\Gamma[\ell, \bar{\ell}]$ is real over $\mathbb{R} \times \mathbb{R}$, it is not at all obvious which extremum one should choose in this subspace, in the case where various extrema are present. The criterion that we identified in the case of an imaginary chemical potential, because it crucially relied on the positivity of the fermion determinant, is not valid anymore. In fact, it is easily seen that the physical point cannot correspond to the absolute minimum of the effective action. This is because, for a vanishing chemical potential, one can choose to work either over Σ or over $\mathbb{R} \times \mathbb{R}$. Due to charge conjugation invariance, the physical point is such that $\ell = \bar{\ell}$ and lies therefore at the intersection of these two subspaces. Seen from the perspective of the subspace Σ, the physical point appears as the absolute minimum, but seen from the perspective of the subspace $\mathbb{R} \times \mathbb{R}$, it appears as a saddle point.

We mention that this version of the sign problem is not as dramatic as the one on the lattice, because one can always try to locate the different extrema (assuming that there is a finite number of them) and select the correct one based on some additional physical insight. However, this entails inevitably a loss of predictive power of continuum approaches when various saddle points are present. We can try to circumvent the problem by inferring a general rule from the case of zero chemical potential. In that case, the physical point in the subspace $\mathbb{R} \times \mathbb{R}$ is not only a saddle point; it is in fact the deepest saddle

(continued)

point.[11] In what follows, we assume that this rule applies even at non-zero chemical potential. We stress however that we found no way to justify this rule from the first principles. The best we can do is to provide a posteriori justifications, based on the relevance of the physical results that we obtain with this rule, in comparison to other strategies.

9.3 Background-Field Gauges

The previous considerations extend of course also to a gauge-fixed setting. In particular, as we now discuss, they have a strong imprint on the way the background field method should be implemented at finite density.

9.3.1 Complex Self-Consistent Backgrounds

To understand this point, recall that the background-field approach at finite temperature relies crucially on the use of self-consistent backgrounds defined by the condition $\bar{A} = \langle A \rangle_{\bar{A}}$. Making the expectation value explicit and projecting along the generators t^a, this condition writes

$$\bar{A}_\nu^a = \frac{\int \mathcal{D}_{\text{gf}}[A; \bar{A}] \, \Delta[A; \mu] \, A_\nu^a \, e^{-S_{YM}[A]}}{\int \mathcal{D}_{\text{gf}}[A; \bar{A}] \, \Delta[A; \mu] \, e^{-S_{YM}[A]}} \equiv \langle A_\nu^a \rangle_{\bar{A}} . \tag{9.30}$$

Since the background is taken constant, temporal, and along the Cartan sub-algebra, this identity is trivially valid for $\nu \neq 0$ or for $a \neq 3$ and $a \neq 8$. Self-consistent backgrounds appear, therefore, as fixed points of the mapping

$$(\bar{A}_0^3, \bar{A}_0^8) \mapsto (\langle A_0^3 \rangle_{\bar{A}}, \langle A_0^8 \rangle_{\bar{A}}) . \tag{9.31}$$

Of course, the existence of fixed points depends crucially on which space the fixed points are searched for. If one views the background as a particular configuration of the gauge field A_μ^a, the natural choice would be to look for fixed points in the space of real background components. As we now discuss, however, this choice is not always the relevant one. The right choice depends on the considered situation and is again intimately related to the properties of the measure under the functional integral.

---

[11] At least among those compatible with the charge conjugation invariance present in the $\mu = 0$ case.

Let us consider the case of an imaginary chemical potential first. In this case, the fermion determinant is real. Then, if we choose the background components to be real, the measure under the functional integral is real. The expectation values $\langle A_0^3 \rangle_{\bar{A}}$ and $\langle A_0^8 \rangle_{\bar{A}}$ are also real, and (9.31) maps $\mathbb{R} \times \mathbb{R}$ into itself. This is certainly a favorable condition for the existence of a fixed point solution and, thus, of self-consistent backgrounds.

The problem occurs when changing to a real chemical potential. Indeed, the fermion determinant being complex in this case, the expectation values $\langle A_0^3 \rangle_{\bar{A}}$ and $\langle A_0^8 \rangle_{\bar{A}}$ are not real anymore if we insist in keeping the background components real, and the previous favorable condition for the existence of a (real) self-consistent background is not met. The way out is to allow for backgrounds with complex components. This may look surprising at first sight if one insists in viewing the background as a particular configuration of the gauge field. However, from the perspective of the gauge-fixing procedure, the background should be rather seen as an infinite collection of gauge-fixing parameters that characterizes the particular gauge under consideration.[12] From this perspective, nothing prevents the background components to be taken complex.

To see how such an extended background allows us to solve the problem, we need the counterpart of the property (9.20) regarding the gauge-fixed measure. One finds

$$\mathcal{D}_{\mathrm{gf}}[A; \bar{A}] = \left(\mathcal{D}_{\mathrm{gf}}[A^{\mathcal{K}}; \bar{A}^{\mathcal{K}}]\right)^*. \tag{9.32}$$

We note that, since we leave open the possibility of complex background components, we have $\bar{A}^{\mathcal{K}} \neq \bar{A}^C$. More precisely, for our constant, temporal, and diagonal background, we have $(\bar{A}_0^{\mathcal{K}})^{3,8} = -(\bar{A}_0^{3,8})^*$, whereas $(\bar{A}_0^C)^{3,8} = -(\bar{A}_0^{3,8})$. The background is thus invariant under $\mathcal{K}$ if its components are taken purely imaginary. In this case, it is readily checked that

$$\langle A_0^{3,8} \rangle_{\bar{A}} = -\langle A_0^{3,8} \rangle_{\bar{A}}^*, \tag{9.33}$$

and therefore (9.31) maps $i\mathbb{R} \times i\mathbb{R}$ into itself. Thus, there is again some chance to find self-consistent backgrounds but with purely imaginary components this time.

Another possibility is to consider $\bar{A}_0^3$ real and $\bar{A}_0^8$ imaginary. Indeed, in this case, the background is invariant under a combination of $\mathcal{K}$ and the Weyl transformation that flips the sign of $\bar{A}_0^3$. Since the Weyl transformation is a symmetry of the

---

[12] Complex backgrounds have also been considered in Refs. [195,196]. There, the search for saddle points in the presence of a complex action forces one to analytically continue the original real gauge field to complex configurations.

problem, it is readily checked in this case that

$$\langle A_0^3 \rangle_{\bar{A}} = \langle A_0^3 \rangle_{\bar{A}}^*, \tag{9.34}$$

$$\langle A_0^8 \rangle_{\bar{A}} = -\langle A_0^8 \rangle_{\bar{A}}^*. \tag{9.35}$$

It follows that (9.31) maps $\mathbb{R} \times i\mathbb{R}$ into itself, opening the possibility for the existence of self-consistent backgrounds with $\bar{A}_0^3$ real and $\bar{A}_0^8$ imaginary.

9.3.2 Background-Field Effective Potential

The previous discussion extends to the background-field effective potential, from which the self-consistent backgrounds should be obtained in principle. It is convenient to introduce the generating functional

$$Z[J; \bar{A}_0, \mu] \equiv \int \mathcal{D}_{\mathrm{gf}}[A; \bar{A}_0] \, \Delta[A; \mu] \exp\left\{ -S_{YM}[A] + \int d^d x \, J^j(x) A_0^j(x) \right\}, \tag{9.36}$$

with $J \equiv (J^3, J^8)$ and $\bar{A}_0 \equiv (\bar{A}_0^3, \bar{A}_0^8)$. We recall that the background-field effective action $\tilde{\Gamma}[\bar{A}_0]$ is constructed from this functional by first Legendre transforming $W[J; \bar{A}_0, \mu] \equiv -\ln Z[A_0; \bar{A}_0, \mu]$ with respect to the sources

$$\Gamma[A_0; \bar{A}_0, \mu] = -W[J; \bar{A}_0, \mu] + \int d^d x \, J^j(x) A_0^j(x), \tag{9.37}$$

with

$$A_0^j(x) \equiv \frac{\delta W}{\delta J^j(x)} = \frac{1}{Z} \frac{\delta Z}{\delta J^j(x)}, \tag{9.38}$$

and then evaluating $\tilde{\Gamma}[\bar{A}_0, \mu] = \Gamma[\bar{A}_0; \bar{A}_0, \mu]$. We can now follow a similar discussion as the one presented in Sect. 9.2.

By using the identities (9.19)–(9.21), it is found that (see Problem 9.5)

$$Z[J; \bar{A}_0, \mu] = Z[-J; -\bar{A}_0, -\mu] \tag{9.39}$$

$$= Z[-J^*; -\bar{A}_0^*, \mu^*]^* \tag{9.40}$$

$$= Z[J^*; \bar{A}_0^*, -\mu^*]^*, \tag{9.41}$$

where the last identity is a combination of the previous two. One can also use a Weyl transformation to flip the sign of both J^3 and $\bar{A}_0^3$ in Eq. (9.40). Using these identities, it is easily checked that, in the case of an imaginary chemical potential, $Z[J; \bar{A}_0, \mu]$ is real if $J \in \mathbb{R} \times \mathbb{R}$ and $\bar{A}_0 \in \mathbb{R} \times \mathbb{R}$. Similarly, in the case of a real chemical potential, $Z[J; \bar{A}_0, \mu]$ is real if $J \in i\mathbb{R} \times i\mathbb{R}$ and

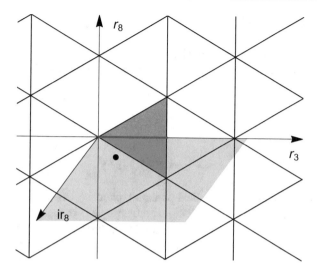

Fig. 9.2 The two different planes where the background-field effective action should be considered depending on whether $\mu \in i\mathbb{R}$ or $\mu \in \mathbb{R}$. This figure is the counterpart of Fig. 9.1

$\bar{A}_0 \in i\mathbb{R} \times i\mathbb{R}$ or if $J \in \mathbb{R} \times i\mathbb{R}$ and $\bar{A}_0 \in \mathbb{R} \times i\mathbb{R}$. From Eq. (9.38), it then follows that these subspaces are stabilized by the Legendre transform and are thus the natural subspaces where to study the background effective action $\tilde{\Gamma}[\bar{A}_0]$. In fact, we will see in the next chapter that the effective action is somewhat ill-defined over $i\mathbb{R} \times i\mathbb{R}$, so we shall restrict to $\mathbb{R} \times i\mathbb{R}$ from now on. The change of subspace for the background components, as one changes from the case of imaginary chemical potential to the case of a real chemical potential, is illustrated in Fig. 9.2.

The properties that we have just discussed are the counterpart of properties (R1) and (R2) discussed in the previous section. The equivalent of the properties (P1), (P2), and (P3) can again be shown to hold true in the case of an imaginary chemical potential using the positivity of the fermion determinant. In particular, the background-field effective action $\tilde{\Gamma}[\bar{A}_0]$ is real if $(\bar{A}_0^3, \bar{A}_0^8) \in \mathbb{R} \times \mathbb{R}$, and self-consistent backgrounds are obtained by minimizing the background-field effective action over this subspace. In the case of a real chemical potential, the proof is jeopardized by the non-positivity of the fermion determinant, but it is reasonable to admit that the background-field effective action $\tilde{\Gamma}[\bar{A}_0]$ is again real if $\bar{A}_0 \in \mathbb{R} \times i\mathbb{R}$, which we shall check on explicit examples in the next chapter. Again, we shall also choose the physical point as the deepest saddle point in this subspace although we lack a derivation of this recipe from first principles.

9.3.3 Background-Dependent Polyakov Loop

The connection with the general discussion of Sect. 9.2 can be done using the background-dependent Polyakov loops. From the identities (9.19)–(9.21), we find that

$$\ell[\bar{A}_0, \mu] = \bar{\ell}[-\bar{A}_0, -\mu] \tag{9.42}$$

$$= \ell[-\bar{A}_0^*, \mu^*]^* \tag{9.43}$$

$$= \bar{\ell}[\bar{A}_0^*, -\mu^*]^* . \tag{9.44}$$

In the case of an imaginary chemical potential, choosing $\bar{A}_0 \in \mathbb{R} \times \mathbb{R}$, one finds from Eq. (9.44) that $\ell[\bar{A}_0, \mu]$ and $\bar{\ell}[\bar{A}_0, \mu]$ are complex conjugate of each other, in agreement with the previous discussion. In the case of a real chemical potential, choosing $\bar{A}_0 \in \mathbb{R} \times i\mathbb{R}$, one finds from Eq. (9.43) that $\ell[\bar{A}_0, \mu]$ and $\bar{\ell}[\bar{A}_0, \mu]$ are both real.

In particular, at leading order, the background-dependent Polyakov loops are not modified by the quark content, and one finds

$$\ell(r_3, r_8, \mu) = \frac{e^{-i\frac{r_8}{\sqrt{3}}} + 2\cos\left(\frac{r_3}{2}\right) e^{i\frac{r_8}{2\sqrt{3}}}}{3} = \bar{\ell}(r_3, -r_8, \mu) , \tag{9.45}$$

which obey the above-mentioned properties.

9.3.4 Other Approaches

Some works propose instead to restrict to $r_8 = 0$ as a way to ensure that the background potential remains real. This is done either directly [193, 197, 198] or effectively by first dropping the imaginary part of the potential and then realizing that the real part has a minimum such that $r_8 = 0$ [190]. However, this is at odds with the fact that a non-vanishing chemical potential breaks charge conjugation symmetry and should then correspond to a non-zero r_8. We will see in the next chapter that approaches that artificially set $r_8 = 0$ miss part of the physical picture by not reproducing the expected behavior of the Polayakov loops as a function of the chemical potential. However, for some other aspects, the $r_8 = 0$ and $r_8 \in i\mathbb{R}$ prescriptions give quantitatively similar results. As an example, we evaluated the thermodynamical observables in the Polyakov-extended Quark-Meson model of [193]. Figure 9.3 shows both the deviation of the pressure p and the trace anomaly $e - 3p$ with respect to the zero-density case, using both prescriptions for r_8.

The prescription $r_8 = 0$ has also been used in Ref. [191] to obtain bubble nucleation rates by computing the barrier between two minima of the potential in the case of a first-order phase transition. We stress again that the analysis should in principle be carried out with a non-zero, imaginary r_8. In this case however, the very

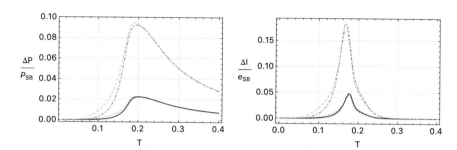

Fig. 9.3 Pressure and trace anomaly deviation with respect to the zero-density case (divided, respectively, by the Stefan-Boltzmann pressure and energy density), as functions of the temperature, for various values of the baryonic chemical potential $\mu_B = 3\mu$, $\mu_B = 200$ MeV (plain and dotted) and $\mu_B = 400$ MeV (dashed and dash-dotted). The thin lines correspond to the results obtained with the $r_8 = 0$ prescription, as in Ref. [193], whereas the thick lines show the results obtained with the $r_8 \in i\mathbb{R}$ presciption

method for extracting nucleation rates has to be revisited before a comparison such as the one in Fig. 9.3 can be even considered.

Problems

9.1 Euclidean Dirac Matrices in the Weyl Basis (⋆)
Consider the Euclidean Dirac matrices in the Weyl basis such that $\gamma^*_{0,2} = \gamma^t_{0,2} = \gamma_{0,2}$ and $\gamma^*_{1,3} = \gamma^t_{1,3} = -\gamma_{1,3}$. Show that

$$\gamma_2\gamma_0\gamma_\mu\gamma_0\gamma_2 = -\gamma^t_\mu \quad \text{and} \quad \gamma_3\gamma_1\gamma_\mu\gamma_1\gamma_3 = \gamma^*_\mu.$$

9.2 Charge Conjugation (⋆⋆)
Consider the transformation defined by $A^C \equiv -A^t$ and $\psi^C \equiv \gamma_0\gamma_2\bar{\psi}^t$.

(a) Show that $\bar{\psi}^C \equiv -\psi^t\gamma_2\gamma_0$. *Tip:* recall that the bar operator works similar to the operation that turns kets into bras.
(b) Evaluate $S[A^C, \psi^C, \bar{\psi}^C; \mu]$, and show that it equals $S[A, \psi, \bar{\psi}; -\mu]$.

9.3 Hermicity (⋆⋆)
Consider the transformation defined by $A^K \equiv A^*$ and $\psi^K \equiv \gamma_1\gamma_3\psi$.

(a) Show that $\bar{\psi}^K \equiv \bar{\psi}\gamma_3\gamma_1$. *Tip:* recall that the bar operator works similar to the operation that turns kets into bras.
(b) Evaluate $S[A^K, \psi^K, \bar{\psi}^K; \mu]$, and show that it equals $S[A, \psi, \bar{\psi}; \mu^*]^*$.

9.4 Dirac Operator ($\star$)

Consider the Dirac operator (9.14). Show that

$$\gamma_2\gamma_0\, M[A;\mu]\,\gamma_0\gamma_2 = M[A^C; -\mu]^t,$$
$$\gamma_3\gamma_1\, M[A;\mu]\,\gamma_1\gamma_3 = M[A^K; \mu^*]^*,$$
$$\gamma_5\, M[A;\mu]\,\gamma_5 = M[A; -\mu^*]^\dagger.$$

Tip: reinterpret the steps in the resolution of the previous two problems.

9.5 Partition Function ($\star$)

Derive Eqs. (9.39)–(9.41).

QCD Deconfinement Transition in the Heavy Quark Regime

10

We now apply the general considerations of the previous chapter to the study of the deconfinement transition in the presence of quark degrees of freedom. To keep the picture relatively simple, we consider a formal regime where the up, down, and strange quarks are considered heavy and the other, even heavier quarks, are neglected. This departure from the physical QCD case has been largely explored in the literature, from lattice simulations [199], first principle continuum methods [198], or models [200], and this for several reasons.

First, it allows to assess the impact of dynamical quarks on the deconfinement transition without the contamination from the breaking of chiral symmetry, the other relevant transition at play in physical QCD.[1] Moreover, in this formal regime, QCD presents a rich phase structure that allows for the comparison and benchmarking of various approaches. For us, it will serve as a further testing ground of the Curci-Ferrari model and the related hypothesis that some of the low-energy properties of QCD can be described with perturbative methods.

10.1 Background Effective Potential

As it was already the case in Yang-Mills theory, the evaluation of the one-loop background effective potential in the presence of quarks requires only the quadratic part of the action. Thus, for the matter part, we only need

$$\delta S_0[\psi, \bar{\psi}; \mu] \equiv \int d^d x \sum_{f=1}^{N_f} \bar{\psi}_f(x)\left(\slashed{D} + M_f - \mu\gamma_0\right)\psi_f(x), \tag{10.1}$$

---

[1] It is interesting, of course, to study the interplay between these two transitions in the physical case. This lies, however, beyond the scope of the present manuscript.

© The Author(s), under exclusive license to Springer Nature Switzerland AG 2022
U. Reinosa, *Perturbative Aspects of the Deconfinement Transition*, Lecture Notes in Physics 1006, https://doi.org/10.1007/978-3-031-11375-8_10

with $\bar{D}_\nu = \partial_\nu - i T \delta_{\nu 0} r^j t^j$. Decomposing the quark fields into a Cartan-Weyl basis $\{|\rho\rangle\}$ that diagonalizes simultaneously the generators t^j, $\psi = \sum_\rho \psi_\rho |\rho\rangle$, with $t^j |\rho\rangle = \rho_j |\rho\rangle$ and $\langle \rho | \sigma \rangle = \delta_{\rho\sigma}$, one finds

$$\delta S_0[\psi, \bar{\psi}; \mu] \equiv \int d^d x \sum_{f=1}^{N_f} \bar{\psi}_{f,\rho}(x) \left(\bar{\mathcal{D}}^\rho + M_f - \mu \gamma_0 \right) \psi_{f,\rho}(x), \qquad (10.2)$$

with $\mathcal{D}_\nu^\rho = \partial_\nu - i T r \cdot \rho \, \delta_{\nu 0}$. In Fourier space, this becomes

$$\delta S_0[\psi, \bar{\psi}; \mu] \equiv \int_P^T \sum_{f=1}^{N_f} \bar{\psi}_{f,\rho}(P) \left(-i \not{P}^\rho + M_f - \mu \gamma_0 \right) \psi_{f,\rho}(P), \qquad (10.3)$$

where $P^\rho \equiv P + T r \cdot \rho n$ is the generalized momentum in the presence of the background, with $P = (\omega, \mathbf{p})$, ω a fermionic Matsubara frequency, and $n = (1, \mathbf{0})$.

10.1.1 General One-Loop Expression

We mention that each shift $T r \cdot \rho n$ of momentum can be interpreted as an imaginary shift of the chemical potential μ: $\mu \to \mu + i r \cdot \rho T$. This illustrates once more the interpretation of the background as an imaginary chemical potential and allows us to derive the one-loop matter contribution to the background effective potential using the well-known one-loop expression for the free-energy density of a colorless fermionic field of flavor f in the absence of background:

$$V_f(T, \mu) = -\frac{T}{\pi^2} \int_0^\infty dq\, q^2 \left\{ \ln\left[1 + e^{-\beta(\varepsilon_{q,f} - \mu)} \right] + \ln\left[1 + e^{-\beta(\varepsilon_{q,f} + \mu)} \right] \right\}. \qquad (10.4)$$

We find

$$\delta V(r; T, \mu) = \sum_{f, \rho} V_f\left(T, \mu + i T r \cdot \rho \right), \qquad (10.5)$$

where ρ runs over the weights (colors) of the defining representation.

10.1.2 Real-Valuedness in the SU(3) Case

In the SU(3) case, we recall that $r = (r_3, r_8)$ and the weights are $(0, -1/\sqrt{3})$, $(1, 1/\sqrt{3})/2$, and $(-1, 1/\sqrt{3})/2$. Therefore

$$\delta V_{\text{1loop}}^{\text{SU(3)}}(r; T, \mu)$$

$$= -\frac{T}{\pi^2} \int_0^\infty dq \, q^2 \left\{ \ln\left[1 + e^{-\beta(\varepsilon_{q,f}-\mu)-i\frac{r_8}{\sqrt{3}}}\right] + \ln\left[1 + e^{-\beta(\varepsilon_{q,f}+\mu)+i\frac{r_8}{\sqrt{3}}}\right] \right.$$

$$+ \ln\left[1 + e^{-\beta(\varepsilon_{q,f}-\mu)+\frac{i}{2}\left(r_3+\frac{r_8}{\sqrt{3}}\right)}\right] + \ln\left[1 + e^{-\beta(\varepsilon_{q,f}+\mu)-\frac{i}{2}\left(r_3+\frac{r_8}{\sqrt{3}}\right)}\right]$$

$$\left. + \ln\left[1 + e^{-\beta(\varepsilon_{q,f}-\mu)+\frac{i}{2}\left(-r_3+\frac{r_8}{\sqrt{3}}\right)}\right] + \ln\left[1 + e^{-\beta(\varepsilon_{q,f}+\mu)-\frac{i}{2}\left(-r_3+\frac{r_8}{\sqrt{3}}\right)}\right] \right\}. \tag{10.6}$$

For completeness, we recall that the pure glue part of the potential reads

$$V_{\text{1loop}}^{\text{SU(3)}}(r; T)$$

$$= \frac{T}{2\pi^2} \int_0^\infty dq \, q^2 \left\{ \ln\frac{\left(1 - e^{-\beta\varepsilon_q+ir_3}\right)^3}{1 - e^{-\beta q+ir_3}} + \ln\frac{\left(1 - e^{-\beta\varepsilon_q-ir_3}\right)^3}{1 - e^{-\beta q-ir_3}} \right.$$

$$+ \ln\frac{\left(1 - e^{-\beta\varepsilon_q+i\frac{r_3+r_8\sqrt{3}}{2}}\right)^3}{1 - e^{-\beta q+i\frac{r_3+r_8\sqrt{3}}{2}}} + \ln\frac{\left(1 - e^{-\beta\varepsilon_q-i\frac{r_3+r_8\sqrt{3}}{2}}\right)^3}{1 - e^{-\beta q-i\frac{r_3+r_8\sqrt{3}}{2}}}$$

$$\left. + \ln\frac{\left(1 - e^{-\beta\varepsilon_q+i\frac{r_3-r_8\sqrt{3}}{2}}\right)^3}{1 - e^{-\beta q+i\frac{r_3-r_8\sqrt{3}}{2}}} + \ln\frac{\left(1 - e^{-\beta\varepsilon_q-i\frac{r_3-r_8\sqrt{3}}{2}}\right)^3}{1 - e^{-\beta q-i\frac{r_3-r_8\sqrt{3}}{2}}} \right\}. \tag{10.7}$$

It is readily checked that the above expressions are real if $(\mu, r_3, r_8) \in i\mathbb{R} \times \mathbb{R} \times \mathbb{R}$, $(\mu, r_3, r_8) \in \mathbb{R} \times i\mathbb{R} \times i\mathbb{R}$, or $(\mu, r_3, r_8) \in \mathbb{R} \times \mathbb{R} \times i\mathbb{R}$, as anticipated in the previous chapter. We mention however that the case $(\mu, r_3, r_8) \in \mathbb{R} \times i\mathbb{R} \times i\mathbb{R}$ is problematic since some of the bosonic integrals become ill-defined. For this reason, for a real chemical potential, we shall only consider the case $(\mu, r_3, r_8) \in \mathbb{R} \times \mathbb{R} \times i\mathbb{R}$.

10.1.3 Polyakov Loop Potential

As we have already mentioned in the previous chapter, at leading order, the background-dependent Polyakov loops take the same expressions as in the pure YM case, namely,

$$\ell(r) = \frac{1}{N} \sum_{\rho} e^{ir \cdot \rho} \quad \text{and} \quad \bar{\ell}(r) = \frac{1}{N} \sum_{\rho} e^{-ir \cdot \rho} . \tag{10.8}$$

In the SU(3) case, these two Polyakov loops are in one-to-one correspondence with the background components r_3 and r_8, and one can consider expressing the latter in terms of the former (see Problem 10.1). Doing so, the background-field effective potential becomes an effective potential for the Polyakov loops.[2] We find

$$\delta V_{1\text{loop}}^{\text{SU(3)}}(\ell, \bar{\ell}; T, \mu)$$

$$= -\frac{T}{\pi^2} \int_0^\infty dq \, q^2 \left\{ \ln \left[1 + 3\ell e^{-\beta \, (\varepsilon_{q,f} - \mu)} + 3\bar{\ell} e^{-2\beta \, (\varepsilon_{q,f} - \mu)} + e^{-3\beta \, (\varepsilon_{q,f} - \mu)} \right] \right.$$

$$\left. + \ln \left[1 + 3\bar{\ell} e^{-\beta \, (\varepsilon_{q,f} + \mu)} \ell + 3\ell e^{-2\beta \, (\varepsilon_{q,f} + \mu)} + e^{-3\beta \, (\varepsilon_{q,f} + \mu)} \right] \right\}. \tag{10.9}$$

The formula can be extended to SU(N), but in this case, one needs to introduce $N - 1$ Polyakov loops to be mapped to the $N - 1$ background components (see Problem 10.2). These are the Polyakov loops associated with the $N - 1$ fundamental SU(N) representations [168].

A similar treatment can be done for the one-loop glue potential. In the SU(3) case, one obtains (see Problem 10.3)

$$V_{1\text{loop}}^{\text{SU(3)}}(\ell, \bar{\ell}; T)$$

$$= \frac{3T}{2\pi^2} \int_0^\infty dq \, q^2 \ln \left[1 + e^{-8\beta\varepsilon_q} - (9\ell\bar{\ell} - 1)(e^{-\beta\varepsilon_q} + e^{-7\beta\varepsilon_q}) \right.$$

$$- (81\ell^2\bar{\ell}^2 - 27\ell\bar{\ell} + 2)(e^{-3\beta\varepsilon_q} + e^{-5\beta\varepsilon_q})$$

$$+ (27\ell^3 + 27\bar{\ell}^3 - 27\ell\bar{\ell} + 1)(e^{-2\beta\varepsilon_q} + e^{-6\beta\varepsilon_q})$$

$$\left. + (162\ell^2\bar{\ell}^2 - 54\ell^3 - 54\bar{\ell}^3 + 18\ell\bar{\ell} - 2)e^{-4\beta\varepsilon_q} \right],$$

$$- \frac{T}{2\pi^2} \int_0^\infty dq \, q^2 \ln \left[1 + e^{-8\beta q} - (9\ell\bar{\ell} - 1)(e^{-\beta q} + e^{-7\beta q}) \right.$$

$$- (81\ell^2\bar{\ell}^2 - 27\ell\bar{\ell} + 2)(e^{-3\beta q} + e^{-5\beta q})$$

$$+ (27\ell^3 + 27\bar{\ell}^3 - 27\ell\bar{\ell} + 1)(e^{-2\beta q} + e^{-6\beta q})$$

$$\left. + (162\ell^2\bar{\ell}^2 - 54\ell^3 - 54\bar{\ell}^3 + 18\ell\bar{\ell} - 2)e^{-4\beta q} \right]. \tag{10.10}$$

---

[2] That this corresponds to the Polyakov loop potential, as it would be obtained from a Legendre transformation with respect to sources coupled to the Polyakov loops, can be shown up to two-loop order (see [94]).

In line with the discussion of the previous chapter, these expressions should be considered for complex-conjugated variables $\bar{\ell} = \ell^*$ in the case of an imaginary chemical potential and for real and independent variables ℓ and $\bar{\ell}$ in the case of a real chemical potential.[3]

10.2 Phase Structure at $\mu = 0$

Let us now combine Eqs. (10.6) and (10.7) to study the phase structure of the model as a function of the (heavy) quark masses. This dependence has been studied in various approaches, including non-gauge-fixed lattice QCD, and offers, therefore, a valuable benchmark for the Curci-Ferrari model. For simplicity, we consider the case of two degenerate flavors with mass $M_u = M_d$ and a third flavor with mass M_s.

We first analyze the phase diagram at $\mu = 0$. As we have discussed above, in this case, the analysis of the background effective potential can be done either over $(r_3, r_8) \in \mathbb{R} \times \mathbb{R}$ or over $(r_3, r_8) \in \mathbb{R} \times i\mathbb{R}$. This is related to the fact that, because charge conjugation is not broken, we expect the physical point to lie along the axis $r_8 = 0$ in the fundamental Weyl chamber. This is indeed what we find numerically. Correspondingly, we set $r_8 = 0$ in what follows. Along this axis, the background-dependent Polyakov loops $\ell(r)$ and $\bar{\ell}(r)$ are guaranteed to be equal and real, in line with their standard interpretation in terms of the free energy of a static quark or anti-quark [37, 159] and the fact that there should be no distinction between the free energy of a quark and that of an anti-quark at $\mu = 0$.

Depending on the values of the quark masses, we find different types of behaviors as the temperature is varied (see Fig. 10.1). For large masses, the absolute minimum presents a finite jump at some transition temperature, signaling a first-order transition. Instead, for small masses, there is always a unique minimum, whose location rapidly changes with temperature in some crossover regime. At the common boundary of these two mass regions, the system presents a critical behavior: there exists a unique minimum of the potential for all temperatures which however behaves as a power law around some critical temperature. At this critical point, the curvature of the potential at the minimum needs to vanish, and, to distinguish it from a mere spinodal in the first-order transition region, we need to require that the third derivative vanishes as well (which is the condition for the merging of spinodals in the first-order transition region). The three conditions

$$0 = V'(r_3) = V''(r_3) = V'''(r_3) \tag{10.11}$$

---

[3] Being the average values of traced unitary matrix, these variables are further constrained to lie in some subregion of Σ or $\mathbb{R} \times \mathbb{R}$, respectively.

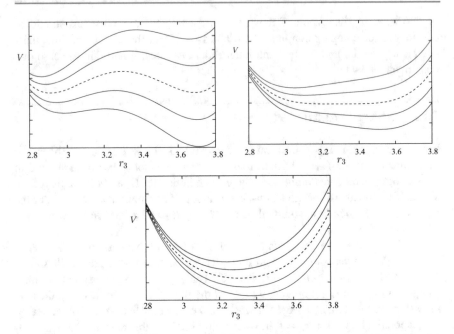

Fig. 10.1 The background-field potential (in arbitrary units) for $\mu = 0$ and $r_8 = 0$, as a function of $x = r_3/(2\pi)$ for different temperatures in the degenerate case $M_u = M_s = M$. Top figure: M slightly larger than the critical value M_c (see Fig. 10.2). Temperature increases for curves from bottom to top. The dashed line corresponds to the transition temperature. Middle figure: $M = M_c$ with the same conventions as for the top figure. Bottom figure: $M < M_c$. The dashed line is for the crossover temperature, where the curvature of the potential at the minimum is the smallest. Reprinted from Ref. [95]

thus determine the critical values T_c and r_3^c for the temperature and the background, together with a critical line in the (M_u, M_s) plane, the so-called Columbia plot.

Our result for the Columbia plot is shown in Fig. 10.2. In the degenerate case $M_u = M_s$, we get, for the critical mass, $M_c/m = 2.867$, and, for the critical temperature, $T_c/m = 0.355$. We thus have $M_c/T_c = 8.07$. This dimensionless ratio does not depend on the value of m and can be directly compared to lattice results. For instance, the calculation of Ref. [199] yields, for three degenerate quarks, $(M_c/T_c)^{\text{latt.}} = 8.32$. We obtain similar good agreement for different numbers of degenerate quark flavors, as summarized in Table 10.1. In fact, at one-loop order, the values for R_{N_f} for different values of N_f are related to each other by a universal relation that does not depend on the modeling of the gauge sector [99] (see Problem 10.5).

We observe that the critical temperature is essentially unaffected by the presence of quarks. It is actually close to the one obtained in the present approach for the pure gauge SU(3) theory [92]. This is due to the fact that, for the typical values of M/T near the critical line, the quark contribution to the potential is Boltzmann suppressed as compared to that of the gauge sector [95].

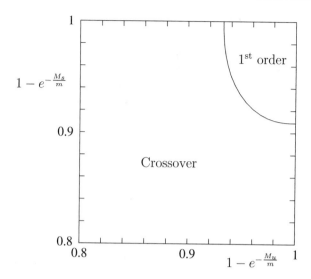

Fig. 10.2 Columbia plot at $\mu = 0$. In the upper right corner of the plane (M_u, M_s), the phase transition is of the first-order type. In the lower left corner, the system presents a crossover. On the plain line, the system has a critical behavior. Reprinted from Ref. [95]

Table 10.1 Values of the critical quark mass and temperature for $N_f = 1, 2, 3$ degenerate quark flavors from the present one-loop calculation. The values of the dimensionless and parameter-independent ratio M_c/T_c are compared to the lattice results of Ref. [199] (previous to last column). For comparison we also show the values from the matrix model of Ref. [200] (last column)

N_f	M_c/m	T_c/m	M_c/T_c	$(M_c/T_c)^{\text{latt.}}$	$(M_c/T_c)^{\text{matr.}}$
1	2.395	0.355	6.74	7.22(5)	8.04
2	2.695	0.355	7.59	7.91(5)	8.85
3	2.867	0.355	8.07	8.32(5)	9.33

We also mention that recent calculations in the Dyson-Schwinger approach [198] yields values of the ratio M_c/T_c that are systematically smaller than the ones obtained on the lattice. The origin of this discrepancy lies in the fact that in these studies a certain renormalized quark mass is used to compute the ratios M_c/T_c, whereas on the lattice and in any one-loop approach such as the one considered here, the bare quark mass is used.

Two-loop corrections to the previous results within the Curci-Ferrari model have been evaluated in Ref. [98] and improve the agreement with lattice results. More recently, another perturbative approach inspired by the Gribov-Zwanziger approach has been considered in Ref. [201] and supports once more the idea that the phase structure of QCD in the top-left corner of the Columbia plot is akin to perturbative methods.

10.3   Phase Structure for $\mu \in i\mathbb{R}$

The case of imaginary chemical potential is interesting in many respects. First, as already mentioned in the previous chapter, the sign problem is under control, allowing for the use of lattice simulations. But more importantly, the corresponding phase structure is quite rich and offers a new source for comparison between the various approaches.

10.3.1 Roberge-Weiss Symmetry

The richness of the phase structure has to do with the fact that, despite the explicit breaking of center symmetry by the defining quarks, a symmetry exists for particular values of the chemical potential.

To see this, consider a center transformation such that $U(\tau + \beta) = e^{i2\pi/3}U(\tau)$. We have seen that this transformation modifies the boundary conditions of the quark field to $\psi(\tau + \beta, \mathbf{x}) = -e^{i2\pi/3}\psi(\tau, \mathbf{x})$ which explicitly breaks center symmetry at the quantum level. However, the usual anti-periodic boundary conditions can be restored by means of an abelian transformation $e^{i2\pi/3T\tau}$, with the effect of shifting the chemical potential by $-i2\pi/3T$. It follows that

$$S[A^U, e^{i2\pi/3T\tau}U\psi, e^{-i2\pi/3T\tau}U^\dagger\bar{\psi}; \mu + i2\pi/3T] = S[A, \psi, \bar{\psi}; \mu]. \qquad (10.12)$$

Now, had we first applied charge conjugation to the system, we would have ended up with the identity

$$S[(A^C)^U, e^{i2\pi/3T\tau}U\psi^C, e^{-i2\pi/3T\tau}U^\dagger\bar{\psi}^C; i2\pi/3T - \mu] = S[A, \psi, \bar{\psi}; \mu]. \qquad (10.13)$$

In particular, if we choose $\mu = i\pi/3T$, we have a symmetry that survives at the quantum level since the boundary conditions are unaffected. This is the so-called Roberge-Weiss symmetry, a subtle combination of center, charge conjugation, and abelian transformations [202].

The Roberge-Weiss symmetry imposes constraints on certain observables that one can then use as order parameters testing the possible spontaneous breaking of the symmetry. In particular, if the Roberge-Weiss symmetry is not spontaneously broken, we must have (at $\mu = i\pi/3T$)

$$\ell = e^{-i2\pi/3}\bar{\ell} = e^{-i2\pi/3}\ell^*, \qquad (10.14)$$

where we used that $\bar{\ell} = \ell^*$ when the chemical potential is imaginary. The argument of the Polyakov loop is then such that

$$\text{Arg}\,\ell = -i\frac{2\pi}{3} - \text{Arg}\,\ell, \qquad (10.15)$$

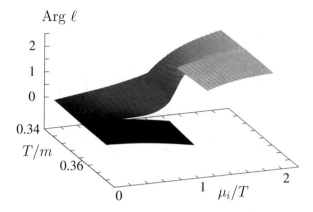

Fig. 10.3 Argument of the Polyakov loop in the $(\mu_i/T, T)$ plane for a degenerate quark mass M. Reprinted from Ref. [95]. Note that the convention for the sign of μ in that reference is opposite to the one considered in this manuscript

and thus Arg $\ell = -i\pi/3$. Any departure from this value signals the breaking of the Roberge-Weiss symmetry.

We mention finally that, as a consequence of the above properties, the system is invariant under $\mu \to \mu + i2\pi/3T$ and $\mu \to i2\pi/3T - \mu$. We shall thus consider chemical potentials $\mu = ixT$ with $x \in [0, \pi/3]$.

10.3.2 Results

We consider for simplicity the case of three degenerate quarks with $M_u = M_s = M$. For a large-enough value of M, we find indeed that the Roberge-Weiss symmetry can be spontaneously broken, with a discontinuity of Arg ℓ as μ_i/T crosses $\pi/3$, if the temperature is large enough (see Fig. 10.3). It is interesting to follow the evolution of the $(\mu_i/T, T)$ phase diagram as the degenerate quark mass M is varied.

For any mass larger than $M_c(\mu = 0) \simeq 2.8m$, such that the transition at vanishing chemical potential is first-order (see Fig. 10.2), we find that the transition persists at non-vanishing μ_i, as depicted in the top panel of Fig. 10.4. This line of first-order transitions, its mirror image by the symmetry $\mu_i/T \to 2\pi/3 - \mu_i/T$, and the line of first-order transitions associated with the Roberge-Weiss symmetry merge into a triple point at $\mu_i/T = \pi/3$.

For $M = M_c(\mu = 0)$, the transition at vanishing chemical potential is second-order, and there appears, in the $(\mu_i/T, T)$ phase diagram, a couple of Z_2 critical points which terminate the lines of first-order transitions described above at $\mu_i/T = 0$ and $\mu_i/T = 2\pi/3$. Decreasing the mass M further, these critical points penetrate deeper in the phase diagram toward $\mu_i/T = \pi/3$, as shown in the middle panel of Fig. 10.4. The critical points are located by generalizing the approach at $\mu = 0$. Since we have one extra variable r_8, we need of course one additional condition. We

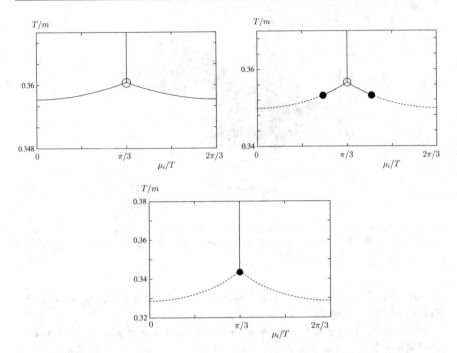

Fig. 10.4 Phase diagram in the plane $(\mu_i/T, T)$ for different values of the degenerate quark mass M. Top panel: $M = 3m$ is larger than the critical mass at vanishing chemical potential $M_c(0) \simeq 2.8m$. Middle panel: $M = 2.6m$ is below $M_c(0)$ but larger than the second critical mass $M_c(i\pi T/3) \simeq 2.2m$ (see text). Bottom panel: $M = 2m$ is smaller than $M_c(i\pi T/3)$. Plain lines correspond to first-order phase transitions, dashed lines to crossovers, black dots represent Z_2 critical points, and the empty circles are triple points. Reprinted from Ref. [95]

require that both equations of motion for r_3 and r_8 are fulfilled:

$$0 = \frac{\partial V}{\partial r_3} \quad \text{and} \quad 0 = \frac{\partial V}{\partial r_8}, \tag{10.16}$$

that the determinant of the Hessian vanishes, meaning that the curvature of the potential vanishes in a certain direction,

$$0 = \frac{\partial^2 V}{\partial r_3^2} \frac{\partial^2 V}{\partial r_8^2} - \left(\frac{\partial^2 V}{\partial r_3 \partial r_8}\right)^2, \tag{10.17}$$

and that the third derivative in this direction vanishes as well

$$0 = a^3 \frac{\partial^3 V}{\partial r_3^3} + 3a^2 b \frac{\partial^3 V}{\partial r_3^2 \partial r_8} + 3ab^2 \frac{\partial^3 V}{\partial r_3 \partial r_8^2} + b^3 \frac{\partial^3 V}{\partial r_3^3}, \tag{10.18}$$

with $a = -\partial^2 V/\partial r_3 \partial r_8$ and $b = \partial^2 V/\partial r_3^2$.

At a critical value $M_c(\mu = i\pi T/3) \simeq 2.2m$, the two Z_2 critical points merge at the symmetric point $\mu_i/T = \pi/3$ and give rise to a tricritical point which terminates the vertical line of first-order transition [203]. The horizontal lines of first-order transitions for $\mu_i/T \neq \pi/3$ have completely disappeared and are replaced by crossovers. For $M < M_c(i\pi T/3)$, the picture is the same with, however, the tricritical point ending the first-order transition line at $\mu_i/T = \pi/3$ replaced by a Z_2 critical point, as shown in the lower panel of Fig. 10.4.

The approach to tricriticality is controlled by the scaling behavior [203]

$$\frac{M_c(\mu)}{T_c(\mu)} = \frac{M_{\text{tric.}}}{T_{\text{tric.}}} + K\left[\left(\frac{\pi}{3}\right)^2 + \left(\frac{\mu}{T}\right)^2\right]^{2/5}. \qquad (10.19)$$

A fit of our results at $\mu = i\mu_i$ yields $M_{\text{tric.}}/T_{\text{tric.}} = 6.15$ and $K = 1.85$,[4] to be compared with the lattice result of Ref. [199], $(M_{\text{tric.}}/T_{\text{tric.}})^{\text{latt.}} = 6.66$ and $K^{\text{latt.}} = 1.55$ for three degenerate quark flavors.

10.4 Phase Structure for $\mu \in \mathbb{R}$

10.4.1 Columbia Plot

Using Eqs. (10.16)–(10.18), we can follow the critical line in the Columbia plot for increasing values of $\mu^2 > 0$. The only subtlety is that we need to solve these equations for $(r_3, r_8) \in \mathbb{R} \times i\mathbb{R}$. We mention however that there is no ambiguity here on the choice of the saddle point since on the critical line there is typically only one saddle point. Our result is shown in Fig. 10.5 and shows that the critical line moves toward the Yang-Mills point, in line with the observations made on the lattice. We can also compare our result at real chemical potential with the extrapolation of the tricritical scaling law. We observe that tricritical scaling survives deep in the $\mu^2 > 0$ region [95], as also observed in other approaches.

10.4.2 T-Dependence of the Polyakov Loops

Our analysis reveals that the location of the saddle point is typically at $r_8 \neq 0$, in line with the fact that charge conjugation invariance is explicitly broken by the presence of a finite chemical potential. A non-vanishing r_8 induces a difference between $\ell(\mu)$ and $\bar{\ell}(\mu)$ and therefore between the associated free energies for quarks and anti-quarks. This is illustrated in Fig. 10.6, which shows the temperature dependence of the averaged Polyakov loops in the region of first-order transition. We observe a significant difference between $\ell(\mu)$ and $\bar{\ell}(\mu)$ below the transition temperature,

---

[4] We mention that the tricritical point can be obtained by setting directly $\mu_i/T = \pi/3$ and using the Roberge-Weiss symmetry (see Ref. [99]).

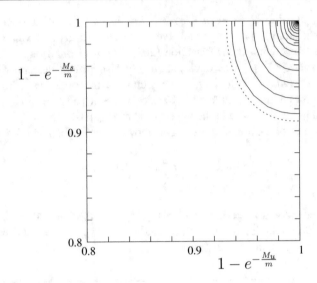

Fig. 10.5 Columbia plot for real chemical potential. The first-order region contracts with increasing μ. The dotted line corresponds to $\mu = 0$. Successive plain lines correspond to a chemical potential increased by steps $\delta\mu = 0.1$ GeV. Reprinted from Ref. [95]

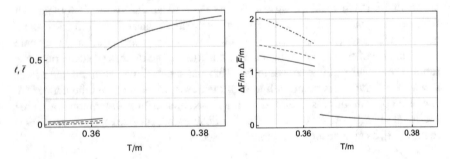

Fig. 10.6 The Polyakov loops ℓ (dot-dashed) and $\bar{\ell}$ (plain) and the corresponding quark and anti-quark free energies computed with a non-zero imaginary background along the r_8 direction, in the case of a positive chemical potential ($\mu = 0.6m$). We also show the same quantities computed with $r_8 = 0$ (dashed) and r_3 at the minimum of the potential along this axis

whereas they essentially agree in the high-temperature phase. In other words, the energetic price to pay for a static quark is much higher than that for an anti-quark (at $\mu > 0$) in the quasi-confined, low-temperature phase, where the Polyakov loops are small, whereas it is essentially the same in the high-temperature deconfined phase. In this latter case, the explanation is that, in a deconfined phase, quarks and anti-quarks are free to roam around and therefore the energetic cost is essentially the same for bringing a quark or an anti-quark. In contrast, in the quasi-confined phase, the capacity of the medium to confine a test color charge depends more notably on the properties of the medium. That anti-quarks are more easily confined (for $\mu > 0$) can be interpreted in terms of screening of a quark by the anti-quarks of the thermal bath [160].

We mention that previous studies of the phase diagram with background-field methods in the functional renormalization group and Dyson-Schwinger approaches of Refs. [193, 197, 198] have employed another criterion than the one used here to determine the physical properties of the system. Instead of searching for a saddle point of the function $V(r_3, ir_8, \mu)$ as we propose here, the authors of Refs. [197, 198] define the physical point as the absolute minimum of the function $V(r_3, 0, \mu)$ as a function of r_3. We have repeated our analysis using this procedure for comparison. A clear artefact of this procedure is that on the axis $r_8 = 0$, the tree-level expressions of the Polyakov loops $\ell(\mu)$ and $\bar{\ell}(\mu)$ are equal, as already mentioned. However, we have found that both criteria give essentially the same critical temperatures in our calculation. This is illustrated in Fig. 10.6.

That the critical temperatures are not significantly modified in these two prescriptions can be traced back to the relative smallness of the values of r_8 obtained by following the saddle points in our procedure. This, in turn, originates from the strong Boltzmann suppression of the (heavy) quark contribution to the potential, which is responsible for the departure of the saddle point from the axis $r_8 = 0$. We point out that the situation might be very different in the case of light quarks [193, 197, 198] and that different procedures for identifying the relevant extremum of the potential may have more dramatic consequences. This needs to be investigated further.

10.4.3 μ-Dependence of the Polyakov Loops

Finally, in Fig. 10.7, we show the Polyakov loops and the corresponding free energies as functions of μ for fixed $T/m = 0.33$ and $M/m = 2.22$, for

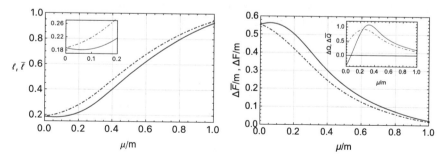

Fig. 10.7 The Polyakov loops ℓ (plain) and $\bar{\ell}$ (dot-dashed) and the corresponding free energies as functions of μ/m. The inset in the first plot shows a close-up view on the small μ region where the change of monotony for ℓ (and therefore ΔF_q) occurs. The second inset shows the difference of average baryonic charge of the bath in the presence of a test quark (plain) or a test anti-quark (dot-dashed), with respect to the average charge in the absence of test charges

$N_f = 3$ degenerate flavors. We observe that the Polyakov loops have a different monotonicity at small μ but then increase together toward one, in line with the observations of Ref. [187]. In this reference, the different monotonicity at small μ was used to question the interpretation of the logarithms of the Polyakov loops as free energies. Here, we show instead that this behavior is perfectly in line with the free-energy interpretation.

The point is that, even though the average charge Q of the thermal bath should vanish for a vanishing chemical potential, this is not so for the average charge Q_q and $\bar{Q}_q$ of the thermal bath in the presence of a test quark/anti-quark. Therefore, we must have

$$\frac{\partial \Delta F}{\partial \mu}\bigg|_{\mu=0} = -\frac{\partial \Delta \bar{F}}{\partial \mu}\bigg|_{\mu=0} \neq 0, \tag{10.20}$$

which explains why the two Polyakov loops have a different monotonicity for small μ. In fact we find that

$$\Delta Q_{\mu=0} \equiv -\frac{\partial \Delta F}{\partial \mu}\bigg|_{\mu=0} < 0 \quad \text{and} \quad \Delta \bar{Q}_{\mu=0} \equiv -\frac{\partial \Delta \bar{F}}{\partial \mu}\bigg|_{\mu=0} > 0, \tag{10.21}$$

so that, in the presence of a test quark (anti-quark) at $\mu = 0$, the thermal bath charges negatively (positively) (see Fig. 10.7). We also show $\Delta Q = Q_q - Q$ and $\Delta \bar{Q} = \bar{Q}_q - Q$ for increasing values of μ. The fact that they both approach 0 is expected since both Q_q and $\bar{Q}_q$ should approach Q at large μ.

Problems

10.1 Matter Contribution to the SU(3) Potential (★★)

(a) Use Eq. (10.5) and the defining weights of SU(3) to derive Eq. (10.6).
(b) Combine the various logarithms that appear in this expression in the form

$$\ln\left[1 + A_{\pm}e^{-\beta(\varepsilon_{q,f}\mp\mu)} + B_{\pm}e^{-2\beta(\varepsilon_{q,f}\mp\mu)} + C_{\pm}e^{-3\beta(\varepsilon_{q,f}\mp\mu)}\right],$$

where $A_{\pm}$, $B_{\pm}$, and $C_{\pm}$ are expressions symmetric in (ρ_1, ρ_2, ρ_3).
(c) Use the constraint $\rho_1 + \rho_2 + \rho_3 = 0$ to relate these expressions with each other, and deduce Eq. (10.9).

10.2 Matter Contribution to the SU(N) Potential (⋆)

Generalize the previous exercise to the SU(N) case. *Tip:* introduce the fundamental Polyakov loops (see Chap. 8)

$$\ell_{C_N^v}(r) = \frac{1}{C_N^v} \sum_{j_1 < \cdots < j_v} e^{i\left(\rho^{(j_1)} + \cdots + \rho^{(j_v)}\right) \cdot r},$$

$$\ell_{\overline{C}_N^v}(r) = \frac{1}{C_N^v} \sum_{j_1 < \cdots < j_v} e^{-i\left(\rho^{(j_1)} + \cdots + \rho^{(j_v)}\right) \cdot r}.$$

Argue that, for a vanishing chemical potential, one can restrict to $\ell_{\overline{C}_N^v} = \ell_{C_N^v}$.

10.3 (Gluonic Contribution to the SU(3) Potential (⋆⋆⋆))

(a) Combine the logarithms in Eq. (10.7) in terms of $1 - e^{-\beta\varepsilon_q + i r \cdot \alpha^{(j)}}$ and its complex conjugate, with $\alpha^{(1)} = \rho^{(2)} - \rho^{(3)}$, $\alpha^{(2)} = \rho^{(3)} - \rho^{(1)}$, $\alpha^{(3)} = \rho^{(1)} - \rho^{(2)}$, and then $\alpha^{(1)} + \alpha^{(2)} + \alpha^{(3)} = 0$.

(b) Express $\sum_{j=1}^{3} e^{i r \cdot \alpha^{(j)}} + \sum_{j=1}^{3} e^{-i r \cdot \alpha^{(j)}}$ in terms of $\ell_3(r)$ and $\ell_{\bar{3}}(r)$, and, similarly,

$\sum_{j=1}^{3} e^{i r \cdot \alpha^{(j)}} \sum_{k=1}^{3} e^{-i r \cdot \alpha^{(k)}}$ and $(\sum_{j=1}^{3} e^{i r \cdot \alpha^{(j)}})^2 + (\sum_{j=1}^{3} e^{-i r \cdot \alpha^{(j)}})^2$ in terms of $\ell_3(3r)$ and $\ell_{\bar{3}}(3r)$. *Tip:* use that $3\ell_3(r) = \sum_{j=1}^{3} e^{i r \cdot \rho^{(j)}}$ and $3\ell_{\bar{3}}(r) = \sum_{j=1}^{3} e^{-i r \cdot \rho^{(j)}}$.

(c) Express $\ell_3(3r)$ and $\ell_{\bar{3}}(3r)$ in terms of $\ell_3(r)$ and $\ell_{\bar{3}}(r)$. *Tip:* start doing the same exercice for $\ell_3(2r)$ and $\ell_{\bar{3}}(2r)$ by expanding $9\ell_3(r)^2$ using Eq. (10.8) and identifying the various terms of the expansions in terms of $\ell_3(r)$. Then, do the same starting from $27\ell_3(r)^3$.

(d) Deduce Eq. (10.7).

10.4 Roberge-Weiss Symmetry and Phase of the Polyakov Loop (⋆)

Show that, for $\mu = i\pi/3T$, and in the case where the Roberge-Weiss symmetry is not spontaneously broken, we must have

$$\ell = e^{-i2\pi/3}\bar{\ell}.$$

10.5 Universality of the Upper Boundary Line in the Columbia Plot (⋆⋆⋆)

(a) At one-loop order, in the heavy quark regime, the Polyakov loop potential for degenerate flavors can be shown to be approximated by

$$V(\ell) = V_{\text{glue}}(\ell) - \frac{6}{\pi^2} N_f \beta^6 M^2 K_2(\beta M)\ell,$$

where K_2 is a the Bessel function of the second kind. Using this expression, write the three conditions $0 = V'(\ell) = V''(\ell) = V'''(\ell)$ that define the upper boundary line in the Columbia plot. Deduce that the critical temperature is essentially constant along this line and that the following "universal" relation holds

$$N_f R_{N_f}^2 K_2(R_{N_f}) = N_f' R_{N_f'}^2 K_2(R_{N_f'}),$$

where R_{N_f} is the critical value of βM.

(b) Show that this result remains true at finite chemical potential.

A Novel Look at the Background-Field Method at Finite Temperature

11

In this final chapter, we revisit once more the rudiments of the background-field method at finite temperature and put forward an alternative formulation of the latter. Although equivalent, in principle, to the one based on self-consistent backgrounds discussed at length in Chaps. 4 and 5,[1] it is different in practical terms and could in fact offer a more robust framework to study the deconfinement transition. After explaining the rationale for this novel proposal and reviewing its various advantages, we implement it within the Curci-Ferrari model. The predictions for the SU(2) and SU(3) transition temperatures are in very good agreement with the lattice results already at one-loop order. Moreover, the Polyakov loop features a slower rise at the transition allowing for a faithful comparison to the lattice data.

This chapter is also the opportunity to revisit the original question of whether the YM two-point correlation functions at finite temperature can be used as probes for the deconfinement transition. We show that, within this new formulation of the background-field method, the gluon two-point correlation function is indeed strongly sensitive to the transition. We compare this behavior to that obtained using self-consistent backgrounds or within the standard Landau gauge.

11.1   Limitations of the Standard Approach

In Chap. 4, we saw that the very reason for extending the Landau gauge in the presence of a background is that this offers a framework where center symmetry is manifest at each computational step. In technical terms, this is encoded in the following symmetry identity for the effective action

$$\Gamma_{\bar{A}^U}[A^U] = \Gamma_{\bar{A}}[A], \quad \forall U \in \mathcal{G}. \tag{11.1}$$

---

[1] See also Chap. 9 for the case of a non-vanishing chemical potential.

© The Author(s), under exclusive license to Springer Nature Switzerland AG 2022
U. Reinosa, *Perturbative Aspects of the Deconfinement Transition*, Lecture Notes in Physics 1006, https://doi.org/10.1007/978-3-031-11375-8_11

We also insisted, however, on the fact that (11.1) makes center symmetry manifest only to the price of connecting two different gauges, labeled by the respective background configurations $\bar{A}$ and $\bar{A}^U$. In particular, the state of the system in the gauge associated with $\bar{A}$, as given by the minimum $A_{\min}[\bar{A}]$ of $\Gamma_{\bar{A}}[A]$, is transformed into $A^U_{\min}[\bar{A}] = A_{\min}[\bar{A}^U]$, that is, the minimum of a different functional $\Gamma_{\bar{A}^U}[A]$. In other words, for a generic $\bar{A}$, the minima of $\Gamma_{\bar{A}}[A]$ do not qualify as order parameters for the center symmetry. Within this framework, it is thus difficult to identify the center-invariant states and then to decide when center symmetry is spontaneously broken.

As discussed in Chaps. 4 and 5, the standard way out is to consider instead the background-field effective action $\tilde{\Gamma}[\bar{A}] \equiv \Gamma_{\bar{A}}[A = \bar{A}]$, a functional of the background-field configuration only that obeys the symmetry identity

$$\tilde{\Gamma}[\bar{A}^U] = \tilde{\Gamma}[\bar{A}], \quad \forall U \in \mathcal{G}. \tag{11.2}$$

The important point here is that the symmetry now connects the absolute minima of the same functional $\tilde{\Gamma}[\bar{A}]$ with each other, making it possible to identify the invariant states. In other words, the minima of $\tilde{\Gamma}[\bar{A}]$ qualify as order parameters for center symmetry. Of course, this strategy makes sense only provided one is able to interpret the absolute minima of $\tilde{\Gamma}[\bar{A}]$ as the actual states of the system, which we did by showing that these minima correspond to self-consistent backgrounds $\bar{A}_s$ such that $A_{\min}[\bar{A}_s] = \bar{A}_s$. This, however, relies on some ideal properties of the gauge-fixing procedure which are not easily met in practice due to the necessary approximations or the degree of modeling of the gauge-fixing procedure in the infrared. One such property is the background independence of the free energy, and, although violations of the latter might get reduced as one refines the level of approximation or modeling, they can be the source of systematic errors in the evaluation of observables using self-consistent backgrounds. For this reason, it is preferable to find an alternative formulation of the background-field method at finite temperature whose rationale does not rely on this property. This is precisely what we shall discuss in this chapter.

11.2   Center-Symmetric Landau Gauge

One possible way of solving the problem of the functional $\Gamma_{\bar{A}}[A]$ without resorting to the use of $\tilde{\Gamma}[\bar{A}]$ would be to find background configurations $\bar{A}$ that are invariant under $\mathcal{G}$. In this way, one could write $\Gamma_{\bar{A}}[A^U] = \Gamma_{\bar{A}}[A]$, for all $U \in \mathcal{G}$, thus connecting absolute minima of the same functional $\Gamma_{\bar{A}}[A]$ via the symmetry, without having to change the gauge in the process. Unfortunately, there is no such configuration that is invariant under $\mathcal{G}$. This relates to the nature of some of the transformations within $\mathcal{G}$.

On the other hand, we have seen that the true physical content of the center-symmetric group is not contained in $\mathcal{G}$ but rather in $\mathcal{G}/\mathcal{G}_0$. We have also seen that there exist backgrounds that are invariant under the action of this quotient group.

These are of course the center-symmetric backgrounds $\bar{A}_c$ identified in Chap. 4 such that

$$\forall U \in \mathcal{G}, \ \exists U_0[U] \in \mathcal{G}_0, \ \bar{A}_c^U = \bar{A}_c^{U_0[U]}. \tag{11.3}$$

Here, we have used the notation $U_0[U]$ to emphasize that the transformation U_0 that is needed for (11.3) to hold true depends in general on the considered transformation U. In fact, if we restrict to constant temporal and diagonal backgrounds as we did in Chap. 5, the transformation $U_0[U]^{-1}$ is nothing, but that transformation of $\mathcal{G}_0$ allows one to bring back the Weyl chamber containing $\bar{A}_c$ into its original location, after being transformed under U. Stated differently given a center transformation $U \in \mathcal{G}$, $U_0[U]^{-1}U$ is an equivalent transformation, in the same class of equivalence than U from the perspective of the quotient group $\mathcal{G}/\mathcal{G}_0$ that leaves the Weyl chamber containing $\bar{A}_c$ globally invariant.

Suppose now that we decide to work within the gauge defined by the choice of background $\bar{A} = \bar{A}_c$. We can then write

$$\begin{aligned} \Gamma[A, \bar{A}_c] &= \Gamma[A^U, \bar{A}_c^U] \\ &= \Gamma[A^U, \bar{A}_c^{U_0[U]}] \\ &= \Gamma[A^{U_0[U]^{-1}U}, \bar{A}_c], \end{aligned} \tag{11.4}$$

where we have successively made use of Eqs. (11.1), (11.3), and again (11.1) this time with U replaced by $U_0[U]^{-1}$. What Eq. (11.4) shows is that the effective action $\Gamma_{\bar{A}_c}[A] \equiv \Gamma[A, \bar{A}_c]$ is invariant under the transformations $U_0[U]^{-1}U$. These are admittely not all possible transformations within $\mathcal{G}$, but they cover all the possible classes of equivalence within $\mathcal{G}/\mathcal{G}_0$ and, thus, carry all the relevant information regarding center symmetry. In summary, we have found a way of making center symmetry manifest without the need to change the gauge. This particular gauge, corresponding to the choice $\bar{A} = \bar{A}_c$, will be referred to in what follows as the *center-symmetric Landau gauge*.[2]

The benefit of using this strategy is clear: the order parameter for the breaking of center symmetry corresponds to the minimum $A_{\min}[\bar{A}_c]$ of $\Gamma_{\bar{A}_c}[A]$, that is, the minimum of a regular Legendre transform, leaving no ambiguity on the identification of the minimum as the state of the system. This also makes certain properties transparent. In particular, a continuous, second-order transition will always be characterized by the appearance of a zero-mode in the inverse propagator $\Gamma_{\bar{A}_c}^{(2)}[A = A_{\min}[\bar{A}_c]]$. Finally, as we shall argue below, the center-symmetric Landau gauge admits a simple implementation on the lattice.

---

[2] In fact, there are as many possible center-symmetric Landau gauges as there are choices for a center-symmetric background. All these gauges are, however, trivially connected to each other.

11.2.1 Constant Backgrounds

In general, the center-symmetric background can be taken constant, temporal, and diagonal

$$\beta g \bar{A}_{c,\mu} = \delta_{\mu 0} \bar{r}_c^j t^j .$$ (11.5)

Thanks to the symmetries in the presence of such type of background, the minimum $A_{\min}[\bar{A}_c]$ of $\Gamma_{\bar{A}_c}[A]$ shares the same properties. This means that one can also restrict the effective action $\Gamma_{\bar{A}_c}[A]$ to

$$\beta g A_\mu = \delta_{\mu 0} r^j t^j ,$$ (11.6)

or better work with the effective potential $V_{\bar{r}_c}(r)$ which we denote $V_c(r)$ in what follows. From this potential, one has access to the zero-frequency/zero-momentum limit of correlation functions along the temporal spacetime dimension and along the neutral color directions:

$$\left(\frac{g}{T}\right)^n \left. \frac{\partial^n V_c(r)}{\partial r^{j_1} \cdots \partial r^{j_n}} \right|_{r=r_{\min}(\bar{r}_c)} ,$$ (11.7)

where the factors g/T have been introduced to restore the relation between r^j and A_0^j. In particular, for $n = 2$, one gets access to the zero-momentum masses.

It is convenient to consider a more general potential $V_{\bar{r}}(r)$ corresponding to the restriction of $\Gamma_{\bar{A}}[A]$ to both $\beta g \bar{A}_\mu = \delta_{\mu 0} \bar{r}^j t^j$ and $\beta g A_\mu = \delta_{\mu 0} r^j t^j$. This potential makes transparent the link between the two implementations of the background-field method since $V_c(r) = V_{\bar{r}_c}(r)$ while $\tilde{V}(\bar{r}) = V_{\bar{r}}(\bar{r})$. In the SU(2) case, for instance, $V_{\bar{r}}(r)$ is a function of two variables. One possible representation of the non-trivial center transformation is

$$V_{2\pi - \bar{r}}(2\pi - r) = V_{\bar{r}}(r) .$$ (11.8)

This implies the symmetry identities $\tilde{V}(2\pi - \bar{r}) = \tilde{V}(\bar{r})$ and $V_c(2\pi - r) = V_c(r)$ which make $\tilde{V}(\bar{r})$ and $V_c(r)$ two alternatives to the study of the deconfinement transition, with the symmetric point located at $r = \bar{r} = \bar{r}_c = \pi$. This is illustrated graphically in Fig. 11.1. We shall explore the connection between the two approaches further below.

Before we continue, however, an important remark on notation is in order. In this chapter, we use letter r to refer to the one-point function (11.6), whereas in previous chapters, this letter was referring to the background. The latter corresponds now to $\bar{r}$, not to be confused with the reduced background introduced in the previous chapters (which we shall not use here).

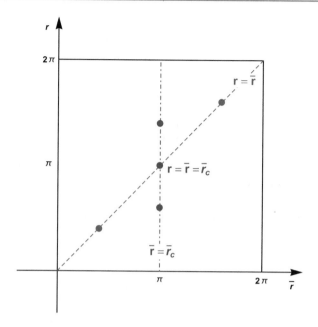

Fig. 11.1 The two possible strategies to discuss the breaking of the center symmetry in background-field gauges (here in the SU(2) case for the purpose of illustration): the standard approach is based on the *background* potential $\tilde{V}(\bar{r})$, defined along the (red) dashed line, whereas the present proposal is based on the *center-symmetric* potential $V_c(r) = V(r, \pi)$, defined along the (blue) dash-dotted line. Both functions have the center symmetry in the form $x \to 2\pi - x$ with $x = r$ or $x = \bar{r}$, as illustrated by the two pairs of points connected by center symmetry. Reprinted from Ref. [100] under CC-BY 4.0 license

11.2.2 Twisted Boundary Conditions and Lattice Implementation

Before closing this section, let us mention another benefit of working in the center-symmetric Landau gauge. The point is that the center-symmetric Landau gauge can be mapped onto the standard Landau gauge but on a different space of gauge-field configurations. That is, the center-symmetric Landau gauge in the space of β-periodic gauge-field configurations is equivalent to the standard Landau gauge in a space of gauge fields obeying modified periodic boundary conditions. This allows one to envisage a simple lattice implementation based on similar algorithms than those for the standard Landau gauge.

For simplicity, let us illustrate this discussion using the SU(2) case.[3] We start writing the gauge condition that defines the center-symmetric Landau gauge on

---

[3] Here, we unveil the connection between the center-symmetric Landau gauge and the standard Landau gauge at a continuum level. Details on the lattice implementation will be given elsewhere.

β-periodic fields. It writes as

$$D_\mu[\bar{A}_c](A_\mu - \bar{A}_{c,\mu}) = 0 \,, \tag{11.9}$$

with

$$\bar{A}_{c,\mu} = i\frac{T}{g}\pi\frac{\sigma^3}{2}\delta_{\mu 0} \,, \tag{11.10}$$

and $D_\mu[\bar{A}] \equiv \partial_\mu - g[\bar{A}_\mu, \ ]$ the background covariant derivative in the presence of a background $\bar{A}$. Let us now use the identity

$$D_\mu[\bar{A}^U](A_\mu^U - \bar{A}_\mu^U) = U D_\mu[\bar{A}](A_\mu - \bar{A}_\mu)U^\dagger \,, \tag{11.11}$$

with

$$A_\mu^U \equiv U A_\mu U^\dagger - \frac{1}{g}U\partial_\mu U^\dagger \,. \tag{11.12}$$

Choosing $\bar{A} = \bar{A}_c$ and

$$U = U_c \equiv e^{-i\tau T\pi\sigma^3/2} \tag{11.13}$$

such that

$$\bar{A}_{c,\mu}^{U_c} = \bar{A}_{c,\mu} - i\frac{T}{g}\pi\frac{\sigma^3}{2}\delta_{\mu 0} = 0 \,, \tag{11.14}$$

we find

$$U_c D_\mu[\bar{A}_c](A_\mu - \bar{A}_{c,\mu})U_c^\dagger = \partial_\mu(A_\mu^{U_c}) \,. \tag{11.15}$$

This shows that the gauge condition (11.9) is equivalent to the standard Landau gauge condition but acting on the space of gauge fields that are the U_c transforms of the periodic gauge fields. It is easily checked that this is nothing but the space of gauge fields whose component along the color direction 3 is periodic and whose components along the color directions 1 and 2 are anti-periodic (see Problem 11.1), which we refer to as *twisted configurations*. These considerations can be easily extended to SU(N).

11.3   Implementation Within the Curci-Ferrari Model

We now apply the above strategy to the Curci-Ferrari model, at one-loop order. The crucial difference with the calculations in the presence of a self-consistent backgroud is that the gauge field A_μ^a that fluctuates under the functional integral needs to be shifted around its expectation value, which we also denote A_μ^a for convenience and which differs from $\bar{A}_\mu^a$.

11.3.1 One-Loop Potential

After adding back the Yang-Mills term to the action (6.5), we thus consider the shift $A_\mu^a \to A_\mu^a + \tilde{a}_\mu^a$. At quadratic order in $\tilde{a}_\mu^a$, we find

$$\sum_\kappa \int_Q^T \left\{ \bar{c}^{-\kappa}(-Q)\, Q_\mu^\kappa \bar{Q}_\mu^\kappa\, c^\kappa(Q) + h^\kappa(Q)^* \bar{Q}_\mu^\kappa \tilde{a}_\mu^\kappa(Q) \right.$$

$$\left. + \frac{1}{2}\tilde{a}_\mu^\kappa(Q)^*(Q_\kappa^2 P_{\mu\nu}^\perp(Q_\kappa) + m^2\delta_{\mu\nu})\,\tilde{a}_\nu^\kappa(Q) \right\}, \qquad (11.16)$$

where $Q_\mu^\kappa \equiv Q_\mu + \kappa r T \delta_{\mu 0}$ and $\bar{Q}_\mu^\kappa \equiv Q_\mu + \kappa \bar{r} T \delta_{\mu 0}$.

The evaluation of the one-loop effective potential in the presence of the background $\bar{A}$ essentially boils down to the evaluation of the determinant of this quadratic form. For the ghost contribution, we find

$$\delta V_{\rm gh}(r, \bar{r}) = -\frac{1}{2}\sum_\kappa \int_Q^T \ln\left(Q_\kappa \cdot \bar{Q}_\kappa\right)^2. \qquad (11.17)$$

As for the gluonic contribution, the corresponding determinant can be evaluated using Schur's complement (see Problem 11.3). We find

$$(Q_\kappa^2 + m^2)^{d-2}\left((Q_\kappa \cdot \bar{Q}_\kappa)^2 + m^2 \bar{Q}_\kappa^2\right). \qquad (11.18)$$

It follows that the gluonic contribution to the one-loop potential reads

$$\delta V_{\rm gl}(r, \bar{r}) = \frac{d-2}{2}\sum_\kappa \int_Q^T \ln\left[Q_\kappa^2 + m^2\right]$$

$$+ \frac{1}{2}\sum_\kappa \int_Q^T \ln\left[(Q_\kappa \cdot \bar{Q}_\kappa)^2 + m^2 \bar{Q}_\kappa^2\right]. \qquad (11.19)$$

Putting all one-loop contributions together and adding the tree-level contribution, we arrive at the following expression for the one-loop potential

$$V(r, \bar{r}) = \frac{m^2 T^2}{2g^2} (r - \bar{r})^2$$

$$+ \frac{d-2}{2} \sum_\kappa \int_Q^T \ln \left[Q_\kappa^2 + m^2 \right]$$

$$+ \frac{1}{2} \sum_\kappa \int_Q^T \ln \left[1 + \frac{m^2 \bar{Q}_\kappa^2}{(Q_\kappa \cdot \bar{Q}_\kappa)^2} \right]. \tag{11.20}$$

It is readily checked that, for $r = \bar{r}$, this general expression gives back the formula (6.15) for the background-field effective potential obtained in Chap. 6.

11.3.2   Practical Evaluation

The presence of $(Q_\kappa \cdot \bar{Q}_\kappa)^2 + m^2 \bar{Q}_\kappa^2$, because it involves a quartic polynomial in the frequencies with no obvious roots, renders the analytical evaluation of the corresponding Matsubara sum cumbersome. A different strategy consists in performing the q-integrals analytically and the Matsubara sums numerically. One subtlety with this approach is, however, how to perform the ϵ-expansion correctly since the latter not always commutes with the Matsubara sum.

One strategy is to add and subtract to $V(r, \bar{r})$ its Taylor expansion $\mathcal{T}_{r=\bar{r}}^{(n)}[V](r, \bar{r})$ in the variable r around $r = \bar{r}$, taken to an appropriate order:

$$V(r, \bar{r}) = \left[V(r, \bar{r}) - \mathcal{T}_{r=\bar{r}}^{(n)}[V](r, \bar{r}) \right] + \mathcal{T}_{r=\bar{r}}^{(n)}[V](r, \bar{r}). \tag{11.21}$$

The function $\mathcal{T}_{r=\bar{r}}^{(n)}[V](r, \bar{r})$ is in fact a polynomial in the variable $\Delta r \equiv r - \bar{r}$ with coefficients depending on $\bar{r}$. They can be evaluated using the standard strategy (see Appendix B) because the frequency dependence of their denominators is quadratic. On the other hand, n should be taken large enough such that Matsubara sum in the bracket of (11.21) is convergent enough not to be affected by the subtleties of the ϵ-expansion. The bracket can then be evaluated by first performing the q-integrals analytically, followed by a numerical Matsubara sum. We refer to [100] for implementation details where it is argued that n can be chosen equal to 2.[4]

---

[4] More precisely, this is true for the integral in the third line of Eq. (11.20). In the case of the integral in the second line, even though the choice $n = 2$ leads to a convergent Matsubara sum, the $\epsilon \to 0$ is still subtle. On the other hand, this integral can be treated using the standard method of performing the Matsubara sum analytically.

11.3.3 Renormalization

The right most term in Eq. (11.21) is also the only one containing the UV divergences. It writes

$$\mathcal{T}_{r=\bar{r}}^{(2)}[V](r,\bar{r}) = \tilde{V}(\bar{r}) + \left.\frac{\partial V}{\partial r_j}\right|_{r=\bar{r}}(r_j - \bar{r}_j) + \frac{1}{2}\left.\frac{\partial^2 V}{\partial r_j \partial r_k}\right|_{r=\bar{r}}(r_j - \bar{r}_j)(r_k - \bar{r}_k).$$

(11.22)

The divergences contained in $\tilde{V}(\bar{r})$ are irrelevant for this is a mere constant that redefines the zero of the potential seen as a function of r. It can be shown that the zero-temperature (and thus the divergent) part of $\partial V/\partial r_j|_{r=\bar{r}}$ vanishes. From this, it follows that the only divergence originates in $\partial^2 V/\partial r_j \partial r_k|_{r=\bar{r}}$.

To absorb this divergence, one should rescale the bare parameters m and g as well as the bare backgrounds r and $\bar{r}$. In turns out that the parameters that rescale multiplicatively are m, g, $\bar{r}$, and $r - \bar{r}$:

$$m \to Z_m m, \quad g \to Z_g g, \quad \bar{r} \to Z_g Z_A^{1/2}\bar{r}, \quad r - \bar{r} \to Z_g Z_a^{1/2}(r - \bar{r}),$$

(11.23)

where a refers to $A - \bar{A}$. The combination $Z_g Z_A^{1/2}$ is finite and can then be set equal to 1 such that $\bar{r}$ does not renormalize. Under this rescaling, the tree-level contribution to $\partial^2 V/\partial r_j \partial r_k|_{r=\bar{r}}$ becomes $Z_a Z_{m^2} m^2 T^2/g^2$ in line with the fact that, up to the factor T^2/g^2, $\partial^2 V/\partial r_j \partial r_k|_{r=\bar{r}}$ is nothing but the inverse propagator, at vanishing frequency and momentum, in the neutral color sector. In what follows, we choose $Z_a Z_{m^2}$ such that this zero-momentum mass strictly equals m^2 in the zero-temperature limit.

11.3.4 Polyakov Loop

The Polyakov loop can be computed by adapting the techniques developed in previous chapters. We introduce a function $\ell(r,\bar{r})$ that gives the Polyakov loop when r is evaluated at the minimum $r_{\min}(\bar{r})$ of the potential. Since the potential is evaluated at next-to-leading order, so should the Polyakov loop

$$\ell(r,\bar{r}) = \ell_0(r,\bar{r}) + g^2\ell_1(r,\bar{r}) + O(g^4).$$

(11.24)

Now, in the present gauge, $\bar{r} = \bar{r}_c$ and $r_{\min}(\bar{r}_c) = \bar{r}_c + O(g^2)$. This last identity follows from the fact that the tree-level term contribution to the potential is proportional to $(r - \bar{r}_c)^2$. Moreover, center symmetry implies that $\ell(\bar{r}_c, \bar{r}_c) = 0$ at all orders and, thus, $\ell_1(\bar{r}_c, \bar{r}_c) = 0$. We then conclude that the next-to-leading-order

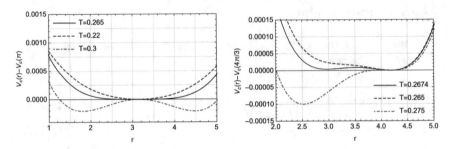

Fig. 11.2 The potential $V_c(r)$ in the SU(2) case (left) and in the SU(3) case (right). Reprinted from Ref. [100] under CC-BY 4.0 license

Polyakov loop is given by its tree-level expression $\ell_0(r_{\min}(\bar{r}_c), \bar{r}_c)$, that is,

$$\ell = \frac{1}{N} \mathrm{tr}\, e^{i r^j_{\min}(\bar{r}_c) t^j} + O(g^4), \qquad (11.25)$$

where $r_{\min}(\bar{r}_c)$ is to be evaluated from the next-to-leading-order potential (11.20).

11.4 Results

Let us now discuss the results obtained with the new implementation of the background-field method at finite temperature.

11.4.1 Transition Temperatures

In Fig. 11.2, the potential $V_c(r)$ is shown for various temperatures in the SU(2) and SU(3) cases. The transition is found to be second-order in the SU(2) case and first-order in the SU(3) case as expected. With the parameters $m = 680\,\mathrm{MeV}$ and $g = 7.5$ obtained from fitting the one-loop zero-temperature CF propagators to lattice data, we find $T_c \simeq 265\,\mathrm{MeV}$ in the SU(2) case, a value much closer to the lattice value $295\,\mathrm{MeV}$ [175] than the previous result $\bar{T}_c \simeq 227\,\mathrm{MeV}$ obtained with the same value of m and $\tilde{V}(\bar{r})$.[5] In the SU(3) case, with the parameters $m = 510\,\mathrm{MeV}$ and $g = 4.9$, we find $T_c \simeq 267\,\mathrm{MeV}$, pretty close to the lattice value $270\,\mathrm{MeV}$ [175] and much better than the estimate using $\tilde{V}(\bar{r})$. We note that the agreement with the lattice value is better in the SU(3) case than in the SU(2) case, as expected since the SU(3) coupling is smaller [141].

---

[5] In this case, because $\tilde{V}(\bar{r})$ does not feature a tree-level term, only the mass parameter is needed. In Ref. [92], $\bar{T}_c \simeq 237\,\mathrm{MeV}$ was determined using the value $m = 710\,\mathrm{MeV}$ obtained by fitting tree-level CF propagators to the lattice data.

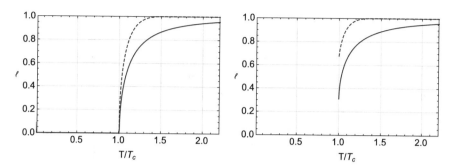

Fig. 11.3 SU(2) (left) and SU(3) (right) Polyakov loops as obtained using the self-consistent background field approach [92] (dashed) or the present center-symmetric background approach (plain). Reprinted from Ref. [100] under CC-BY 4.0 license

11.4.2 Polyakov Loop

We can also evaluate the Polyakov loops both in the SU(2) case and in the SU(3) case and compare them with the results previously obtained using the self-consistent background-field approach [92]. This is shown in Fig. 11.3. We observe that the rise of the Polyakov loop in the deconfined phase toward its maximal value is systematically slower with the alternative method, in line with what is observed in lattice simulations. A faithful comparison with lattice data requires, however, to take into account the renormalization of the Polyakov loop. Here, we have computed the bare Polyakov loop in dimensional regularization which should be interpreted as the renormalized loop in a particular scheme. Since this is not the same scheme as the one on the lattice, we need to allow for an overall (temperature-independent) normalization. In the SU(3) case (where the perturbative CF approach works best), we find that the best fit is obtained over the interval $[T_c, 1.5T_c]$ (see Fig. 11.4). We note that the relative error remains rather small up to $2T_c$. Above this temperature, our result departs from the lattice one. This is expected as various effects not included here, such as the resummation of hard thermal loops [205] or renormalization group running [180, 206], are known to play an important role in this regime.

11.4.3 Masses

As already mentioned, the potential $V_c(r)$ gives access from Eq. (11.7) to the zero-momentum mass in the temporal direction and in the diagonal color direction. In particular, in the SU(2) case, it is obvious that this neutral electric mass needs to vanish at the transition or, in other words, that the propagator should diverge, which the results in Fig. 11.5 confirm. In the SU(3) case, the mass decreases substantially when approaching the transition from below but never vanishes, leading instead to a sharp peak in the propagator. We conclude that the propagator in the center-

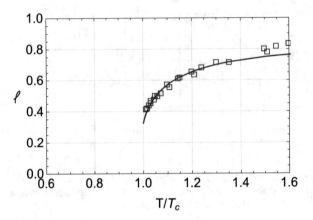

Fig. 11.4 Comparison of our SU(3) Polyakov to the lattice data of Refs. [186, 204]. Reprinted from Ref. [100] under CC-BY 4.0 license

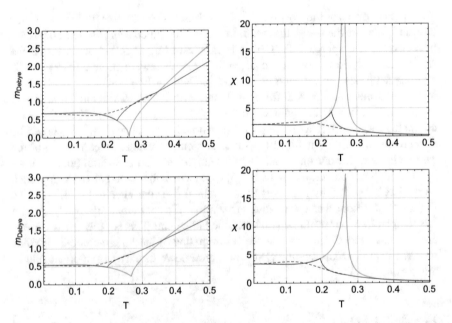

Fig. 11.5 The neutral electric mass (left) and its inverse, the neutral electric susceptibility or longitudinal propagator $D_L(0)$ at vanishing frequency and momentum (right) as functions of the temperature for the SU(2) theory (upper panel) and for the SU(3) theory (lower panel). The dashed lines show the corresponding results obtained in the self-consistent background approach [96]. T and m are given in GeV. Reprinted from Ref. [100] under CC-BY 4.0 license

symmetric Landau gauge can be used as a probe to the deconfinement transition. A thourough study of the gluon propagator in the center symmetry Landau gauge has been recently considered in [207], both in the SU(2) case and in the SU(3) case, with, in particular, the identification of new order parameters for the deconfinement transition. These results need to be contrasted to the results obtained within the standard Landau gauge as reviewed in Chap. 1. We now compare them with those obtained within the self-consistent approach.

11.5 Connection to the Self-Consistent Backgrounds

We have now at our disposal two alternative methods to study the deconfinement transition. Let us here discuss to which extent they are equivalent in principle and why they differ in practice.

11.5.1 Case of an Ideal Gauge Fixing

In the absence of approximations or modeling of the gauge-fixing procedure in the infrared, the two approaches should be strictly equivalent. Indeed, the free energy of the system is strictly background independent, and, as we saw in Chap. 4, this ensures the equivalence between the absolute minima of $\tilde{V}(\bar{r})$ and the self-consistent backgrounds. In turn, this means that, as long as the minimum of $V_c(r)$ is located at $r = \bar{r}_c$, $\bar{r}_c$ is a self-consistent background and thus a minimum of $\tilde{V}(\bar{r})$. Reciprocally, as soon the minimum of $V_c(r)$ departs from $\bar{r}_c$, $\bar{r}_c$ is not anymore a self-consistent background and neither a minimum of $\tilde{V}(\bar{r})$. In summary, for an ideal gauge fixing, the breaking of center symmetry occurs at the same temperature regardless of whether we use $\tilde{V}(\bar{r})$ or $V_c(r)$.

In the case of a continuous transition, one also deduces that the mass in the self-consistent approach, that one can define similarly to (11.7) as

$$\left(\frac{g}{T}\right)^2 \frac{\partial^2 V(r, \bar{r})}{\partial r^{j_1} \partial r^{j_2}}\Bigg|_{r=\bar{r}=\bar{r}_s}, \tag{11.26}$$

where $\bar{r}_s$ is an absolute minimum of $\tilde{V}(\bar{r})$, needs to vanish at the transition. This is because, in the low-temperature phase, $\bar{r}_s = \bar{r}_c$ and, therefore, (11.26) coincides with the mass as computed in the center-symmetric Landau gauge.

To illustrate this point further, let us note that, from the definition of $r_{\min}(\bar{r})$ and the background independence of the free energy $V(r_{\min}(\bar{r}), \bar{r})$, one can write

$$0 = \frac{\partial V}{\partial r}\Bigg|_{r_{\min}(\bar{r}),\bar{r}} \quad \text{and} \quad 0 = \frac{\partial V}{\partial \bar{r}}\Bigg|_{r_{\min}(\bar{r}),\bar{r}}. \tag{11.27}$$

Taking a derivative with respect to $\bar{r}$, this leads to the system of equations

$$0 = \frac{\partial^2 V}{\partial r^2}\bigg|_{r_{\min}(\bar{r}),\bar{r}} r'_{\min}(\bar{r}) + \frac{\partial^2 V}{\partial r \partial \bar{r}}\bigg|_{r_{\min}(\bar{r}),\bar{r}}, \tag{11.28}$$

$$0 = \frac{\partial^2 V}{\partial r \partial \bar{r}}\bigg|_{r_{\min}(\bar{r}),\bar{r}} r'_{\min}(\bar{r}) + \frac{\partial^2 V}{\partial \bar{r}^2}\bigg|_{r_{\min}(\bar{r}),\bar{r}}. \tag{11.29}$$

On the other hand, the second derivative of $\tilde{V}(\bar{r})$ reads

$$\tilde{V}''(\bar{r}) = \frac{\partial^2 V}{\partial r^2}\bigg|_{\bar{r},\bar{r}} + 2 \frac{\partial^2 V}{\partial r \partial \bar{r}}\bigg|_{\bar{r},\bar{r}} + \frac{\partial^2 V}{\partial \bar{r}^2}\bigg|_{\bar{r},\bar{r}}. \tag{11.30}$$

If we evaluate this second derivative at a self-consistent background $\bar{r}_s = r_{\min}(\bar{r}_s)$, we can use (11.28)–(11.29), and we arrive at

$$\tilde{V}''(\bar{r}_s) = (1 - r'_{\min}(\bar{r}_s))^2 \frac{\partial^2 V}{\partial r^2}\bigg|_{\bar{r}_s,\bar{r}_s}. \tag{11.31}$$

If we want the continuous SU(2) transition to occur at the same temperature when seen from the perspective of $\tilde{V}(\bar{r})$ or from that of $V_c(r)$, $\tilde{V}''(\bar{r}_s)$ and $\partial^2 V / \partial r^2|_{\bar{r}_s,\bar{r}_s}$ should vanish simultaneously, and we should then expect $r'_{\min}(\bar{r}_c) - 1 \neq 0$ at the transition.[6] This, in turn, implies that the departure of $r_{\min}(\bar{r}_c)$ from the value $\bar{r}_c$ should occur as illustrated in Fig. 11.6, where we represent the function $r_{\min}(\bar{r}) - \bar{r}$. This function should be odd under $\bar{r} \rightarrow 2\pi - \bar{r}$, and, because its slope at $\bar{r}_c$ is assumed not to vanish at the transition, the only possibility for $r_{\min}(\bar{r}_c)$ to depart from $\bar{r}_c$ is that the function becomes discontinuous at $\bar{r} = \bar{r}_c$ above T_c.

Let us also mention that the zeros of this function are nothing but the self-consistent backgrounds. In addition to the ones that emerge to the left and the right of $\bar{r}_c$ at the transition, we should have a self-consistent background, and thus a zero, at $\bar{r} = 0$ (and also at $\bar{r} = 2\pi$ by symmetry). This is the color-symmetric solution which we have discussed in Chap. 4. In the case of an ideal gauge fixing, as any other self-consistent background, the color-symmetric solution should correspond to a minimum of $\tilde{V}(\bar{r})$. There is no contradiction here since this point corresponds to the Landau gauge, and the free energy should be the same at this minimum or at the ones that stem from $\bar{r}_c$. In fact, in this ideal situation, the free energy should be constant along the curve $r_{\min}(\bar{r})$.

---

[6] Of course, it could be that we also have $r'_{\min}(\bar{r}_c) - 1 = 0$ at the transition, but this would require the simultaneous vanishing of three quantity, whereas we only have one identity to connect them.

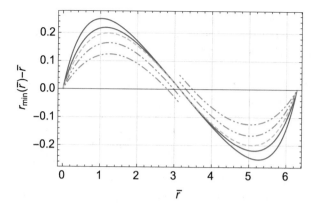

Fig. 11.6 Speculated behavior of the function $r_{\min}(\bar{r}) - \bar{r}$ for various temperatures in the case of an idealized gauge fixing: $T < T_c$ (dash-dotted, violet), $T = T_c$ (dashed, orange), and $T > T_c$ (plain, dotted)

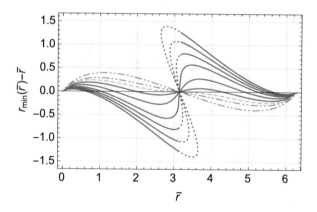

Fig. 11.7 Function $r_{\min}(\bar{r}) - \bar{r}$ for various temperatures: $T < \bar{T}_c$ (dash-dotted, violet), $T = \bar{T}_c$ (dashed, orange), $\bar{T}_c < T < T_c$ (plain, red), and $T > T_c$ (plain, dotted)

11.5.2 In Practice

As we have already discussed, the correspondence between the minima of $\tilde{V}(r)$ and the self-consistent backgrounds is broken in practice. Thus, it is not true anymore a priori that the transition temperature T_c as obtained from $V_c(r)$ coincides with the transition temperature $\bar{T}_c$ as obtained from $\tilde{V}(\bar{r})$. In particular, the mass as computed in the self-consistent approach does not need to vanish at the transition temperature as defined from $\tilde{V}(\bar{r})$. We now illustrate this more precisely in the case of the Curci-Ferrari model.

In Fig. 11.7, we plot once again the function $r_{\min}(\bar{r}) - \bar{r}$ for various temperatures, but this time in the case of the Curci-Ferrari model. First, we observe that $\bar{r} = 0$ (and equivalently $\bar{r} = 2\pi$) is always a self-consistent background. However, unlike the case of an ideal gauge fixing, we find that $\bar{r} = 0$ corresponds to a maximum of $\tilde{V}(\bar{r})$. This is one first illustration of the breaking of the equivalence between the notion of self-consistent backgrounds and the minima of $\tilde{V}(\bar{r})$, due to the non-fulfillment of Eq. (4.5).

For $T < \bar{T}_c$, that is, in the confined phase from the point of view of $\tilde{V}(\bar{r})$, there is another self-consistent background at $\bar{r} = \pi$ which is also the minimum of $\tilde{V}(\bar{r})$, so in this case the equivalence between self-consistent background and minimum of $\tilde{V}(\bar{r})$ holds. At $T = \bar{T}_c$, the minimum of $\tilde{V}(\bar{r})$ moves away from $\bar{r} = \pi$ leaving a maximum instead at $\bar{r} = \pi$. However, in the range $\bar{T}_c < T < T_c$, the background $\bar{r} = \pi$ remains self-consistent despite the fact that it is not anymore the minimum of $\tilde{V}(\bar{r})$. In this range of temperatures, we then have three different types of self-consistent backgrounds (within the interval $[0, \pi]$), $\bar{r} = 0$, $\bar{r} = \pi$, and the true minimum of $\tilde{V}(\bar{r})$. For $T = T_c$, that is, at the deconfinement transition from the point of view of $V_c(r)$, the background $\bar{r} = \pi$ ceases to be self-consistent, and we are left with only two self-consistent backgrounds $\bar{r} = 0$ and the minimum of $\tilde{V}(\bar{r})$.

If we define the zero-momentum mass in the self-consistent case as in (11.26), it is then no doubt that it differs from the mass in the center-symmetric Landau gauge in the range $[\bar{T}_c, T_c]$. Moreover it does not vanish at the associated transition temperature $\bar{T}_c$, as anticipated. This is also illustrated in Fig. 11.5.

To illustrate these features further, we can proceed as in the previous section by deriving a relation between $\tilde{V}''(\bar{r})$ and $\partial^2 V(r, \bar{r})/\partial r^2|_{r=\bar{r}}$. We cannot use the background independence of the free energy in this case. However, one can show that, in a similar fashion as we derived Eq. (6.7),[7]

$$\frac{m^2}{g^2}(\bar{r} - r_{\min}(\bar{r})) = \left.\frac{\partial V}{\partial \bar{r}}\right|_{r_{\min}(\bar{r}),\bar{r}}, \qquad (11.32)$$

which implies

$$\frac{m^2}{g^2}(1 - r'_{\min}(\bar{r})) = \left.\frac{\partial^2 V}{\partial r \partial \bar{r}}\right|_{r_{\min}(\bar{r}),\bar{r}} r'_{\min}(\bar{r}) + \left.\frac{\partial^2 V}{\partial \bar{r}^2}\right|_{r_{\min}(\bar{r}),\bar{r}}, \qquad (11.33)$$

in place of Eq. (11.29). From this, we deduce that

$$\tilde{V}''(\bar{r}_s) = (1 - r'_{\min}(\bar{r}_s))^2 \left.\frac{\partial^2 V}{\partial r^2}\right|_{\bar{r}_s,\bar{r}_s} + \frac{m^2}{g^2}(1 - r'_{\min}(\bar{r}_s)), \qquad (11.34)$$

---

[7] It can also be checked that the one-loop expression for $V(r, \bar{r})$ is such that $\partial V/\partial \bar{r}|_{r,\bar{r}} = 0$ for any $\bar{r}$, from which it follows that $\tilde{V}'(\bar{r}) = \partial V/\partial r|_{r=\bar{r},\bar{r}}$ and, therefore, $\tilde{V}''(\bar{r}) = \partial^2 V/\partial r^2|_{r=\bar{r},\bar{r}} + \partial^2 V/\partial r \partial \bar{r}|_{r=\bar{r},\bar{r}}$. Combining this identity with (11.28) which is still valid, we arrive, for a self-consistent background $\bar{r}$, at $\tilde{V}''(\bar{r}) = (1 - r'_{\min}(\bar{r}))\partial^2 V/\partial r^2|_{r=\bar{r},\bar{r}}$. This formula is exact at one-loop order. It is seen to be compatible with (11.34) after expanding to leading order in g^2, with $1 - r'_{\min}(\bar{r}) \sim g^2$ and $g^2 \partial^2 V/\partial r^2|_{r=\bar{r},\bar{r}} = m^2 + O(g^2)$.

at a self-consistent background $\bar{r}_s$. We see here that, contrary to the case of an ideal gauge fixing, and because there is no reason that self-consistent backgrounds always coincide with minima of $\tilde{V}$, the vanishing of $\tilde{V}''(\bar{r}_c)$ at $\bar{T}_c$ has no reason to imply that of $\partial^2 V/\partial r^2|_{r=\bar{r}=\bar{r}_c}$. Then $1 - r'_{min}(\bar{r}_c)$ can vanish at $\bar{T}_c$ without representing an extra constraint. This is what we find in Fig. 11.5 and explains how $\bar{r}_c$ can remain self-consistent in the range $[\bar{T}_c, T_c]$ while not being a minimum of $\tilde{V}(\bar{r})$ anymore.

Problems

11.1 Twisted Gauge Fields (⋆⋆)

(a) Consider the transformation $U_c \equiv e^{-i\tau T \pi \sigma^3/2}$, and explicitate how it transforms a given gauge-field configuration. *Tip:* decompose the field into a Cartan-Weyl basis.

(b) Assume that the original configurations is β-periodic. What boundary conditions satisfy the color components of the transformed field?

(c) Generalize to the SU(3) case.

11.2 Center-Symmetric Landau Gauge and Gribov Copies (⋆⋆)

(a) Show that the set of gauge configurations obeying the center-symmetric Landau gauge condition $F[A] \equiv \bar{D}_\mu[\bar{A}_c](A_\mu - \bar{A}_{c,\mu}) = 0$ is invariant under the transformations that leave the center-symmetric background invariant.

(b) Show that for any $U_0^{(i)}(A) \in \mathcal{G}_0$ such that $F[A^{U_0^{(i)}(A)}] = 0$ and any $U \in \mathcal{G}$ such that $\bar{A}_c^U = \bar{A}_c$

$$\frac{\left| \det \frac{\delta F[(A^{U_0})^U]}{\delta U_0} \Big|_{U_0^{(i)}(A)} \right|^{-1}}{\left| \det \frac{\delta F[A^{U_0}]}{\delta U_0} \Big|_{U_0^{(i)}(A)} \right|^{-1}}$$

does not depend on i. This shows that this particular conditional gauge fixing obeys the two conditions discussed in Prob. 3.5 which ensure that the gauge-fixing functional $z[A]$ (see Chap. 3) is invariant under the transformations that leave $\bar{A}_c$ invariant.

11.3 Gluonic Determinant (⋆⋆)

(a) Given a block square matrix

$$M = \begin{pmatrix} A & B \\ C & D \end{pmatrix},$$

with A a square invertible matrix, show that

$$\begin{pmatrix} A & B \\ C & D \end{pmatrix} = \begin{pmatrix} A & 0 \\ C & 1 \end{pmatrix} \begin{pmatrix} 1 & A^{-1}B \\ 0 & D - CA^{-1}B \end{pmatrix}.$$

Deduce that $\det M = \det A \times \det(D - CA^{-1}B)$.

(b) Apply this to the gluonic quadratic form, whose matrix is

$$M = \begin{pmatrix} Q^2 P_{\mu\nu}^{\perp}(Q) + m^2 \delta_{\mu\nu} & -\bar{Q}_{\mu} \\ \bar{Q}_{\nu} & 0 \end{pmatrix}.$$

Tip: start inverting the upper left block after expressing it in terms of $P_{\mu\nu}^{\perp}(Q)$ and $P_{\mu\nu}^{\parallel}(Q) \equiv \delta_{\mu\nu} - P_{\mu\nu}^{\perp}(Q)$.

Conclusions and Outlook

12

Tackling the most intriguing properties of quantum chromodynamics and Yang-Mills theories in the continuum requires one to go beyond the standard (but incomplete) Faddeev-Popov gauge-fixing procedure which, although very efficient at high energies, is known not to be valid at low energies. In this respect, the Curci-Ferrari model has received a certain attention lately as a possible candidate for such an extension in the Landau gauge. Part of that attention is rooted in the impressive agreement between one-loop vacuum correlation functions computed within the model and the most accurate determinations of the Landau gauge correlation functions on the lattice. We know now that this agreement extends and even improves at two-loop accuracy. Moreover, the Curci-Ferrari running coupling needed in these comparisons remains moderate over the whole range of scales, as opposed to the running coupling in the standard Faddeev-Popov approach which diverges at a finite scale. These two surprising results open the exciting possibility to tackle some of the low-energy properties from perturbative means.

Of course, ideally one should aim at generating the Curci-Ferrari model from first principles, starting from the QCD/YM actions, and applying a *bona fide* gauge-fixing procedure that deals with the Gribov ambiguity. Such a clear-cut mechanism is not known at present in the Landau gauge, but various lines of investigation are being pursued in this direction. In the meantime, a more pragmatic approach is to test the perturbative predictions of the model further and confront them to existing results, in particular those from the lattice simulations. In this manuscript, we have reviewed some of these predictions with regard to the finite temperature properties of the QCD/YM system, in particular the confinement/deconfinement transition.

We have first reviewed (in Chap. 1) the results for the YM correlators at finite temperature, as computed from the Curci-Ferrari model, with an emphasis on the so-called longitudinal gluon propagator. The latter was indeed foreseen for some time as providing a direct probe onto the deconfinement transition. Although the Curci-Ferrari model describes qualitatively well the lattice results for the various correlators, it fails in describing the behavior of the zero-momentum longitudinal propagator (inverse susceptibility). However, this negative result does not necessarily signal a failure of the model because the same limitations are observed within other approaches that do not rely on the Curci-Ferrari model. Rather it has been argued that the limitations originate in the use of the Landau gauge which fails in capturing the order parameter associated with the deconfinement transition. It is known that a more appropriate gauge at finite temperature is the so-called Landau-DeWitt gauge, the background generalization of the Landau gauge.

We have devoted Chaps. 3, 4, and 5 to a thorough review of the rationale behind the use of background-field gauges at finite temperature and why they are indeed the *good* gauges to discuss the deconfinement transition. Although the standard recipe based on the use of self-consistent backgrounds is a well-known result, we have tried to give a self-contained and original derivation, with special emphasis and discussion of the underlying assumptions that are usually left implicit. The relevance of this discussion is that some of these assumptions are usually violated by practical implementations of the gauge fixing (including the Curci-Ferrari modeling), with a potential impact on the interpretation of the results. Some of these assumptions can also be violated by the physical set-up, as in the presence of quarks at finite density. A different implementation of the background-field method at finite temperature which does not rely on the use of self-consistent backgrounds and which does not suffer from the limitations of the latter has been discussed in Chap. 11.

Based on this knowledge, we have extended the Curci-Ferrari model in the presence of a background, which needs to be seen as the finite temperature counterpart of the standard Curci-Ferrari model in the vacuum. We have investigated various perturbative predictions of this model with regard to the deconfinement transition. In particular, in Chap. 6, we have analyzed the one-loop predictions in the YM case. We find that, already at one-loop order, the model predicts a deconfinement transition, with transition temperatures in reasonably good agreement with the results of other approaches, including lattice. Chapter 7 is devoted to testing the convergence properties of the perturbative expansion within the model. We find in particular improved values for the transition temperatures that get closer to the values determined on the lattice. Chapter 8 further deepens the connection between center symmetry and the deconfinement transition by providing characterizations of center symmetry in terms of the vanishing of certain collections of Polyakov loops and by analyzing the symmetry-breaking pattern in the SU(4) case. In Chaps. 9 and 10, we have extended our analysis in the presence of quarks, in the formal but interesting regime where all quarks are considered heavy. We find that most of the known qualitative and quantitative features in this case can be reproduced by using our model approach at one-loop order. Higher-order corrections have also been investigated in this case and shown to improve the results [98]. In Chap. 11, we have casted a new light on the results of Chaps. 4 and 5, through our novel implementation of background-field techniques at finite temperature. Within this new setting, we have also revisited the question of whether the gluon susceptibility can probe the deconfinement transition.

In parallel to these studies, we have discussed various open questions which concern not only the Curci-Ferrari model but also other continuum approaches.

In particular, it seems that a common mechanism for confinement in all these approaches is the ghost dominance observed at low temperatures. At first sight, a ghost-dominated phase at low temperature seems worrisome for ghost degrees of freedom which are unphysical and come with negative thermal distribution functions, leading potentially to inconsistent thermodynamics. We have shown, however, that the presence of a surrounding confining background leads to a transmutation of some of these negative distribution functions into positive distribution functions, with a net-positive contribution to, say, the entropy density at low temperatures. The same mechanism leads to a slightly negative entropy density right below the deconfinement transition at one-loop order. However, we have shown that two-loop corrections cure this inconsistent behavior.

A problem that found no solution so far is that, in most continuum approaches in the Landau/Landau-DeWitt gauges, there remain massless degrees of freedom at low temperature that contribute polynomially to the thermodynamical observables in this limit, at odds with the observations made on the lattice. This is certainly a future challenge for continuum approaches that could relate to the use of more exotic background configurations than the ones used in this work, to the way we envisage and implement gauge fixing, or to the ability of the approach to generate the low-lying bound states, in particular glueball states in the case of YM theory. All these aspects are currently under investigation.

These problems aside, the natural question that comes next is whether the applicability of the perturbative Curci-Ferrari model (be it in the Landau gauge in the vacuum or in the Landau-DeWitt gauge at finite temperature) extends to the physical QCD case, that is, in the presence of light quarks, in which case the relevant symmetry is chiral symmetry and its spontaneous breaking. We know already that the answer to this question is negative, the reason being that, in the presence of light quarks, the strength of the interaction between quarks and gluons in the infrared is two to three times larger than that in the YM sector of the theory, preventing a full perturbative analysis. We note, however, that this does not necessarily point to a limitation of the Curci-Ferrari model but rather to the use of a strict perturbative expansion within the model when applied to investigate QCD phenomena. Interestingly enough, one can use the knowledge that the Curci-Ferrari model in the YM sector is essentially perturbative to construct a systematic expansion for the Curci-Ferrari model in the QCD case, controlled by two small parameters, the YM coupling and the inverse of the number of colors. This strategy has already been used in the vacuum where it has been shown to capture the physics of chiral symmetry breaking [85, 86, 207]. Within this setting it has been possible to perform a calculation of the pion decay constant in the chiral limit [209] which opens interesting perspectives on the ab initio determination of hadronic observables within the Curci-Ferrari model. A first investigation within this double expansion scheme has also been done at finite temperature and finite chemical potential [87]. In particular, with a very crude implementation of the leading order in the double expansion, a critical end-point is found in the phase diagram, in the same ballpark as certain low-energy phenomenological approaches. A challenging question for the future is whether such critical end-point persists within a full implementation of the double expansion scheme and, if so, where it is exactly located.

The SU(N) Lie Algebra

A

The special unitary group SU(N) is the group of complex $N \times N$ unitary matrices with unit determinant. The corresponding SU(N) Lie algebra is obtained by considering infinitesimal elements $U = \mathbb{1} + X$ and imposing the two conditions $UU^\dagger = \mathbb{1}$ and $\det U = 1$. One finds $X + X^\dagger = 0$ and $\operatorname{tr} X = 0$, which means that SU(N) is the space of traceless anti-Hermitian matrices. Its dimension is $N^2 - 1$, and a commonly used basis is provided by the matrices $i H_j$ ($1 \le j \le N-1$), $i X^{jj'}$, and $i Y^{jj'}$ ($1 \le j < j' \le N$), with

$$H_j \equiv \frac{1}{\sqrt{2j(j+1)}} \begin{pmatrix} \mathbb{1}_j & & \\ & -j & \\ & & 0_{N-1-j} \end{pmatrix},\tag{A.1}$$

and

$$X^{jj'}_{kk'} = \frac{1}{2}\left(\delta_{jk}\delta_{j'k'} + \delta_{jk'}\delta_{j'k}\right),\tag{A.2}$$

$$Y^{jj'}_{kk'} = -\frac{i}{2}\left(\delta_{jk}\delta_{j'k'} - \delta_{jk'}\delta_{j'k}\right).\tag{A.3}$$

With this choice, the trace of the square of any basis element is $-1/2$, and the trace of the product of two distinct basis elements is 0. In what follows, we review various important notions related to the SU(N) algebra.

© The Author(s), under exclusive license to Springer Nature Switzerland AG 2022
U. Reinosa, *Perturbative Aspects of the Deconfinement Transition*, Lecture Notes
in Physics 1006, https://doi.org/10.1007/978-3-031-11375-8

A.1   Defining Weights of SU(N)

The above choice of basis shows explicitly that some of the generators of the SU(N) algebra can be diagonalized simultaneously. This can be rewritten formally as

$$H_j|\rho^{(k)}\rangle = \rho_j^{(k)}|\rho^{(k)}\rangle \,, \quad \forall 1 \le k \le N \,, \quad \forall 1 \le j \le N-1 \,, \tag{A.4}$$

where the N kets $|\rho^{(k)}\rangle$ define a basis of eigenstates in the space of the defining representation of the algebra. For each eigenstate $|\rho^{(k)}\rangle$, the various eigenvalues $\rho_j^{(k)}$ are conveniently gathered into a $(N-1)$-dimensional real vector $\rho^{(k)}$ that also serves labeling the eigenstate. The $\rho^{(k)}$ are referred to as the weights of the defining representation or *defining weights* for short. We mention that these vectors sum up to one since

$$\sum_{k=1}^{N} \rho_j^{(k)} = \sum_{k=1}^{N} \left(H_j\right)_{kk} = \operatorname{tr} H_j = 0 \,. \tag{A.5}$$

Similarly,

$$\sum_{k=1}^{N} \rho_j^{(k)} \rho_{j'}^{(k)} = \operatorname{tr} H_j H_{j'} = \frac{1}{2}\delta_{jj'} \,, \tag{A.6}$$

from our normalization convention.

It is easy to compute the norms of the defining weights and also the angles between various such weights. To this purpose we note that

$$\rho_j^{(k)} = (H_j)_{kk} = \frac{1}{\sqrt{2j(j+1)}} \times \begin{cases} 0 \,, & \text{for } k > j+1 \\ -j \,, & \text{for } k = j+1 \\ 1 \,, & \text{for } k \le j \end{cases} \tag{A.7}$$

and write

$$\rho^{(k)} \cdot \rho^{(k)} = \sum_{j=1}^{k-1} \left(H_j\right)_{kk}^2 + \sum_{j=k}^{N-1} \left(H_j\right)_{kk}^2 \tag{A.8}$$

$$= \frac{(k-1)^2}{2(k-1)k} + \sum_{j=k}^{N-1} \frac{1}{2j(j+1)} = \frac{1}{2}\left(1 - \frac{1}{N}\right).$$

Similarly, for $k < k'$,

$$\rho^{(k)} \cdot \rho^{(k')} = \sum_{j=1}^{k-1} (H_j)_{kk} (H_j)_{k'k'}$$ (A.9)

$$+ \sum_{j=k}^{k'-1} (H_j)_{kk} (H_j)_{k'k'}$$

$$+ \sum_{j=k'}^{N-1} (H_j)_{kk} (H_j)_{k'k'}$$

$$= -\frac{(k'-1)}{2(k'-1)k'} + \sum_{j=k'}^{N-1} \frac{1}{2j(j+1)} = -\frac{1}{2N} .$$

The defining weights of the SU(N) algebra are thus N equal norm vectors of $\mathbb{R}^{N-1}$ such that the angles between any two of these vectors are also all equal.

Finally, it is easily seen that, except for the full set, any subset of defining weights is linearly independent. To prove this, we choose, without loss of generality, the first n weights $\rho^{(1)}, \ldots, \rho^{(n)}$ and evaluate the Gram matrix

$$G_n \equiv \left(\rho^{(k)} \cdot \rho^{(l)}\right)_{(k,l)} = \frac{1}{2} \left(\mathbb{1}_n - \frac{1_n}{N}\right),$$ (A.10)

where 1_n denotes the $n \times n$ matrix with all components equal to 1. The chosen set of weights is linearly independent iff the Gram matrix is invertible. We then just need to evaluate the determinant of the Gram matrix as

$$\ln \det (2G_n) = \operatorname{tr} \ln(2G_n) = -\operatorname{tr} \sum_{k=1}^{\infty} \frac{1}{k} \left(\frac{1_n}{N}\right)^k$$ (A.11)

$$= -\sum_{k=1}^{\infty} \frac{1}{k} \left(\frac{n}{N}\right)^k = \ln \left(1 - \frac{n}{N}\right),$$

where we have used that $1_n^k = n^{k-1} 1_n$ for $k \geq 1$. Then, $\det (2G_n) = 1 - n/N$, and, as announced, except for the full set $(n = N)$, any subset of defining weights is linearly independent. In particular, any set of $N - 1$ defining weights forms a basis of $\mathbb{R}^{N-1}$.

A.2   Roots of SU(N)

The notion of weight is not restricted to the defining representation but applies in fact to any representation of the algebra, in particular to the adjoint representation $X \mapsto \mathrm{ad}_X \equiv [X, \_]$. Since $[\mathrm{ad}_{H_j}, \mathrm{ad}_{H_k}] = \mathrm{ad}_{[H_j, H_k]} = 0$, it is again meaningful to try to diagonalize simultaneously all the generators ad_{H_j}. In fact, we know already $N - 1$ eigenstates since

$$\mathrm{ad}_{H_j} H_k = [H_j, H_k] = 0 \,. \tag{A.12}$$

The corresponding weights are all equal to the nul vector of $\mathbb{R}^{N-1}$. We are thus left with determining the remaining eigenstates E_α such that

$$[H_j, E_\alpha] = \alpha_j E_\alpha \,. \tag{A.13}$$

The non-zero weights α are referred to as the *roots* of the algebra, and the diagonalizing basis $\{i\,H_j, i\,E_\alpha\}$ is known as a Cartan-Weyl basis.

To find the elements E_α, consider the matrices $Z_\pm^{jj'} \equiv X^{jj'} \pm iY^{jj'}$, with $j < j'$. They are such that

$$\left(Z_+\right)_{kk'}^{jj'} = \delta_{jk}\delta_{j'k'} \quad \text{and} \quad \left(Z_-\right)_{kk'}^{jj'} = \delta_{jk'}\delta_{j'k} \,. \tag{A.14}$$

Therefore,

$$\left(H_j\right)_{\alpha\beta}\left(Z_+\right)_{\beta\gamma}^{kk'} - \left(Z_+\right)_{\alpha\beta}^{kk'}\left(H_j\right)_{\beta\gamma} \tag{A.15}$$

$$= \left[\left(H_j\right)_{\alpha\alpha} - \left(H_j\right)_{\gamma\gamma}\right]\left(Z_+\right)_{\alpha\gamma}^{kk'}$$

$$= \left[\left(H_j\right)_{kk} - \left(H_j\right)_{k'k'}\right]\left(Z_+\right)_{\alpha\gamma}^{kk'} \,,$$

and

$$\left(H_j\right)_{\alpha\beta}\left(Z_-\right)_{\beta\gamma}^{kk'} - \left(Z_-\right)_{\alpha\beta}^{kk'}\left(H_j\right)_{\beta\gamma} \tag{A.16}$$

$$= \left[\left(H_j\right)_{\alpha\alpha} - \left(H_j\right)_{\gamma\gamma}\right]\left(Z_-\right)_{\alpha\gamma}^{kk'}$$

$$= \left[\left(H_j\right)_{k'k'} - \left(H_j\right)_{kk}\right]\left(Z_-\right)_{\alpha\gamma}^{kk'} \,.$$

This shows that the $Z_\pm^{kk'}$ (with $k < k'$) are nothing but the sought-after E_α, and the corresponding roots appear as differences of the defining weights $\alpha^{(kk')} = \rho^{(k)} - \rho^{(k')}$.[1]

We have

$$\alpha^{(kk')} \cdot \alpha^{(kk')} = (\rho^{(k)})^2 + (\rho^{(k')})^2 - 2\rho^{(k)} \cdot \rho^{(k')} = 1 - \frac{1}{N} + \frac{1}{N} = 1, \qquad (A.17)$$

where we have used that $k \neq k'$. This means that the roots are all of norm unity. Scalar products between roots can be computed in a similar way, but, contrary to the defining weights, they depend on the chosen pair of roots. Finally, we have

$$\sum_{k \neq k'} \alpha_j^{(kk')} \alpha_{j'}^{(kk')} = \sum_{k,k'} \left(\rho_j^{(k)} \rho_{j'}^{(k)} - 2\rho_j^{(k)} \rho_{j'}^{(k')} + \rho_j^{(k')} \rho_{j'}^{(k')} \right) = N\delta_{jj'}, \qquad (A.18)$$

which is nothing but the Casimir of the adjoint representation.

A.3   Relations Between Roots and Weights

Next, we consider the scalar product between a defining weight $\rho^{(k)}$ and a root $\alpha^{(k'k'')}$, with $k' < k''$. If $k = k'$, we have

$$\rho^{(k)} \cdot \alpha^{(k'k'')} = (\rho^{(k)})^2 - \rho^{(k)} \cdot \rho^{(k'')} = \frac{1}{2}\left(1 - \frac{1}{N}\right) + \frac{1}{2N} = \frac{1}{2}, \qquad (A.19)$$

while, if $k \neq k'$ and $k \neq k''$, we have

$$\rho^{(k)} \cdot \alpha^{(k'k'')} = \rho^{(k)} \cdot \rho^{(k')} - \rho^{(k)} \cdot \rho^{(k'')} = 0. \qquad (A.20)$$

Consider then $N - 1$ of the defining weights. Without loss of generality, we can choose $\rho^{(1)}, \ldots, \rho^{(N-1)}$. As we have seen above, they form a basis of the space and generate the remaining weight $\rho^{(N)}$ as a linear combination with integer coefficients. Next, consider the $N - 1$ roots $\alpha^{(1)} \equiv \alpha^{1N}, \ldots, \alpha^{N-1} \equiv \alpha^{(N-1)N}$ which also generate the other roots as linear combinations with integer coefficients. From the above equations, it is easily seen that $4\pi\alpha^{(1)}, \ldots, 4\pi\alpha^{(N-1)}$ is a basis dual to $\rho^{(1)}, \ldots, \rho^{(N-1)}$ in the sense defined in Chap. 5. Similarly $4\pi\rho^{(1)}, \ldots, 4\pi\rho^{(N-1)}$ is a basis dual to $\alpha^{(1)}, \ldots, \alpha^{(N-1)}$.

---

[1] This comes as no surprise since the adjoint representation is found when decomposing the tensor product of the defining representation and the corresponding contragredient representation, after symmetrization and elimination of the singlet trace. Since the weights of the contragredient representation are opposite to the weights of the defining representation, it is no doubt that the roots appear as differences of the defining weights.

These remarks are very useful because, as we have seen in Chap. 5, generic (resp. periodic) winding transformations are generated by vectors that generate a lattice dual to the one generated by the roots (resp. by the defining weights). Equivalently, we can now say that, up to a factor 4π, generic (resp. periodic) winding transformations are generated by the defining weights (resp. by the roots). Moreover, the center element associated with a given weight $4\pi\rho^{(k)}$ is $e^{4\pi i\rho^{(k)}\cdot\rho^{(l)}} = e^{-i\frac{2\pi}{N}}$ and depends neither on l nor on k. It can be convenient instead to find a generating set of winding transformations that covers all the center elements. This set is provided by $4\pi\rho^{(1)}$, $4\pi(\rho^{(1)} + \rho^{(2)}), \ldots, 4\pi(\rho^{(1)} + \cdots \rho^{(N-1)}) = -4\pi\rho^{(N)}$.

A.4   Complexified Algebra and Killing Form

We mention that the eigenstates $i E_\alpha$ do not belong to the original (real) Lie algebra, but rather to the complexified algebra. The reason why the complexification is needed is that the operators $\mathrm{ad}_{i H_j}$ are not symmetric with respect to the Killing form $(X_1; X_2) \mapsto -2\mathrm{tr}\, X_1 X_2$ and thus not necessarily diagonalizable over the original algebra. They become however anti-Hermitian with respect to an extended version of the Killing form and therefore diagonalizable over the complexified algebra.

To see this in more details, consider first the Killing form over the original Lie algebra. First, because

$$(X; X) = 2\mathrm{tr}\, XX^\dagger = 2\sum_a |X_a|^2, \tag{A.21}$$

the Killing form defines a non-degenerate, positive-definite symmetric form over the original Lie algebra. Moreover, it obeys the cyclicality property

$$(X_1; \mathrm{ad}_{X_2} X_3) = (X_2; \mathrm{ad}_{X_3} X_1) = (X_3; \mathrm{ad}_{X_1} X_2), \tag{A.22}$$

which implies in particular $(X_1; \mathrm{ad}_{X_2} X_3) = -(\mathrm{ad}_{X_2} X_1; X_3)$. This means that the operator ad_{X_2} is antisymmetric, so not diagonalizable a priori over the original Lie algebra.

To turn this antisymmetric operator into an anti-Hermitian one, we now define the complexified algebra as the space of pairs $Z = (X, Y)$, which we denote also as $Z = X + IY$, equipped with the extended Lie bracket $[X_1 + IY_1, X_2 + IY_2] \equiv [X_1, X_2] - [Y_1, Y_2] + I([X_1, Y_2] + [Y_1, X_2])$. We also define a complex conjugation over the complexified algebra as $\overline{X + IY} \equiv X - IY$. This complex conjugation should not be mistaken with the complex conjugation of matrices. We have in fact $\bar{Z} = -Z^\dagger$. Finally we define

$$\langle Z_1; Z_2 \rangle \equiv -2\mathrm{tr}\, \bar{Z}_1 Z_2 = \langle \bar{Z}_2; \bar{Z}_1 \rangle = \langle Z_2; Z_1 \rangle^*, \tag{A.23}$$

which extends the Killing form over the complexified algebra (since $\bar{X} = X$ for elements of the original algebra). We have

$$\langle Z; Z \rangle = (X; X) + (Y; Y) \,, \tag{A.24}$$

so the extended Killing form is now Hermitian positive definite. Moreover

$$\langle X_1; \mathrm{ad}_{X_2} X_3 \rangle = \langle \bar{X}_2; \mathrm{ad}_{X_3} \bar{X}_1 \rangle = \langle \bar{X}_3; \mathrm{ad}_{\bar{X}_1} X_2 \rangle \,. \tag{A.25}$$

For an element $X_2 = \bar{X}_2$ of the original Lie algebra, such as iH_j, this implies in particular $\langle X_1; \mathrm{ad}_{X_2} X_3 \rangle = -\langle \mathrm{ad}_{X_2} X_1; X_3 \rangle$, which means that the operator ad_{X_2} is anti-Hermitian and thus diagonalizable, with imaginary eigenvalues.

This is why the ad_{H_j} can be diagonalized simultanously and the roots are real vectors. Moreover, by conjugating $[H_j, E_\alpha] = \alpha_j E_\alpha$, we find $[H_j, \bar{E}_\alpha] = -\alpha_j \bar{E}_\alpha$ which shows why roots come by pairs and tells us that, with an appropriate choice of normalization we have $\bar{E}_\alpha = -E_\alpha^\dagger = -E_{-\alpha}$ (this is precisely the choice made above for the $Z_\pm^{kk'}$). Finally, this shows that the diagonalization basis can be chosen such that $\langle iH_j; iH_k \rangle = \delta_{jk}$, $\langle iH_j, iE_\alpha \rangle = 0$, and $\langle E_\alpha; E_\beta \rangle = \delta_{\alpha\beta}$, which implies $(iH_j; iH_k) = \delta_{jk}$, $(iH_j, iE_\alpha)) = 0$, and $(iE_\alpha; iE_\beta) = \delta_{-\alpha,\beta}$, owing to $\bar{H}_j = -H_j$ and our phase convention between $\bar{E}_\alpha$ and $E_{-\alpha}$.

We mention finally that the components of an element X of the original algebra in the Cartan-Weyl basis are not necessarily real. However since $\bar{H}_j = -H_j$ and $\bar{E}_\alpha = -E_{-\alpha}$, we find $X = \bar{X} = i(X_j^* H_j + X_\alpha^* E_{-\alpha})$ and, thus, $X_j^* = X_j$ and $X_\alpha^* = X_{-\alpha}$. This conclusion can also be reached using the Hermitian conjugation which is nothing but $Z^\dagger = -\bar{Z}$.

Evaluating Matsubara Sums

B

Consider a complex function $f(z)$ with a finite number of poles z_* (no branch cuts) distinct from any $i\omega_q$ with $\omega_q = 2\pi q/\beta$ a Matsubara frequency. Suppose also that $f(z)$ goes to zero rapidly enough as $|z| \to \infty$ (and also uniformly enough as a function of $\mathrm{Arg}\, z$). The goal of this Appendix is to recall how to evaluate the Matsubara sum

$$\frac{1}{\beta} \sum_{q \in \mathbb{Z}} f(i\omega_q), \tag{B.1}$$

in this case.

B.1   Basic Result

Let us consider the contour integral

$$I_N \equiv \int_{C_N} \frac{dz}{2\pi i} f(z)\, n(z), \tag{B.2}$$

where C_N is the circular path centered around $z = 0$ and of radius $R_N \equiv 2\pi(N + 1/2)/\beta$ and $n(z)$ is the function $1/(e^{\beta z} - 1)$ whose poles are located at $i\omega_q$ and thus distinct from the poles of $f(z)$ y assumption, with residue $1/\beta$.

If N is taken large enough, all the poles of $f(z)$ are within the circle. Using the residue theorem, we can then write

$$I_N = \frac{1}{\beta} \sum_{q=-N}^{N} f(i\omega_q) + \sum_{z_*} \mathrm{Res}\, f(z)|_{z=z_*} n(z_*). \tag{B.3}$$

© The Author(s), under exclusive license to Springer Nature Switzerland AG 2022
U. Reinosa, *Perturbative Aspects of the Deconfinement Transition*, Lecture Notes in Physics 1006, https://doi.org/10.1007/978-3-031-11375-8

On the other hand, under the assumption that $f(z)$ goes to zero rapidly and uniformly enough as $|z| \to \infty$, one can expect that $I_N \to 0$ as $N \to \infty$.[1] Using this result in Eq. (B.3), one arrives eventually at

$$\frac{1}{\beta} \sum_{q \in \mathbb{Z}} f(i\omega_q) = - \sum_{z_*} \mathrm{Res}\, f(z)|_{z=z_*} n(z_*),$$ (B.4)

where z_* refers to the poles of $f(z)$.

B.2   Application to Sum-Integrals

B.2.1   One-Loop Tadpole Sum-Integrals of Type I

Consider the one-loop sum-integral

$$\int_Q^T \frac{q^n}{Q_\kappa^2 + m^2} = \int \frac{d^{d-1}q}{(2\pi)^{d-1}} q^n \frac{1}{\beta} \sum_{q \in \mathbb{Z}} \frac{1}{(\omega_q + \kappa \cdot \hat{r})^2 + \varepsilon_q^2},$$ (B.5)

with $\varepsilon_q \equiv \sqrt{q^2 + m^2}$. We can write

$$\frac{1}{(\omega_q + \kappa \cdot \hat{r})^2 + \varepsilon_q^2} = \frac{1}{2\varepsilon_q} \left[-\frac{1}{i(\omega_q + \kappa \cdot \hat{r}) - \varepsilon_q} + \frac{1}{i(\omega_q + \kappa \cdot \hat{r}) + \varepsilon_q} \right]$$
(B.6)

so the relevant poles in the complex plane are located at $z = \pm\varepsilon_q - i\kappa \cdot \hat{r}$ with residues $\mp 1/2\varepsilon_q$. It follows that

$$\int_Q^T \frac{q^n}{Q_\kappa^2 + m^2} = \int \frac{d^{d-1}q}{(2\pi)^{d-1}} q^n \frac{n(\varepsilon_q - i\kappa \cdot \hat{r}) - n(-\varepsilon_q - i\kappa \cdot \hat{r})}{2\varepsilon_q}.$$ (B.7)

Using the identity $n(-z) = -1 - n(z)$, this rewrites

$$\int_Q^T \frac{q^n}{Q_\kappa^2 + m^2} = \int \frac{d^{d-1}q}{(2\pi)^{d-1}} q^n \frac{1 + n(\varepsilon_q - i\kappa \cdot \hat{r}) + n(\varepsilon_q + i\kappa \cdot \hat{r})}{2\varepsilon_q},$$ (B.8)

where the term with 1 corresponds to the zero-temperature limit since $n(\varepsilon_q \pm i\kappa \cdot \hat{r})$ is exponentially suppressed as $T \to 0$.

---

[1] Of course, this needs to be checked in more detail for each specific case, although we shall not do it here.

B.2.2 One-Loop Tadpole Sum-Integrals of Type II

We also find one-loop sum-integrals of the form

$$
\int_Q^T \frac{q^n \omega_q^\kappa}{Q_\kappa^2 + m^2} = \int \frac{d^{d-1}q}{(2\pi)^{d-1}} q^n \frac{1}{\beta} \sum_{q \in \mathbb{Z}} \frac{\omega_q + \kappa \cdot \hat{r}}{(\omega_q + \kappa \cdot \hat{r})^2 + \varepsilon_q^2} \,. \tag{B.9}
$$

We note that the corresponding Matsubara sum is here not absolutely convergent, but it admits nonetheless a limit if we consider the sum as the limit of $\sum_{q=-N}^N$ as $N \to \infty$. In fact, one can subtract 0 in the form

$$
\frac{1}{\beta} \sum_{q=-N}^N \frac{\omega_q + \kappa \cdot \hat{r}}{(\omega_q + \kappa \cdot \hat{r})^2 + \varepsilon_q^2} = \frac{1}{\beta} \sum_{q=-N}^N \frac{\omega_q + \kappa \cdot \hat{r}}{(\omega_q + \kappa \cdot \hat{r})^2 + \varepsilon_q^2} - \frac{1}{\beta} \sum_{q=-N}^N \frac{\omega_q}{(\omega_q)^2 + \varepsilon_q^2}
$$

$$
= \frac{1}{\beta} \sum_{q=-N}^N \left[\frac{\omega_q + \kappa \cdot \hat{r}}{(\omega_q + \kappa \cdot \hat{r})^2 + \varepsilon_q^2} - \frac{\omega_q}{(\omega_q)^2 + \varepsilon_q^2} \right], \tag{B.10}
$$

so that we now have to deal with an absolutely convergent sum. We can now write

$$
\frac{\omega_q + \kappa \cdot \hat{r}}{(\omega_q + \kappa \cdot \hat{r})^2 + \varepsilon_q^2} = -\frac{1}{2i} \left[\frac{1}{i(\omega_q + \kappa \cdot \hat{r}) - \varepsilon_q} + \frac{1}{i(\omega_q + \kappa \cdot \hat{r}) + \varepsilon_q} \right] \tag{B.11}
$$

so that the relevant poles in the complex plane are located at $z = \pm\varepsilon_q - i\kappa \cdot \hat{r}$ as well as $z = \pm\varepsilon_q$, all with residue $-1/2i$. It follows that

$$
\int_Q^T \frac{q^n \omega_q^\kappa}{Q_\kappa^2 + m^2} = \int \frac{d^{d-1}q}{(2\pi)^{d-1}} q^n \frac{n(\varepsilon_q - i\kappa \cdot \hat{r}) + n(-\varepsilon_q - i\kappa \cdot \hat{r}) - n(\varepsilon_q) - n(-\varepsilon_q)}{2i} \,. \tag{B.12}
$$

Making use of $n(-z) = -1 - n(z)$, this rewrites

$$
\int_Q^T \frac{q^n \omega_q^\kappa}{Q_\kappa^2 + m^2} = \int \frac{d^{d-1}q}{(2\pi)^{d-1}} q^n \frac{n(\varepsilon_q - i\kappa \cdot \hat{r}) - n(\varepsilon_q + i\kappa \cdot \hat{r})}{2i} \,, \tag{B.13}
$$

and we note that there is no zero-temperature contribution this time.

B.2.3 The Two-Loop Sunset Sum-Integral

We now treat the scalar two-loop integral

$$
S_{\alpha\beta\gamma}^{\kappa\lambda\tau} \equiv \int_{Q,K} G_\alpha(Q_\kappa) G_\beta(K_\lambda) G_\gamma(L_\tau) \,, \tag{B.14}
$$

where $Q + K + L = 0$ and $\kappa + \lambda + \tau = 0$ such that $Q_\kappa + K_\lambda + L_\tau = 0$ and where $G_\alpha(Q_\kappa)$ stands for the free propagator $1/(Q_\kappa^2 + \alpha^2)$. It proves useful to introduce the spectral representation of the latter

$$G_\alpha(Q_\kappa) = \tilde{G}_\alpha(i\omega_q^\kappa; q) \equiv \int_{q_0} \frac{\rho_\alpha(q_0, q)}{q_0 - i\omega_q^\kappa}, \tag{B.15}$$

where $\int_{q_0} = \int dq_0/(2\pi)$ and

$$\rho_\alpha(q_0, q) = 2\pi \, \mathrm{sign}(q_0) \, \delta\left(q_0^2 - \varepsilon_{\alpha,q}^2\right), \tag{B.16}$$

with $\varepsilon_{\alpha,q} = \sqrt{q^2 + \alpha^2}$. The double Matsubara sum in Eq. (B.14) yields

$$T^2 \sum_{q,k} \frac{1}{(q_0 - i\omega_q^\kappa)(k_0 - i\omega_k^\lambda)(l_0 + i\omega_q^\kappa + i\omega_k^\lambda)}.$$

$$= \frac{(n_{k_0 - i\lambda \cdot \hat{r}} - n_{-l_0 + i\tau \cdot \hat{r}})(n_{q_0 - i\kappa \cdot \hat{r}} - n_{-l_0 - k_0 - i\kappa \cdot \hat{r}})}{l_0 + k_0 + q_0}$$

$$= \frac{n_{k_0 - i\lambda \cdot \hat{r}} n_{l_0 - i\tau \cdot \hat{r}} - n_{-q_0 + i\kappa \cdot \hat{r}}(n_{k_0 - i\lambda \cdot \hat{r}} - n_{-l_0 + i\tau \cdot \hat{r}})}{l_0 + k_0 + q_0}, \tag{B.17}$$

where we have used the identity $(1 + n_x + n_y)n_{x+y} = n_x n_y$ and $\kappa + \lambda + \tau = 0$. The second line of this equation makes it clear that Eq. (B.17) is well defined for all k_0, q_0, and l_0, including the limiting case $l_0 + k_0 + q_0 \to 0$, for which both the numerator and the denominator vanish linearly. In the following, we shall decompose the fraction in the third line of Eq. (B.17) in different pieces whose numerators do not vanish at $l_0 + k_0 + q_0 = 0$, thus making the corresponding contribution to Eq. (B.17) formally divergent in this limit. To avoid this problem, we regulate the denominator in Eq. (B.17) as

$$\frac{1}{l_0 + k_0 + q_0} \to \mathrm{Re}\left(\frac{1}{l_0 + k_0 + q_0 + i0^+}\right). \tag{B.18}$$

Now, we use the identities

$$n_{k_0 - i\kappa \cdot \hat{r}} = -\theta(-k_0) + \mathrm{sign}(k_0)n_{|k_0| - i \, \mathrm{sign}(k_0)\kappa \cdot \hat{r}} \tag{B.19}$$

$$n_{-k_0 + i\kappa \cdot \hat{r}} = -\theta(k_0) - \mathrm{sign}(k_0)n_{|k_0| - i \, \mathrm{sign}(k_0)\kappa \cdot \hat{r}} \tag{B.20}$$

in the third line of Eq. (B.17) to rewrite Eq. (B.14) as

$$
S^{\kappa\lambda\tau}_{\alpha\beta\gamma} = S^{\text{vac}}_{\alpha\beta\gamma} + \mu^{2\epsilon} \int_{q_0,\mathbf{q}} \sigma^\kappa_\alpha(q_0, q) \operatorname{Re} \tilde{I}_{\beta\gamma}(q_0 + i0^+; q)
$$

$$
+ \int_{q_0,k_0,\mathbf{q},\mathbf{k}} \sigma^\kappa_\alpha(q_0, q)\sigma^\lambda_\beta(k_0, k) \operatorname{Re} \tilde{G}_\gamma(\ell_0 + i0^+; \ell)
$$

$$
+ \text{perm.,} \tag{B.21}
$$

where $S^{\text{vac}}_{\alpha\beta\gamma}$ is an unimportant vacuum contribution, independent of the temperature and of the background field, $\ell_0 = q_0 + k_0$, $\ell = |\mathbf{q} + \mathbf{k}|$,

$$
\sigma^\kappa_\alpha(q_0, q) = \rho_\alpha(q_0, q) \operatorname{sign}(q_0) \, n_{|q_0| - i\operatorname{sign}(q_0)\kappa\cdot\hat{r}}, \tag{B.22}
$$

and "perm." denotes the circular permutations of the pairs of indices (α, κ), (β, λ), and (γ, τ) in the two integrals that appear explicitly in (B.21). The function $\tilde{G}_\alpha(z; q)$ has been defined in Eq. (B.15), and the function $\tilde{I}_{\alpha\beta}(z; q)$ is related to the vacuum one-loop integral [here $Q = (\omega, \mathbf{q})$]

$$
I^{\text{vac}}_{\alpha\beta}(Q) = \tilde{I}_{\alpha\beta}(i\omega; q) \equiv \mu^{2\epsilon} \int \frac{d^d K}{(2\pi)^d} G_\alpha(K) G_\beta(Q + K)
$$

$$
= \mu^{2\epsilon} \int_{k_0,l_0,\mathbf{k}} \rho_\alpha(k_0, k)\rho_\beta(l_0, \ell) \frac{\theta(l_0) - \theta(-k_0)}{i\omega + l_0 + k_0}. \tag{B.23}
$$

In obtaining Eq. (B.21), we have used $\tilde{G}_\alpha(x; q) = \tilde{G}_\alpha(-x; q)$ and $\tilde{I}_{\alpha\beta}(x; q) = \tilde{I}_{\alpha\beta}(-x; q)$, and we have set $d \to 4$ in the second, UV-finite line. In contrast, one has to keep d arbitrary in the second term on the right-hand side of Eq. (B.21) since it contains UV-divergent contributions, arising from the zero-temperature loop in Eq. (B.23).

It is now an easy matter to perform explicitly the frequency and, for the double integral on the second line, the angular integrations. As before, we decompose the result according to the number of thermal factors n in each contribution as

$$
S^{\kappa\lambda\tau}_{\alpha\beta\gamma} = S^{\kappa\lambda\tau}_{\alpha\beta\gamma}(0n) + S^{\kappa\lambda\tau}_{\alpha\beta\gamma}(1n) + S^{\kappa\lambda\tau}_{\alpha\beta\gamma}(2n). \tag{B.24}
$$

We obtain

$$
S^{\kappa\lambda\tau}_{\alpha\beta\gamma}(1n) = \mu^{2\epsilon} \int_{\mathbf{q}} \operatorname{Re} \frac{n_{\varepsilon_{\alpha,q} - i\kappa\cdot\hat{r}}}{\varepsilon_{\alpha,q}} \operatorname{Re} \tilde{I}_{\beta\gamma}(\varepsilon_{\alpha,q} + i0^+; q) + \text{perm.} \tag{B.25}
$$

$$
S^{\kappa\lambda\tau}_{\alpha\beta\gamma}(2n) = \frac{1}{32\pi^4} \int_0^\infty dq \, q \int_0^\infty dk \, k \operatorname{Re} \frac{n_{\varepsilon_{\alpha,q} - i\kappa\cdot\hat{r}} \, n_{\varepsilon_{\beta,k} - i\lambda\cdot\hat{r}}}{\varepsilon_{\alpha,q} \, \varepsilon_{\beta,k}}
$$

$$
\times \operatorname{Re} \ln \frac{(\varepsilon_{\alpha,q} + \varepsilon_{\beta,k} + i0^+)^2 - (\varepsilon_{\gamma,k+q})^2}{(\varepsilon_{\alpha,q} + \varepsilon_{\beta,k} + i0^+)^2 - (\varepsilon_{\gamma,k-q})^2}
$$

$$+ \frac{1}{32\pi^4} \int_0^\infty dq\, q \int_0^\infty dk\, k \operatorname{Re} \frac{n_{\varepsilon_{\alpha,q} - i\kappa\cdot\hat{r}}\, n_{\varepsilon_{\beta,k} + i\lambda\cdot\hat{r}}}{\varepsilon_{\alpha,q}\, \varepsilon_{\beta,k}}$$

$$\times \operatorname{Re} \ln \frac{(\varepsilon_{\alpha,q} - \varepsilon_{\beta,k} + i0^+)^2 - (\varepsilon_{\gamma,k+q})^2}{(\varepsilon_{\alpha,q} - \varepsilon_{\beta,k} + i0^+)^2 - (\varepsilon_{\gamma,k-q})^2} .$$

$$+ \text{ perm.,} \tag{B.26}$$

and $S_{\alpha\beta\gamma}^{\kappa\lambda\tau}(0n) = S_{\alpha\beta\gamma}^{\text{vac}}$. As mentioned before, the contribution $S_{\alpha\beta\gamma}^{\kappa\lambda\tau}(2n)$ is UV finite, but the contribution $S_{\alpha\beta\gamma}^{\kappa\lambda\tau}(1n)$ contains UV-divergent terms which explicitly depend on the temperature and on the background field. We note that Eq. (B.25) rewrites

$$S_{\alpha\beta\gamma}^{\kappa\lambda\tau}(1n) = J_\alpha^\kappa(1n) \operatorname{Re} \tilde{I}_{\beta\gamma}(\varepsilon_{\alpha,q} + i0^+; q) + \text{perm.,} \tag{B.27}$$

where $J_\alpha^\kappa(1n)$ is defined[2] in Eq. (7.16) and where we used the fact, owing to the $O(d)$ invariance of the Euclidean integral Eq. (B.23), $\operatorname{Re} \tilde{I}_{\beta\gamma}(\varepsilon_{\alpha,q} + i0^+; q)$ depends only on $\varepsilon_{\alpha,q}^2 - q^2 = \alpha^2$. The expression for $I_{\alpha\beta}^{\text{vac}}(Q) = \tilde{I}_{\alpha\beta}(i\omega; q)$ can be found, for instance, in [91] and reads, up to corrections of $O(\epsilon)$,[3]

$$I_{\alpha\beta}^{\text{vac}}(Q) = \frac{1}{16\pi^2} \left\{ \frac{1}{2\epsilon} + 1 + \frac{1}{2} \ln \frac{\bar{\mu}^2}{Q^2} - \frac{1}{2} \ln \left(C_{\alpha\beta}^2(Q) - \frac{1}{4} \right) \right.$$

$$\left. + C_{\alpha\beta}(Q) \ln \frac{C_{\alpha\beta}(Q) - \frac{1}{2}}{C_{\alpha\beta}(Q) + \frac{1}{2}} \right\} + (\alpha \leftrightarrow \beta), \tag{B.28}$$

where $C_{\alpha\beta}(Q)$ (which should not be mistaken with $C_{\kappa\lambda\tau}$) is given by

$$C_{\alpha\beta}(Q) = \frac{B_{\alpha\beta}(Q) + \alpha^2 - \beta^2}{2Q^2}, \tag{B.29}$$

with

$$B_{\alpha\beta}(Q) = \sqrt{Q^4 + 2Q^2 \left(\alpha^2 + \beta^2 \right) + \left(\alpha^2 - \beta^2 \right)^2}. \tag{B.30}$$

---

[2] Strictly speaking, we should use the expression of $J(1n)$ for arbitrary d in Eq. (B.27) since it multiplies a divergent integral. However, as we discuss in the main text, this makes no difference upon renormalization

[3] A typo in the formula (B2) of Ref. [91] is corrected here: the first three terms on the right-hand side receive a factor $1/2$ due to the symmetrization $(\alpha \leftrightarrow \beta)$.

Using this formula and the definition Eq. (B.23), we get

$$\text{Re } \tilde{I}_{00}(\varepsilon_q + i0^+; q) = \frac{1}{16\pi^2}\left[\frac{1}{\epsilon} + \ln\frac{\bar{\mu}^2}{m^2} + 2\right], \tag{B.31}$$

$$\text{Re } \tilde{I}_{m0}(q + i0^+; q) = \frac{1}{16\pi^2}\left[\frac{1}{\epsilon} + \ln\frac{\bar{\mu}^2}{m^2} + 1\right], \tag{B.32}$$

$$\text{Re } \tilde{I}_{mm}(\varepsilon_q + i0^+; q) = \frac{1}{16\pi^2}\left[\frac{1}{\epsilon} + \ln\frac{\bar{\mu}^2}{m^2} + 2 - \frac{\pi}{\sqrt{3}}\right]. \tag{B.33}$$

Solutions

Problems of Chap. 2

2.1 Faddeev-Popov Determinant (⋆)

(a) We have

$$A_\mu^U = U A_\mu U^\dagger - U \partial_\mu U^\dagger$$
$$= (1 + \theta) A_\mu (1 - \theta) - (1 + \theta) \partial_\mu (1 - \theta)$$
$$= A_\mu + [\theta, A_\mu] + \partial_\mu \theta ,$$

and thus $A_\mu^U - A_\mu = \partial_\mu \theta - [A_\mu, \theta] = D_\mu \theta$.

(b) First, we recall that $\delta F[A^U]/\delta U$ depends only on A^U since the variation of $F[A^U]$ with respect to U is evaluated by considering a variation $F[A^{\theta U}] = F[(A^U)^\theta]$ with θ infinitesimal. Now, assuming that A is such that $F[A] = 0$, we can take $U = 1$. The chain rule then gives

$$\Delta[A] = \det \left. \frac{\delta F^a[A^U](x)}{\delta \theta^b(y)} \right|_{\theta=0} = \det \int d^d z \left. \frac{\delta F^a[A](x)}{\delta A_\mu^c(z)} \frac{\delta (A^U)_\mu^c(z)}{\delta \theta^b(y)} \right|_{\theta=0} .$$

Combining this with the previous result, we eventually arrive at

$$\Delta[A] = \det \left(\int d^d z \, \frac{\delta F^a[A](x)}{\delta A_\mu^c(z)} D_\mu^{cb} \delta(z - y) \right) .$$

© The Author(s), under exclusive license to Springer Nature Switzerland AG 2022
U. Reinosa, *Perturbative Aspects of the Deconfinement Transition*, Lecture Notes
in Physics 1006, https://doi.org/10.1007/978-3-031-11375-8

2.2 BRST Symmetry (⋆⋆)

(a) Under the considered transformation, the Yang-Mills part of the action is of course invariant by construction, so let us focus on the gauge-fixing part δS_{FP}. The variation of the ghost contribution writes

$$-\int d^d x \int d^d y \, s\bar{c}^a(x) \frac{\delta F^a[A](x)}{\delta A_\mu^b(y)} D_\mu c^b(y)$$

$$+\int d^d x \int d^d y \, \bar{c}^a(x) \frac{\delta F^a[A](x)}{\delta A_\mu^b(y)} s^2 A_\mu^b(y)$$

$$+\int d^d x \, \bar{c}^a(x) \int d^d y \int d^d z \, \frac{\delta F^a[A](x)}{\delta A_\mu^b(y)\delta A_\nu^c(z)} D_\nu c^c(z) D_\mu c^b(y),$$

but we notice that the last line vanishes because

$$\int d^d y \int d^d z \, \frac{\delta F^a[A](x)}{\delta A_\mu^b(y)\delta A_\nu^c(z)} D_\nu c^c(z) D_\mu c^b(y)$$

$$= -\int d^d z \int d^d y \, \frac{\delta F^a[A](x)}{\delta A_\nu^c(z)\delta A_\mu^b(y)} D_\mu c^b(y) D_\nu c^c(z)$$

$$= -\int d^d y \int d^d z \, \frac{\delta F^a[A](x)}{\delta A_\mu^b(y)\delta A_\nu^c(z)} D_\nu c^c(z) D_\mu c^b(y) = 0.$$

On the other hand, the variation of the Nakanishi-Lautrup field contribution writes

$$\int d^d x \int d^d y \, ih^a(x) \frac{\delta F^a[A](x)}{\delta A_\mu^b(y)} D_\mu c^b(y)$$

$$+\int d^d x \, sih^a(x) \, F^a[A](x).$$

Putting all the pieces together, we arrive at

$$s\delta S_{\mathrm{FP}} = \int d^d x \int d^d y \, (ih^a(x) - s\bar{c}^a(x)) \frac{\delta F^a[A](x)}{\delta A_\mu^b(y)} D_\mu c^b(y)$$

$$+\int d^d x \int d^d y \, \bar{c}^a(x) \frac{\delta F^a[A](x)}{\delta A_\mu^b(y)} s^2 A_\mu^b(y)$$

$$+\int d^d x \, sih^a(x) \, F^a[A](x),$$

We thus see that the gauge-fixing contribution is invariant if we set $s\bar{c}^a = ih^a$ and $sih^a = 0$ and choose sc^a such that $s^2 A_\mu^a = 0$.

(b) We have

$$s^2 A_\mu^a = s D_\mu c^a$$
$$= D_\mu s c^a + f^{abc}(s A_\mu^b) c^c$$
$$= D_\mu s c^a + f^{abc}(D_\mu c^b) c^c$$
$$= D_\mu s c^a + f^{abc}(\partial_\mu c^b) c^c + f^{abc} f^{bde} A_\mu^d c^e c^c$$

Using the Jacobi identity and the Grassmannian nature of the field c, we find

$$f^{abc} f^{bde} A_\mu^d c^e c^c = -f^{abd} f^{bec} A_\mu^d c^e c^c - f^{abe} f^{bcd} A_\mu^d c^e c^c$$
$$= -f^{abd} f^{bec} A_\mu^d c^e c^c - f^{abc} f^{bde} A_\mu^d c^e c^c$$
$$= -\frac{1}{2} f^{abd} f^{bec} A_\mu^d c^e c^c .$$

Similarly, we can write

$$f^{abc}(\partial_\mu c^b) c^c = \frac{1}{2} f^{abc} \partial_\mu (c^b c^c) .$$

Altogether, we arrive at

$$s^2 A_\mu^a = D_\mu s c^a + \partial_\mu \left(\frac{1}{2} f^{abc} c^b c^c \right) - \frac{1}{2} f^{abd} f^{bec} A_\mu^d c^e c^c$$
$$= D_\mu \left(s c^a + \frac{1}{2} f^{abc} c^b c^c \right)$$

from which it follows that the choice $s c^a = -(1/2) f^{abc} c^b c^c$ ensures that $s^2 A_\mu^a = 0$.

2.3 Nilpotency (⋆⋆)
(a) We have $s^2 i h = s0 = 0$, $s^2 \bar{c} = s i h = 0$. Moreover

$$s^2 c^a = -\frac{1}{2} f^{abc}(s c^b) c^c + \frac{1}{2} f^{abc} c^b (s c^c)$$
$$= -\frac{1}{2} f^{abc}(s c^b) c^c$$
$$= \frac{1}{2} f^{abc} f^{bde} c^d c^e c^c .$$

Making again use of the Jacobi identity, this becomes

$$s^2 c^a = -\frac{1}{2} f^{abd} f^{bec} c^d c^e c^c - \frac{1}{2} f^{abe} f^{bcd} c^d c^e c^c$$

$$= -f^{abd} f^{bec} c^d c^e c^c$$

$$= -f^{abc} f^{bde} c^d c^e c^c .$$

These two expressions are compatible only if $s^2 c^a = 0$.

(b) We have $s(\phi \chi) = (s\phi)\chi + (-1)^q \phi(s\chi)$ with $q = 1$ if ϕ is a Grassmannian field and 0 otherwise. Then

$$s^2(\phi \chi) = s((s\phi)\chi + (-1)^q \phi(s\chi))$$

$$= (s^2 \phi)\chi + (-1)^{q+1}(s\phi)(s\chi) + (-1)^q (s\phi)(s\chi) + \phi(s^2 \phi)$$

$$= (s^2 \phi)\chi + \phi(s^2 \phi) .$$

(c) Since any field configuration can be seen as a linear combination of products of primary fields, we conclude that $s^2 = 0$ over all possible field configurations.

2.4 Curci-Ferrari Model and Modified BRST Symmetry (⋆)

(a) We have

$$s_m \int d^d x \, \frac{1}{2} m^2 A^a_\mu(x) A^a_\mu(x)$$

$$= \int d^d x \, m^2 D_\mu c^a(x) A^a_\mu(x)$$

$$+ \int d^d x \, m^2 \partial_\mu c^a(x) A^a_\mu(x) + \int d^d x \, m^2 f^{abc} A^b_\mu(x) c^c(x) A^a_\mu(x)$$

$$= \int d^d x \, m^2 \partial_\mu c^a(x) A^a_\mu(x) = -\int d^d x \, m^2 c^a(x) \partial_\mu A^a_\mu(x) .$$

(b) The previous variation combines with part of the variation of $\int d^d x \, i h^a \partial_\mu A^a_\mu$. As compared to the derivation in Problem 2.2, we now find a term

$$\int d^d x \, (s_m i h^a(x) - m^2 c^a(x)) \partial_\mu A^a_\mu(x) ,$$

thus leading this time to $s_m i h^a = m^2 c^a$.

2.5 Curci-Ferrari Model and Gluon Propagator (★★★)

(a) The quadratic part of the action reads

$$\int d^d x\, A_\mu^a(x) \Big[\delta_{\mu\nu}(-\partial^2 + m^2) + \partial_\mu \partial_\nu \Big] \delta^{ab} A_\mu^b(x)$$

$$+ \frac{1}{2} \int d^d x\, i h^a(x)\, \delta^{ab} \partial_\mu A_\mu^b(x) - \frac{1}{2} \int d^d x\, i A_\mu^a(x)\, \delta^{ab} \partial_\mu h^b(x).$$

In Fourier space, this becomes

$$\int d^d Q\, A_\mu^a(-Q) \Big[\delta_{\mu\nu}(Q^2 + m^2) - Q_\mu Q_\nu \Big] \delta^{ab} A_\mu^b(Q)$$

$$- \frac{1}{2} \int d^d x\, h^a(-Q)\, \delta^{ab} Q_\mu A_\mu^b(Q) + \frac{1}{2} \int d^d x\, A_\mu^a(-Q)\, \delta^{ab} Q_\mu h^b(Q)$$

and thus to the matrix (the color structure is trivial)

$$\begin{pmatrix} \delta_{\mu\nu}(Q^2 + m^2) - Q_\mu Q_\nu & Q_\mu \\ -Q_\nu & 0 \end{pmatrix}.$$

Writing the inverse as

$$\begin{pmatrix} A_{\mu\nu} & -B_\mu \\ B_\nu & C \end{pmatrix},$$

we find

$$(Q^2 + m^2) A_{\mu\nu} - A_{\mu\rho} Q_\rho Q_\nu + B_\mu Q_\nu = \delta_{\mu\nu},$$

$$A_{\mu\rho} Q_\rho = 0,$$

$$(Q^2 + m^2) B_\nu - B_\rho Q_\rho Q_\nu - C Q_\nu = 0,$$

$$B_\rho Q_\rho = 1.$$

The second and fourth equations allow one to simplify the other two as

$$(Q^2 + m^2) A_{\mu\nu} + B_\mu Q_\nu = \delta_{\mu\nu},$$

$$(Q^2 + m^2) B_\nu - (C + 1) Q_\nu = 0.$$

Contracting the second one with Q_ν, we arrive at $C = m^2/Q^2$ and thus $B_\mu = Q_\mu/Q^2$. Eventually, this leads to

$$A_{\mu\nu} = \frac{1}{Q^2 + m^2} \Big[\delta_{\mu\nu} - \frac{Q_\mu Q_\nu}{Q^2} \Big].$$

(b) This result could have been anticipated since the Nakanishi-Lautrup field imposes the field A_μ to be transverse, $\partial_\mu A_\mu = 0$. It follows immediately that the propagator, which is nothing but the (connected) correlation function $\langle A_\mu^a A_\nu^b \rangle$, is transverse, $\partial_\mu \langle A_\mu^a A_\nu^b \rangle = 0$. In Fourier space, this means that $Q_\mu G_{\mu\nu}^{ab}(Q) = 0$, as we have indeed found.

2.6 Curci-Ferrari Model and Taylor's Non-renormalization Theorem (★★★)

(a) The interaction term between the ghost, antighost, and gluon fields reads

$$\int d^d x \, f^{abc} \partial_\mu \bar{c}^a(x) A_\mu^b(x) c^c(x) .$$

In Fourier space, this reads

$$\int d^d Q \int d^d K \int d^d L \, \delta^{(d)}(Q + K + L) \, i f^{abc} K_\mu \bar{c}^a(K) A_\mu^b(Q) c^c(L) .$$

Assuming momentum conservation at the vertices, the associated Feynman rule is then $i K_\mu f^{abc}$, where K_μ denotes the momentum of the antighost and a its color label, while b and c are those of the gluon and the ghost, respectively.

(b) The external ghost leg of momentum K is attached to an interaction vertex from which emanates an internal gluon propagator. Denoting by Q the momentum of the gluon, the combination of the gluon propagator and of the interaction vertex produces a factor

$$(K + Q)_\mu \left[\delta_{\mu\nu} - \frac{Q_\mu Q_\nu}{Q^2} \right] = K_\mu \left[\delta_{\mu\nu} - \frac{Q_\mu Q_\nu}{Q^2} \right] + Q_\mu \left[\delta_{\mu\nu} - \frac{Q_\mu Q_\nu}{Q^2} \right]$$

$$= K_\mu \left[\delta_{\mu\nu} - \frac{Q_\mu Q_\nu}{Q^2} \right] + \left[Q_\nu - \frac{Q^2 Q_\nu}{Q^2} \right]$$

$$= K_\mu \left[\delta_{\mu\nu} - \frac{Q_\mu Q_\nu}{Q^2} \right] .$$

We deduce that the loop correction vanishes in the limit $K \to 0$.

(c) The renormalized ghost-antighost-gluon vertex equals the bare vertex times the combination $Z_A^{1/2} Z_c$ of renormalization factors (one has $Z_c = Z_{\bar{c}}$). Since the loop corrections to the bare vertex vanish in the limit of external ghost momentum, and since the tree-level contribution is just the bare coupling g_{bare}, up to tensorial decorations, we deduce that $Z_A^{1/2} Z_c g_{\text{bare}}$ needs to be finite. In terms of the coupling renormalization factor Z_g, this means that the combination $Z_A^{1/2} Z_c Z_g$ is necessarily finite.

Problems of Chap. 3

3.1 Gauge Transformation of the Wilson Line (⋆)

(a) Thanks to the path ordering, we can write

$$\frac{d}{d\sigma_f} L_A(\sigma_f, \sigma_i, \mathbf{x}) = A_0(\sigma_f, \mathbf{x}) L_A(\sigma_f, \sigma_i, \mathbf{x}).$$

This is a linear, first-order differential equation whose solution is entirely determined once an initial condition is given. Here, this condition is simply

$$L_A(\sigma_f = \sigma_i, \sigma_i, \mathbf{x}) = \mathcal{P} \exp\left\{\int_{\sigma_i}^{\sigma_i} d\tau \, A_0(\tau, \mathbf{x})\right\} = \mathbb{1}.$$

(b) We have

$$\frac{d}{d\sigma_f} \left(U(\sigma_f, \mathbf{x}) \, L_A(\sigma_f, \sigma_i, \mathbf{x}) \, U^\dagger(\sigma_i, \mathbf{x}) \right)$$

$$= \frac{dU(\sigma_f, \mathbf{x})}{d\sigma_f} L_A(\sigma_f, \sigma_i, \mathbf{x}) \, U^\dagger(\sigma_i, \mathbf{x}) + U(\sigma_f, \mathbf{x}) \, A_0(\sigma_f, \mathbf{x}) L_A(\sigma_f, \sigma_i, \mathbf{x}) \, U^\dagger(\sigma_i, \mathbf{x})$$

$$= \left(\frac{dU(\sigma_f, \mathbf{x})}{d\sigma_f} U^\dagger(\sigma_f, \mathbf{x}) + U(\sigma_f, \mathbf{x}) \, A_0(\sigma_f, \mathbf{x}) U^\dagger(\sigma_f, \mathbf{x}) \right)$$

$$\times U(\sigma_f, \mathbf{x}) L_A(\sigma_f, \sigma_i, \mathbf{x}) \, U^\dagger(\sigma_i, \mathbf{x})$$

$$= \left(-U(\sigma_f, \mathbf{x}) \frac{dU^\dagger(\sigma_f, \mathbf{x})}{d\sigma_f} + U(\sigma_f, \mathbf{x}) \, A_0(\sigma_f, \mathbf{x}) U^\dagger(\sigma_f, \mathbf{x}) \right)$$

$$\times U(\sigma_f, \mathbf{x}) L_A(\sigma_f, \sigma_i, \mathbf{x}) \, U^\dagger(\sigma_i, \mathbf{x})$$

$$= A_0^U(\sigma_f, \mathbf{x}) \, U(\sigma_f, \mathbf{x}) L_A(\sigma_f, \sigma_i, \mathbf{x}) \, U^\dagger(\sigma_i, \mathbf{x}).$$

It follows that $U(\sigma_f, \mathbf{x}) \, L_A(\sigma_f, \sigma_i, \mathbf{x}) \, U^\dagger(\sigma_i, \mathbf{x})$ obeys the same differential equation
as $L_{A^U}(\sigma_f, \sigma_i, \mathbf{x})$. Since the initial condition is also the same:

$$U(\sigma_i, \mathbf{x}) \, L_A(\sigma_i, \sigma_i, \mathbf{x}) \, U^\dagger(\sigma_i, \mathbf{x}) = U(\sigma_i, \mathbf{x}) \, U^\dagger(\sigma_i, \mathbf{x}) = \mathbb{1},$$

we conclude that $U(\sigma_f, \mathbf{x}) \, L_A(\sigma_f, \sigma_i, \mathbf{x}) \, U^\dagger(\sigma_i, \mathbf{x})$ and $L_{A^U}(\sigma_f, \sigma_i, \mathbf{x})$ coincide.

3.2 Periodic Boundary Conditions and Gauge Transformations (⋆⋆)

(a) That the periodicity conditions are preserved by the transformation U means that

$$U(\tau + \beta, \mathbf{x}) A_\mu(\tau + \beta, \mathbf{x}) U^\dagger(\tau + \beta, \mathbf{x}) - U(\tau + \beta, \mathbf{x}) \partial_\mu U^\dagger(\tau + \beta, \mathbf{x})$$

$$= U(\tau, \mathbf{x}) A_\mu(\tau, \mathbf{x}) U^\dagger(\tau, \mathbf{x}) - U(\tau, \mathbf{x}) \partial_\mu U^\dagger(\tau, \mathbf{x}).$$

Using $A_\mu(\tau + \beta, \mathbf{x}) = A_\mu(\tau, \mathbf{x})$ and the unitarity of $U(\tau + \beta, \mathbf{x})$, we rewrite this conveniently as

$$U(\tau + \beta, \mathbf{x})A_\mu(\tau, \mathbf{x})U^\dagger(\tau + \beta, \mathbf{x}) + (\partial_\mu U(\tau + \beta, \mathbf{x}))U^\dagger(\tau + \beta, \mathbf{x})$$
$$= U(\tau, \mathbf{x})A_\mu(\tau, \mathbf{x})U^\dagger(\tau, \mathbf{x}) - U(\tau, \mathbf{x})\partial_\mu U^\dagger(\tau, \mathbf{x}).$$

Then, upon left multiplication by $U^\dagger(\tau, \mathbf{x})$ and right multiplication by $U(\tau + \beta, \mathbf{x})$, this rewrites

$$Z(\tau, \mathbf{x})A_\mu(\tau + \beta, \mathbf{x}) + U^\dagger(\tau, \mathbf{x})\partial_\mu U(\tau + \beta, \mathbf{x})$$
$$= A_\mu(\tau, \mathbf{x})Z(\tau, \mathbf{x}) - (\partial_\mu U^\dagger(\tau, \mathbf{x}))U(\tau + \beta, \mathbf{x}),$$

which rewrites solely in terms of $Z(\tau, \mathbf{x})$ as $\partial_\mu Z - [A_\mu, Z] = 0$.

(b) Since the previous identity is valid for any periodic gauge field, it should be valid for $A_\mu = 0$. From this we deduce that Z is constant and that $[A_\mu, Z] = 0$ for any periodic gauge field configuration. In particular this means that Z should commute with any generator it^a of the algebra and, therefore, with any element of the group (which all take the form $\exp i\theta^a t^a$ in the case of SU(N)). In other words, Z is an element of the center of the group.

3.3 Center Symmetry Group (⋆)

(a) Let X be a matrix commuting with any special unitary matrix. It should then commute with any element of the corresponding Lie algebra, that is, any anti-Hermitian matrix of vanishing trace. Since it also commutes with the identity and since the latter together with the anti-Hermitian matrices generate (upon allowing for complex coefficients) all possible matrices of size N, we deduce that X should commute with any matrix of size N, in particular the matrices $M^{(h\ell)}$ of elements $M_{ij}^{(h\ell)} = \delta_{hi}\delta_{\ell j}$. This means that $X_{ij}M_{jk}^{(h\ell)} = X_{ih}\delta_{k\ell}$ should be equal to $M_{ij}^{(h\ell)}X_{jk} = \delta_{hi}X_{\ell k}$.

Choosing $k = \ell$ and $i \neq h$, we find that $X_{ih} = 0$ for $i \neq h$, while $k = \ell$ and $i = h$, implies that X_{ii} does not depend on i. It follows that X is proportional to the identity, $X = a\mathbb{1}$. Since X is a special unitary matrix, we should have $a^N = 1$ and then $a = e^{i2\pi k/N}$.

(b) If U_0 is periodic and U periodic modulo an element of the center, we have

$$U(\tau + \beta, \mathbf{x})U_0(\tau + \beta, \mathbf{x})U^\dagger(\tau + \beta, \mathbf{x}) = ZU(\tau, \mathbf{x})U_0(\tau, \mathbf{x})U^\dagger(\tau, \mathbf{x})Z^\dagger$$
$$= (ZZ^\dagger)U(\tau, \mathbf{x})U_0(\tau, \mathbf{x})U^\dagger(\tau, \mathbf{x})$$
$$= U(\tau, \mathbf{x})U_0(\tau, \mathbf{x})U^\dagger(\tau, \mathbf{x}),$$

so that $UU_0U^\dagger$ is periodic. In other words, $\mathcal{G}_0$ is a normal subgroup within $\mathcal{G}$, and therefore the quotient set $\mathcal{G}/\mathcal{G}_0$ inherits a group structure from that pre-existing within $\mathcal{G}$.

3.4 Decomposition of Unity ($\star$)

(a) We write

$$\int_{\mathcal{G}_0} \mathcal{D}U_0\, z[A^{U_0}] = \int_{\mathcal{G}_0} \mathcal{D}U_0\, \frac{\rho[A^{U_0}]}{\int_{\mathcal{G}_0} \mathcal{D}U_0'\, \rho[(A^{U_0})^{U_0'}]}$$

$$= \int_{\mathcal{G}_0} \mathcal{D}U_0\, \frac{\rho[A^{U_0}]}{\int_{\mathcal{G}_0} \mathcal{D}U_0'\, \rho[A^{U_0'U_0}]}\,.$$

The Haar measure being right-invariant, we can consider the change of variables $U_0' \to U_0'U_0^{\dagger}$ which does not change the measure and neither the integration domain. We then arrive at

$$\int_{\mathcal{G}_0} \mathcal{D}U_0\, z[A^{U_0}] = \int_{\mathcal{G}_0} \mathcal{D}U_0\, \frac{\rho[A^{U_0}]}{\int_{\mathcal{G}_0} \mathcal{D}U_0'\, \rho[A^{U_0'}]}$$

$$= \frac{\int_{\mathcal{G}_0} \mathcal{D}U_0\, \rho[A^{U_0}]}{\int_{\mathcal{G}_0} \mathcal{D}U_0'\, \rho[A^{U_0'}]} = 1\,.$$

(b) We write

$$\int_{\mathcal{G}_0} dU_0\, z_U[A^{U_0}] = \int_{\mathcal{G}_0} dU_0\, z[(A^{U_0})^U]$$

$$= \int_{\mathcal{G}_0} dU_0\, z[A^{UU_0}]$$

$$= \int_{\mathcal{G}_0} dU_0\, z[A^{UU_0U^{\dagger}U}]$$

$$= \int_{\mathcal{G}_0} dU_0\, z[(A^U)^{UU_0U^{\dagger}}]\,.$$

The benefit of this rewriting is that we can now consider the change of variables $U_0 \to U^{\dagger}U_0U$ that changes neither the integration domain nor the integration measure since we have assumed the Haar measure to be left- and right-invariant. We then arrive at

$$\int_{\mathcal{G}_0} dU_0\, z_U[A^{U_0}] = \int_{\mathcal{G}_0} dU_0\, z[(A^U)^{U_0}] = 1\,.$$

3.5 Conditional Gauge Fixing (★★★)

(a) If $\delta(F[A^{U_k}]) = \alpha[\mathcal{A}]\,\delta(F[A])$, then

$$z[A^{U_k}] = \frac{\delta(F[A^{U_k}])}{\int_{\mathcal{G}_0} \mathcal{D}U_0\,\delta(F[(A^{U_k})^{U_0}])}$$

$$= \frac{\delta(F[A^{U_k}])}{\int_{\mathcal{G}_0} \mathcal{D}U_0\,\delta(F[(A^{U_k^\dagger U_0 U_k})^{U_k}])}$$

$$= \frac{\alpha[\mathcal{A}]\delta(F[A])}{\alpha[\mathcal{A}]\int_{\mathcal{G}_0} \mathcal{D}U_0\,\delta(F[(A^{U_k^\dagger U_0 U_k})])}$$

$$= \frac{\delta(F[A])}{\int_{\mathcal{G}_0} \mathcal{D}U_0\,\delta(F[(A^{U_0})])} = z[A],$$

where we have used that $U_k^\dagger U_0 U_k \in \mathcal{G}_0$ and performed the change of variables $U_0 \to U_k U_0 U_k^\dagger$. Reciprocally, assuming $z[A^{U_k}] = z[A]$, we have

$$\frac{\delta(F[A^{U_k}])}{\int_{\mathcal{G}_0} \mathcal{D}U_0\,\delta(F[(A^{U_k})^{U_0}])} = \frac{\delta(F[A])}{\int_{\mathcal{G}_0} \mathcal{D}U_0\,\delta(F[A^{U_0}])},$$

and thus

$$\delta(F[A^{U_k}]) = \alpha[A]\,\delta(F[A^{U_0}]),$$

with

$$\alpha[A] \equiv \frac{\int_{\mathcal{G}_0} \mathcal{D}U_0\,\delta(F[(A^{U_k})^{U_0}])}{\int_{\mathcal{G}_0} \mathcal{D}U_0\,\delta(F[A^{U_0}])}.$$

We just need to show that $\alpha[A]$ depends only on the orbit of A. To this purpose, we write

$$\alpha[A^{U_0}] = \frac{\int_{\mathcal{G}_0} \mathcal{D}U_0'\,\delta(F[((A^{U_0})^{U_k})^{U_0'}])}{\int_{\mathcal{G}_0} \mathcal{D}U_0'\,\delta(F[(A^{U_0})^{U_0'}])}$$

$$= \frac{\int_{\mathcal{G}_0} \mathcal{D}U_0'\,\delta(F[A^{U_0'U_kU_0'}])}{\int_{\mathcal{G}_0} \mathcal{D}U_0'\,\delta(F[A^{U_0'U_0'}])}$$

$$= \frac{\int_{\mathcal{G}_0} \mathcal{D}U_0'\,\delta(F[A^{U_0'U_kU_0U_k^\dagger U_k}])}{\int_{\mathcal{G}_0} \mathcal{D}U_0'\,\delta(F[A^{U_0'U_0'}])}$$

$$= \frac{\int_{\mathcal{G}_0} \mathcal{D}U_0' \, \delta(F[A^{U_0'U_k}])}{\int_{\mathcal{G}_0} \mathcal{D}U_0' \, \delta(F[A^{U_0'}])} = \alpha[A] \,,$$

where we have used the change of variables $U_0' \to U_0' U_k U_0^\dagger U_k^\dagger$ in the numerator, and the change of variables $U_0' \to U_0' U_0^\dagger$ in the denominator.

(b) The support of the distributions $\delta(F[A^{U_k}])$ and $\delta(F[A])$ is, respectively, located on the solutions to $F[A^{U_k}] = 0$ and $F[A] = 0$. Thus for these distributions to coincide modulo a factor, their supports should be the same, and thus the set of solutions to $F[A] = 0$ should be invariant under U^k.

(c) We have

$$\delta(F[A^{U_0}]) = \sum_i \left| \det \frac{\delta F[A^{U_0}]}{\delta U_0} \bigg|_{U_0^{(i)}(A)} \right|^{-1} \delta(U_0 - U_0^{(i)}(A)) \,.$$

Since the set of solutions to $F[A] = 0$ is invariant under the U^k, we also have

$$\delta(F[(A^{U_0})^{U_k}]) = \sum_i \left| \det \frac{\delta F[(A^{U_0})^{U_k}]}{\delta U_0} \bigg|_{U_0^{(i)}(A)} \right|^{-1} \delta(U_0 - U_0^{(i)}(A)) \,.$$

Clearly, for $\delta(F[A^{U_k}]) = \alpha[\mathcal{A}] \, \delta(F[A])$ to hold true, we need that

$$\frac{\left| \det \frac{\delta F[(A^{U_0})^{U_k}]}{\delta U_0} \big|_{U_0^{(i)}(A)} \right|^{-1}}{\left| \det \frac{\delta F[A^{U_0}]}{\delta U_0} \big|_{U_0^{(i)}(A)} \right|^{-1}} = \alpha[\mathcal{A}]$$

which depends then only on the considered orbit and not on the particular copy.

(d) If the first condition holds true, we can decompose $\delta(F[A^{U_0}])$ and $\delta(F[(A^{U_0})^{U_k}])$ as in (c), and since the second condition holds true, the determinants are related by a copy-independent factor which then implies $\delta(F[A^{U_k}]) = \alpha[\mathcal{A}] \, \delta(F[A])$.

(e) Since there is only one solution to $F[A] = 0$ along each orbit, it is clear that the condition in (c) is always met. As for the condition in (b), we first note that, in the case with copies, this condition is not easily met because in general the various solutions to $F[A] = 0$ along two orbits connected by a center transformation of index k are connected to each other not by the same U_k but rather by U_k modulo transformations in $\mathcal{G}_0$. Now, in the particular case where there are no copies, this problem although still present is less prominent. Indeed, since there is only one solution to $F[A] = 0$ per orbit, one can define an orbit-dependent transformation that connects the solution along a given orbit to the solution along the transformed orbit.

(f) By construction these orbit-dependent transformations leave the support of $\delta(F[A])$ and thus $\delta(F[A^{U_k}]) = \alpha[\mathcal{A}] \delta(F[A])$ from which it follows that $\Delta[A] \delta(F[A])$

is invariant under these transformations. We note that, because these transformations are orbit-dependent, the associated Jacobian could be non-trivial. Indeed, if the gauge field is varied along the orbit, the transformation is not changed, and one recovers the standard Jacobian. In contrast, if the field is varied transversally to the orbits, the transformation is modified, which leads to new contributions in the evaluation of the Jacobian.

3.6 Effective Action (★★★)

(a) We write

$$\text{tr}\, J_\mu A_\mu^U = \text{tr}\, J_\mu U A_\mu U^\dagger - \text{tr}\, J_\mu U \partial_\mu U^\dagger$$
$$= \text{tr}\, U^\dagger J_\mu U A_\mu - \text{tr}\, J_\mu U \partial_\mu U^\dagger = \text{tr}\, J_\mu^{U^\dagger} A_\mu - \text{tr}\, J_\mu U \partial_\mu U^\dagger.$$

(b) We write

$$\ln \int \mathcal{D}A\, z[A]\, e^{-S_{YM}[A]-2\text{tr}\, J_\mu A_\mu} = \ln \int \mathcal{D}A^U\, z[A^U]\, e^{-S_{YM}[A^U]-2\text{tr}\, J_\mu A_\mu^U}$$
$$= 2\text{tr}\, J_\mu U \partial_\mu U^\dagger + \ln \int \mathcal{D}A\, z[A^U]\, e^{-S_{YM}[A]-2\text{tr}\, J_\mu^{U^\dagger} A_\mu}$$
$$= 2\text{tr}\, J_\mu U \partial_\mu U^\dagger + \ln \int \mathcal{D}A\, z_U[A]\, e^{-S_{YM}[A]-2\text{tr}\, J_\mu^{U^\dagger} A_\mu}.$$

from which it follows that $W_z[J] = W_{z_U}[J^{U^\dagger}] + 2\text{tr}\, J_\mu U \partial_\mu U^\dagger$.

(c) It follows that

$$A_z[J] \equiv -\frac{1}{2}\frac{\partial W_z[J]}{\partial J^t} = U A_{z_U}[J^{U^\dagger}]U^\dagger - U \partial U^\dagger = A_{z_U}^U[J^{U^\dagger}].$$

(d) Upon inversion of $A_z[J]$, one finds $J_z[A] = J_{z_U}^U[A^{U^\dagger}]$. We can check this by evaluating $A_z[J_{z_U}^U[A^{U^\dagger}]]$ and checking that it is equal to A. To this purpose, we use the previous result and write

$$A_z[J_{z_U}^U[A^{U^\dagger}]] = A_{z_U}^U[(J_{z_U}^U)^{U^\dagger}[A^{U^\dagger}]]$$
$$= A_{z_U}^U[J_{z_U}[A^{U^\dagger}]]$$
$$= (A^{U^\dagger})^U = A.$$

(e) By combining the previous results, we can finally write

$$
\begin{aligned}
\Gamma_z[A] &= -W_z[J_z[A]] - 2\mathrm{tr}\, J_{z,\mu}[A]A_\mu \\
&= -W_{z_U}[J_z^{U^\dagger}[A]] - 2\mathrm{tr}\, J_{z,\mu}[A]U\partial_\mu U^\dagger - 2\mathrm{tr}\, J_{z,\mu}[A]A_\mu \\
&= -W_{z_U}[J_{z_U}[A^{U^\dagger}]] - 2\mathrm{tr}\, J_{z_U,\mu}^U[A^{U^\dagger}]U\partial_\mu U^\dagger - 2\mathrm{tr}\, J_{z_U,\mu}^U[A^{U^\dagger}]A_\mu \\
&= -W_{z_U}[J_{z_U}[A^{U^\dagger}]] - 2\mathrm{tr}\, J_{z_U,\mu}[A^{U^\dagger}](\partial_\mu U^\dagger)U - 2\mathrm{tr}\, J_{z_U,\mu}[A^{U^\dagger}]U^\dagger A_\mu U \\
&= -W_{z_U}[J_{z_U}[A^{U^\dagger}]] + 2\mathrm{tr}\, J_{z_U,\mu}[A^{U^\dagger}]U^\dagger \partial_\mu U - 2\mathrm{tr}\, J_{z_U,\mu}[A^{U^\dagger}]U^\dagger A_\mu U \\
&= -W_{z_U}[J_{z_U}[A^{U^\dagger}]] - 2\mathrm{tr}\, J_{z_U,\mu}[A^{U^\dagger}]A_\mu^{U^\dagger} ,
\end{aligned}
$$

that is, $\Gamma_z[A] = \Gamma_{z_U}[A^{U^\dagger}]$.

Problems of Chap. 4

4.1 Background Gauge Symmetry (★★)

This problem can be seen as a particular case of the last problem of the previous chapter.

(a) Performing a change of variables under the functional integral defining $\hat{W}[J^U, \bar{A}^U]$, in the form $A \to A^U$, we find

$$
\hat{W}_{\bar{A}^U}[J^U] \equiv \ln \int \mathcal{D}_{\mathrm{gf}}[A^U; \bar{A}^U] \exp\left\{ -S_{YM}[A^U] + \int d^d x \; (J_\mu^U; a_\mu^U) \right\},
$$

where $a_\mu^U \equiv U a_\mu U^\dagger$. Using $S_{YM}[A^U] = S_{YM}[A]$, $\mathcal{D}_{\mathrm{gf}}[A^U; \bar{A}^U] = \mathcal{D}_{\mathrm{gf}}[A; A]$, and $(J_\mu^U; a_\mu^U) = (J_\mu; a_\mu)$, we arrive at

$$
\hat{W}_{\bar{A}^U}[J^U] \equiv \ln \int \mathcal{D}_{\mathrm{gf}}[A; \bar{A}] \exp\left\{ -S_{YM}[A] + \int d^d x \; (J_\mu; a_\mu) \right\} = \hat{W}[J; \bar{A}].
$$

The relevance of restricting to $U \in \mathcal{G}$ is that only in this case the integration domain is preserved by the considered change of variables, and we can identify the various functional integrals.

(b) We have

$$
a_{\bar{A}}[J] = -2\frac{\delta \hat{W}_{\bar{A}}[J]}{\delta J^\iota} = -2\frac{\delta \hat{W}_{\bar{A}^U}[J^U]}{\delta J^\iota} = -2U^\dagger \left.\frac{\delta \hat{W}_{\bar{A}^U}[J]}{\delta J^\iota}\right|_{J^U} U = U^\dagger a_{\bar{A}^U}[J^U] U ,
$$

and thus $a_{\bar{A}^U}[J^U] = U a_{\bar{A}}[J] U^\dagger = a_{\bar{A}}^U[J]$.

(c) We just need to verify that $J_{\bar{A}}^U[a^{U^\dagger}]$ is the functional inverse of $a_{\bar{A}^U}[J]$. To this purpose, we write

$$a_{\bar{A}U}[J_{\bar{A}}^U[a^{U^\dagger}]] = a_{\bar{A}}^U[J_{\bar{A}}[a^{U^\dagger}]] = (a^{U^\dagger})^U = a\,.$$

4.2 Charge Conjugation Invariance and Polyakov Loop (⋆)
We have

$$\bar{\ell} = \langle \bar{\Phi}_A \rangle = \langle \Phi_{-A^t} \rangle = \langle \Phi_{A^C} \rangle = \langle \Phi_A \rangle = \ell\,.$$

4.3 Convexity and Effective Action (⋆⋆)
(a) That $W_{\bar{A}}[J]$ lies always above any of its tangential planes writes

$$W_{\bar{A}}[J] \geq W_{\bar{A}}[J_0] + \int d^d x \left(\left. \frac{\delta W}{\delta J(x)} \right|_{J_0} ; J(x) - J_0(x) \right).$$

(b) From this, we deduce that

$$\Gamma_{\bar{A}}\left[\frac{\delta W_{\bar{A}}}{\delta J} \right] \equiv -W_{\bar{A}}[J] + \int d^d x \left(\frac{\delta W_{\bar{A}}}{\delta J(x)} ; J(x) \right)$$

$$\leq -W_{\bar{A}}[J_0] + \int d^d x \left(\left. \frac{\delta W_{\bar{A}}}{\delta J(x)} \right|_{J_0} ; J_0(x) \right)$$

$$+ \int d^d x \left(\frac{\delta W_{\bar{A}}}{\delta J(x)} - \left. \frac{\delta W_{\bar{A}}}{\delta J(x)} \right|_{J_0} ; J(x) \right),$$

that is,

$$\Gamma_{\bar{A}}\left[\frac{\delta W_{\bar{A}}}{\delta J} \right] \leq \Gamma_{\bar{A}}\left[\left. \frac{\delta W_{\bar{A}}}{\delta J} \right|_{J_0} \right] + \int d^d x \left(\frac{\delta W_{\bar{A}}}{\delta J(x)} - \left. \frac{\delta W_{\bar{A}}}{\delta J(x)} \right|_{J_0} ; J(x) \right).$$

4.4 Positivity of the Functional Integral Measure and Convexity (⋆⋆)
(a) We can then apply the previous inequality to the functions $h^{1/p} f$ et $h^{1/q} g$. It follows that hfg is integrable (this uses the assumption $1/p + 1/q = 1$) and

$$\int h|fg| \leq \left(\int h|f|^p \right)^{1/p} \left(\int h|g|^q \right)^{1/q}.$$

(b) We can apply the previous result with

$$\mathcal{D}_+ A = \mathcal{D}_{\text{gf}}[A; \bar{A}] e^{-S_{YM}[A]},$$

and

$$f[A] = e^{\alpha_1 J_1 \cdot A}, \quad g[A] = e^{\alpha_2 J_2 \cdot A},$$

as well as $p = 1/\alpha_1$ and $q = 1/\alpha_2$. We find

$$e^{W_{\bar{A}}[\alpha_1 J_1 + \alpha_2 J_2]} = \int \mathcal{D}_{\text{gf}}[A; \bar{A}] e^{-S_{YM}[A]} e^{\alpha_1 J_1 \cdot A} e^{\alpha_2 J_2 \cdot A}$$

$$\leq \underbrace{\left(\int \mathcal{D}_{\text{gf}}[A; \bar{A}] e^{-S_{YM}[A]} e^{J_1 \cdot A} \right)^{\alpha_1}}_{\left(e^{W_{\bar{A}}[J_1]} \right)^{\alpha_1}} \underbrace{\left(\int \mathcal{D}_{\text{gf}}[A; \bar{A}] e^{-S_{YM}[A]} e^{J_2 \cdot A} \right)^{\alpha_2}}_{\left(e^{W_{\bar{A}}[J_2]} \right)^{\alpha_2}}.$$

This implies

$$W_{\bar{A}}[\alpha_1 J_1 + \alpha_2 J_2] \leq \alpha_1 W_{\bar{A}}[J_1] + \alpha_2 W_{\bar{A}}[J_2],$$

which expresses the convexity of $W_{\bar{A}}[J]$ with respect to J.

Problems of Chap. 5

5.1 Restricted Twisted Gauge Transformations (⋆)
(a) A restricted twisted transformation $U(x)$ should be such that

$$U(x) i r_j H_j \delta_{\mu 0} U^\dagger(x) - U(x) \beta \partial_\mu U^\dagger(x) = i s_j H_j \delta_{\mu 0},$$

for any choice of r and for a certain s that depends on r. Choosing $\mu = i$, we deduce that $\partial_i U = 0$ and thus that U can only depend on the Euclidean time τ. Moreover, choosing $\mu = 0$ and $r = 0$, we find that

$$-U(\tau) \beta \partial_\tau U^\dagger(\tau) = i s_j H_j,$$

for a certain s or, in other words

$$\beta \frac{dU(\tau)}{d\tau} = i s_j H_j U(\tau).$$

(b) The previous ordinary differential equation integrates to

$$U(\tau) = e^{i \frac{\tau}{\beta} s_j H_j} W,$$

for some SU(N) color rotation W. Since both U and $e^{i \frac{\tau}{\beta} s_j H_j}$ leave the Cartan sub-algebra invariant, then $W = e^{-i \frac{\tau}{\beta} s_j H_j} U$ has to leave the Cartan sub-algebra globally invariant.

5.2 SU(3) Root Diagram (⋆)
The Gell-Mann matrices write

$$\lambda_1 = \begin{pmatrix} 0 & 1 & 0 \\ 1 & 0 & 0 \\ 0 & 0 & 0 \end{pmatrix}, \quad \lambda_2 = \begin{pmatrix} 0 & -i & 0 \\ i & 0 & 0 \\ 0 & 0 & 0 \end{pmatrix}, \quad \lambda_3 = \begin{pmatrix} 1 & 0 & 0 \\ 0 & -1 & 0 \\ 0 & 0 & 0 \end{pmatrix},$$

$$\lambda_4 = \begin{pmatrix} 0 & 0 & 1 \\ 0 & 0 & 0 \\ 1 & 0 & 0 \end{pmatrix}, \quad \lambda_5 = \begin{pmatrix} 0 & 0 & -i \\ 0 & 0 & 0 \\ i & 0 & 0 \end{pmatrix},$$

$$\lambda_6 = \begin{pmatrix} 0 & 0 & 0 \\ 0 & 0 & 1 \\ 0 & 1 & 0 \end{pmatrix}, \quad \lambda_7 = \begin{pmatrix} 0 & 0 & 0 \\ 0 & 0 & -i \\ 0 & i & 0 \end{pmatrix}, \quad \lambda_8 = \frac{1}{\sqrt{3}} \begin{pmatrix} 1 & 0 & 0 \\ 0 & 1 & 0 \\ 0 & 0 & -2 \end{pmatrix}.$$

From looking at the first line, one recognizes the Pauli sigma matrices in the upper 2×2 block. This suggest introducing $\lambda_1 \pm i\lambda_2$ which diagonalize the adjoint action of λ_3 with respective charges ± 1. Moreover, since the upper 2×2 block of λ_8 is the identity matrix, they also diagonalize the adjoint action of λ_8 with both a vanishing charge.

One can proceed similarly with λ_4 and λ_5 by noting that $\lambda_4 \pm i\lambda_5$ diagonalize the adjoint action of $\lambda_3/2 + \sqrt{3}\lambda_8/2$ and $3\lambda_3/2 - \sqrt{3}\lambda_8/2$ with respective charges ± 1 and 0. In other words, $\lambda_4 \pm i\lambda_5$ diagonalize the adjoint actions of λ_3 and λ_8 with respective charges $\pm 1/2$ and $\pm\sqrt{3}/2$. Finally, $\lambda_6 \pm i\lambda_7$ diagonalize the adjoint action of $-\lambda_3/2 + \sqrt{3}\lambda_8/2$ and $-3\lambda_3/2 - \sqrt{3}\lambda_8/2$ with respective charges ± 1 and 0. In other words, $\lambda_6 \pm i\lambda_7$ diagonalize the adjoint actions of λ_3 and λ_8 with respective charges $\mp 1/2$ and $\pm\sqrt{3}/2$.

Introducing $\alpha^{(1)} = (1/2, -\sqrt{3}/2)$, $\alpha^{(2)} = (1/2, \sqrt{3}/2)$, and $\alpha^{(3)} = (-1, 0)$, as well as $\lambda_{\pm\alpha^{(1)}} = \lambda_4 \mp i\lambda_5$, $\lambda_{\pm\alpha^{(2)}} = \lambda_6 \mp i\lambda_7$, and $\lambda_{\pm\alpha^{(3)}} = \lambda_1 \mp i\lambda_2$, we verify that $[\lambda_j, \lambda_\alpha] = \alpha_j \lambda_\alpha$ with $j = 3, 8$.

5.3 Winding Transformations (⋆)

We first write

$$e^{i\frac{\tau}{\beta}s_j H_j} A_\mu e^{-i\frac{\tau}{\beta}s_j H_j} = e^{i\frac{\tau}{\beta}s_j H_j} i A_\mu^{j'} H_{j'} e^{-i\frac{\tau}{\beta}s_j H_j} + e^{i\frac{\tau}{\beta}s_j H_j} i A_\mu^\alpha E_\alpha e^{-i\frac{\tau}{\beta}s_j H_j}$$

$$= A_\mu^{j'} H_{j'} + e^{i\frac{\tau}{\beta}s_j H_j} i A_\mu^\alpha E_\alpha e^{-i\frac{\tau}{\beta}s_j H_j}.$$

Next, we write

$$\frac{d}{d\tau}e^{i\frac{\tau}{\beta}s_jH_j}A_\mu^\alpha E_\alpha e^{-i\frac{\tau}{\beta}s_jH_j} = \frac{i}{\beta}e^{i\frac{\tau}{\beta}s_jH_j}[s_jH_j, A_\mu^\alpha E_\alpha]e^{-i\frac{\tau}{\beta}s_jH_j}$$

$$= \frac{i}{\beta}e^{i\frac{\tau}{\beta}s_jH_j}s_jA_\mu^\alpha[H_j, E_\alpha]e^{-i\frac{\tau}{\beta}s_jH_j}$$

$$= \frac{i}{\beta}e^{i\frac{\tau}{\beta}s_jH_j}s_j\alpha_j A_\mu^\alpha E_\alpha e^{-i\frac{\tau}{\beta}s_jH_j}$$

$$= \frac{i}{\beta}s_j\alpha_j e^{i\frac{\tau}{\beta}s_jH_j}A_\mu^\alpha E_\alpha e^{-i\frac{\tau}{\beta}s_jH_j}$$

and thus

$$e^{i\frac{\tau}{\beta}s_jH_j}A_\mu^\alpha E_\alpha e^{-i\frac{\tau}{\beta}s_jH_j} = e^{i\frac{\tau}{\beta}s_j\alpha_j}A_\mu^\alpha E_\alpha .$$

5.4 Weyl Rotations (★★)
(a) We have

$$\text{ad}_{w_\alpha}(H_j) = i\theta_\alpha[E_\alpha, H_j] + i\theta_{-\alpha}[E_{-\alpha}, H_j]$$

$$= -i(\theta_\alpha E_\alpha - \theta_\alpha^* E_{-\alpha})\alpha_j ,$$

and then

$$\text{ad}_{w_\alpha}^2(H_j) = [\theta_\alpha E_\alpha + \theta_\alpha^* E_{-\alpha}, \theta_\alpha E_\alpha - \theta_\alpha^* E_{-\alpha}]\alpha_j$$

$$= -2|\theta_\alpha|^2[E_\alpha, E_{-\alpha}]\alpha_j = -2|\theta_\alpha|^2(\alpha_k H_k)\alpha_j ,$$

which we rewrite for convenience as

$$\text{ad}_{w_\alpha}^2(\alpha_k H_k) = -2|\theta_\alpha|^2\alpha^2(\alpha_k H_k) .$$

Subsequent iterations of ad_{w_α} will alternate between $\theta_\alpha E_\alpha - \theta_\alpha^* E_{-\alpha}$ and $\alpha_k H_k$. This suggests to also look into

$$\text{ad}_{w_\alpha}^2(\theta_\alpha E_\alpha - \theta_\alpha^* E_{-\alpha}) = \frac{i}{\alpha^2}\text{ad}_{w_\alpha}^3(\alpha_k H_k)$$

$$= -2i|\theta_\alpha|^2\text{ad}_{w_\alpha}(\alpha_k H_k)$$

$$= -2|\theta_\alpha|^2\alpha^2(\theta_\alpha E_\alpha - \theta_\alpha^* E_{-\alpha}) .$$

We can now apply these formulas repeatedly to the first and second iterations of ad_{w_α} on H_j, as given above. We find

$$\text{ad}_{w_\alpha}^{2p+1}(H_j) = -i\left(-2|\theta_\alpha|^2\alpha^2\right)^p(\theta_\alpha E_\alpha - \theta_{-\alpha}E_{-\alpha})\alpha_j ,$$

for $p \geq 0$, and

$$\mathrm{ad}_{w_\alpha}^{2p}(H_j) = \left(-2|\theta_\alpha|^2\alpha^2\right)^p \frac{\alpha_k H_k}{\alpha^2}\alpha_j\,,$$

for $p > 0$.

(b) We have

$$
\begin{aligned}
e^{\mathrm{ad}_{w_\alpha}} H_j &= \left(1 + \sum_{p=1}^{\infty} \frac{\mathrm{ad}_{w_\alpha}^{2p}}{(2p)!} + \sum_{p=0}^{\infty} \frac{\mathrm{ad}_{w_\alpha}^{2p+1}}{(2p+1)!}\right) H_j \\
&= H_j + \sum_{p=1}^{\infty} \frac{\left(-2|\theta_\alpha|^2\alpha^2\right)^p}{(2p)!} \frac{\alpha_k H_k}{\alpha^2}\alpha_j \\
&\quad -i\sum_{p=0}^{\infty} \frac{\left(-2|\theta_\alpha|^2\alpha^2\right)^p}{(2p+1)!}(\theta_\alpha E_\alpha - \theta_{-\alpha}E_{-\alpha})\,\alpha_j \\
&= H_j + \left(\cos\left(\sqrt{2\alpha^2}|\theta_\alpha|\right) - 1\right)\frac{\alpha_k H_k}{\alpha^2}\alpha_j \\
&\quad - \frac{i\alpha_j}{\sqrt{2\alpha^2}|\theta_\alpha|}\sin\left(\sqrt{2\alpha^2}|\theta_\alpha|\right)(\theta_\alpha E_\alpha - \theta_{-\alpha}E_{-\alpha})\,.
\end{aligned}
$$

5.5 SU(2) Charge Conjugation as a Weyl Rotation (⋆)

Consider $i\sigma_2$. We have $(i\sigma_2)(i\sigma_2)^\dagger = i(-i)\sigma_2\sigma_2^\dagger = \sigma_2^2 = 1$. Thus $i\sigma_2$ is unitary. Moreover $\sigma_2\sigma_{1/3}\sigma_2 = -\sigma_{1/3}$, while $\sigma_2\sigma_2\sigma_2 = \sigma_2$.

5.6 Homogeneity and Isotropy Modulo Gauge Transformations (⋆⋆⋆)

(a) Assume that $\bar{A}_\mu(x)$ fulfils the condition (Hom), and consider transforming this configuration by $U_0 \in \mathcal{G}_0$. We can write

$$
\begin{aligned}
\bar{A}_\mu^{U_0}(x+u) &= U_0(x+u)\,\bar{A}_\mu(x+u)\,U_0^\dagger(x+u) - U_0(x+u)\,\partial_\mu U_0^\dagger(x+u) \\
&= U_0(x+u)U_u(x)\,\bar{A}_\mu(x)\,U_u^\dagger(x)U_0^\dagger(x+u) \\
&\quad - U_0(x+u)\,(U_u(x)\,\partial_\mu U_u^\dagger(x))\,U_0^\dagger(x+u) - U(x+u)\,\partial_\mu U_0^\dagger(x+u) \\
&= U_0(x+u)U_u(x)\,\bar{A}_\mu(x)\,U_u^\dagger(x)U^\dagger(x+u) \\
&\quad - U_0(x+u)U_u(x)\,\partial_\mu(U_u^\dagger(x)U_0^\dagger(x+u)) \\
&= \bar{A}_\mu^{V_0}(x) = (\bar{A}_\mu^{U_0})^{V_0 U_0^{-1}}(x)\,,
\end{aligned}
$$

with $V_0(x) \equiv U_0(x+u)U_u(x) \in \mathcal{G}_0$. We can treat similarly the case of the spatial rotations. It follows that, in order to find configurations obeying the properties (Hom) and (Iso), it is enough to look for configurations in a given gauge. We shall consider a convenient choice of gauge below.

(b) We have

$$\forall x, u \in [0, \beta] \times \mathbb{R}^3, \ \exists U_u(x) \in \mathcal{G}_0, \ \bar{F}_{\mu\nu}(x+u) = U_u(x)\, \bar{F}_{\mu\nu}(x)\, U_u^\dagger(x).$$

In particular, choosing $x = 0$ and $u \to x$, we find that

$$\forall x \in [0, \beta] \times \mathbb{R}^3, \ \exists U(x) \in \mathcal{G}_0, \ \bar{F}_{\mu\nu}(x) = U(x)\, \bar{F}_{\mu\nu}(0)\, U^\dagger(x).$$

It is not clear at this point whether $U(x) \in \mathcal{G}_0$.
(c) From $\bar{F}_{\mu\nu}(\tau + \beta, \mathbf{x}) = \bar{F}_{\mu\nu}(\tau, \mathbf{x})$, we deduce that

$$U(\tau + \beta, \mathbf{x})\, \bar{F}_{\mu\nu}(0)\, U^\dagger(\tau + \beta, \mathbf{x}) = U(\tau, \mathbf{x})\, \bar{F}_{\mu\nu}(0)\, U^\dagger(\tau, \mathbf{x}),$$

that is,

$$U^\dagger(\tau, \mathbf{x})U(\tau + \beta, \mathbf{x})\, \bar{F}_{\mu\nu}(0)\, U^\dagger(\tau + \beta, \mathbf{x})U(\tau, \mathbf{x}) = \bar{F}_{\mu\nu}(0).$$

In the SU(2) case, the left-hand side corresponds to an SO(3) color rotation of $\bar{F}_{\mu\nu}^a(0)$, and the identity expresses the invariance of the six color vectors ($0 \le \mu < \nu \le 3$) under this rotation. Assuming that at least two of these vectors point in different color directions, the only possibility is that the rotation in question is the identity, thus implying that $U(\tau + \beta, \mathbf{x}) = U(\tau, \mathbf{x})$. In this case, which we assume later on, $U(x) \in \mathcal{G}_0$, and it can then be gauged away. In what follows we then assume that $\bar{F}_{\mu\nu}(x) = \bar{F}_{\mu\nu}(0) \equiv \bar{F}_{\mu\nu}$.
(d) We have

$$\forall x \in [0, \beta] \times \mathbb{R}^3, \ \forall R \in SO(3), \ \exists U_R(x) \in \mathcal{G}_0,$$

$$\begin{cases} R_{ij}\bar{F}_{j0} = U_R(x)\, \bar{F}_{i0}\, U_R^\dagger(x) \\ R_{ij}R_{kl}\bar{F}_{jl} = U_R(x)\, \bar{F}_{ik}\, U_R^\dagger(x) \end{cases},$$

where we have already used the result derived in the previous question. Since the right-hand side is an SO(3) color rotation, we can also write

$$\forall x \in [0, \beta] \times \mathbb{R}^3, \ \forall R \in SO(3), \ \exists \mathcal{R}(R, x) \in SO(3),$$

$$\begin{cases} R_{ij}\bar{F}_{j0}^a = \mathcal{R}_{ab}^{-1}(R, x)\bar{F}_{i0}^b \\ R_{ij}R_{kl}\bar{F}_{jl}^a = \mathcal{R}_{ab}^{-1}(R, x)\bar{F}_{ik}^b \end{cases}.$$

Combining two rotations R and S, we find

$$S_{ij}R_{jk}\bar{F}_{k0}^a(R^{-1}S^{-1}x) = \mathcal{R}_{ab}^{-1}(SR, x)\bar{F}_{i0}^b(x),$$

but also

$$S_{ij} R_{jk} \bar{F}^a_{k0}(R^{-1}S^{-1}x) = R^{-1}_{ab}(R, S^{-1}x) S_{ij} \bar{F}^b_{j0}(S^{-1}x)$$
$$= R^{-1}_{ab}(R, S^{-1}x) R^{-1}_{bc}(S, x) \bar{F}^c_{i0}(x).$$

Applying the same argument to the magnetic sector, we arrive at

$$R^{-1}_{ab}(SR, x) \bar{F}^b_{\mu\nu}(x) = R^{-1}_{ab}(R, S^{-1}x) R^{-1}_{bc}(S, x) \bar{F}^c_{\mu\nu}(x).$$

In particular, for $x = 0$, we find

$$R^{-1}_{ab}(SR, 0) \bar{F}^b_{\mu\nu}(0) = R^{-1}_{ab}(R, 0) R^{-1}_{bc}(S, 0) \bar{F}^c_{\mu\nu}(0).$$

Under the same assumption as above concerning the existence of at least two independent color vectors among the $\bar{F}^a_{\mu\nu}(0)$ (see the next Problem for the other cases), this implies

$$R_{ab}(SR, 0) = R_{ab}(S, 0) R_{bc}(R, 0),$$

which means that $R \to R(R, 0)$ is a (real) dimension 3 representation of SO(3). This representation can be either made of the trivial representation $R \to 1$ or, upon an appropriate choice of basis, be the defining representation $R \to R$. In the first case, we should have, $\forall R$,

$$R_{ij} \bar{F}^a_{j0} = \bar{F}^a_{i0} \quad \text{and} \quad R_{ij} R_{kl} \bar{F}^a_{jl} = \bar{F}^a_{ij},$$

but this implies that $\bar{F}^a_{i0} = 0$ and $\bar{F}^a_{ij} \propto \delta_{ij}$ and thus $\bar{F}^a_{ij} = 0$. This does not fit the assumption made earlier about the existence of at least two independent color direction, and we shall treat it in the next problem. In the second case, we should have, $\forall R$,

$$R_{ij} \bar{F}^a_{j0} = (R^{-1})^{ab} \bar{F}^b_{i0} \quad \text{and} \quad R_{ij} R_{kl} \bar{F}^a_{jl} = (R^{-1})^{ab} \bar{F}^b_{ij}.$$

In other words, $\forall R$,

$$R_{ij} R^{ab} \bar{F}^b_{j0} = \bar{F}^a_{i0} \quad \text{and} \quad R_{ij} R_{kl} R^{ab} \bar{F}^b_{jl} = \bar{F}^a_{ij},$$

from which it follows that $\bar{F}^a_{i0} = \alpha \delta_{ia}$ and $\bar{F}^a_{ij} = \beta \varepsilon_{aij}$. We note that α and β cannot vanish simultaneously for otherwise there would not be two independent color directions.

(f) The Bianchi identities read $D_\mu \bar{F}_{\nu\rho} + D_\nu \bar{F}_{\rho\mu} + D_\rho \bar{F}_{\mu\nu} = 0$. Since $\bar{F}_{\mu\nu}$ is constant, only the non-abelian terms survive

$$0 = \varepsilon_{abc}(\beta \bar{A}_0^b \varepsilon_{ijc} + \alpha \bar{A}_i^b \delta_{cj} - \alpha \bar{A}_j^b \delta_{ci}),$$

$$0 = \beta \varepsilon_{abc}(\bar{A}_i^b \varepsilon_{jkc} + \bar{A}_j^b \varepsilon_{kic} + \bar{A}_k^b \varepsilon_{ijc}).$$

The first identity rewrites

$$0 = \beta \bar{A}_0^b (\delta_{ai}\delta_{bj} - \delta_{aj}\delta_{bi}) + \alpha \bar{A}_i^b \varepsilon_{abj} - \alpha \bar{A}_j^b \varepsilon_{abi}.$$

If $i = j$, there is no information. If $i \neq j$, we choose $a = i$ and $b = j$ and find

$$0 = \beta \bar{A}_0^b.$$

If we choose $a = i$ and $b \neq j$, we find

$$\forall i \neq j, \ 0 = \alpha \bar{A}_i^j.$$

Finally, if we choose $b = i$ and $a \neq j$, we find

$$\forall i, \ 0 = \alpha \bar{A}_i^i.$$

The second identity rewrites

$$0 = \beta ((\bar{A}_i^k - \bar{A}_k^i)\delta_{aj} + (\bar{A}_j^i - \bar{A}_i^j)\delta_{ak} + (\bar{A}_k^j - \bar{A}_j^k)\delta_{ai}),$$

that is,

$$\forall i \neq j, \ 0 = \beta (\bar{A}_i^j - \bar{A}_j^i).$$

(g) As already mentioned, the case $\alpha = \beta = 0$ will be treated below. In the case $\alpha \neq 0$ and $\beta = 0$, we have $\bar{A}_i^a = 0$. It follows that $\bar{F}_{i0}^a = \partial_i \bar{A}_0^a = \alpha \delta_i^a$ and then that $\bar{A}_0^a(x) = \alpha x^a + \gamma^a(\tau)$. Because $\alpha \neq 0$, these configurations cannot be in the same orbits as the constant temporal backgrounds. Moreover, they obey the property (Iso). However, it is easy to convince oneself that they do not obey the property (Hom). To see this, let us consider an infinitesimal spatial translation. The property (Hom) would read

$$\alpha u^a = \partial_0 \theta^a + \varepsilon^{abc}\theta^b(\alpha x^c + \gamma^c(\tau)),$$

$$0 = \partial_i \theta^a.$$

Taking a derivative with respect to x_i in the first equation, and using the second, we find

$$0 = \alpha \varepsilon^{abi} \theta^b,$$

which implies in any case $\alpha = 0$ (since $\theta = 0$ also implies $\alpha = 0$). Let us mention also that one can use the constraints from charge conjugation (in the pure YM case) and parity invariance. In the SU(2) case, charge conjugation does not impose any constraint because it is tantamount to a color rotation (Weyl transformation). On the other hand, it is easily checked that parity invariance (modulo color rotations) implies that $\alpha = 0$.
(h) In the case where $\alpha = 0$ and $\beta \neq 0$, the previous equations imply that $\bar{A}_0^a = 0$ and $\bar{A}_i^a = \bar{A}_a^i$ with $\bar{A}_i^a$ not depending on τ and obeying the equation

$$\beta \varepsilon_{aij} = \partial_i \bar{A}_j^a - \partial_j \bar{A}_i^a + \varepsilon^{abc} \bar{A}_i^b \bar{A}_j^c.$$

A particular solution is provided by $\bar{A}_i^a = \sqrt{\beta} \delta_i^a$. It trivially obeys (Hom), and it also obeys (Iso) because, due to the locking of space and color, a spatial color rotation can be absorbed into a color rotation:

$$R_{ij} \bar{A}_j^a = \sqrt{\beta} R_{ij} \delta_j^a = \sqrt{\beta} R_{ia} = \sqrt{\beta} R^{ab} \delta_i^b = R^{ab} \bar{A}_i^b.$$

5.7 Degenerate Case (★★★)
We now investigate the case where the Lorentz components $\bar{F}_{\mu\nu}(0)$ are all aligned along the same color direction, that is, $\bar{F}_{\mu\nu}^a(0) = \bar{f}_{\mu\nu} n^a$.
(a) We have, for any R,

$$R_{ij} \bar{f}_{j0} n^a = \mathcal{R}_{ab}^{-1}(R, 0) \bar{f}_{i0} n^b \quad \text{and} \quad R_{ij} R_{kl} \bar{f}_{jl} n^a = \mathcal{R}_{ab}^{-1}(R, 0) \bar{f}_{ik} n^b.$$

This implies that, for any R,

$$R_{ij} \bar{f}_{j0} = \bar{f}_{i0}, \quad R_{ij} R_{kl} \bar{f}_{jl} = \bar{f}_{ik} \quad \text{and} \quad \mathcal{R}_{ab}(R, 0) n^b = n^a,$$

which is only possible if $\bar{f}_{\mu\nu} = 0$.
(b) We have, for instance,

$$\begin{aligned}
\bar{A}_\mu(x+u) &= -U(x+u) \, \partial_\mu U^\dagger(x+u) \\
&= -U(x+u) \, U^\dagger(x) \, U(x) \, \partial_\mu (U^\dagger(x) \, U(x) \, U^\dagger(x+u)) \\
&= -U(x+u) \, U^\dagger(x) \, [U(x) \, \partial_\mu U^\dagger(x)] \, U(x) \, U^\dagger(x+u) \\
&\quad - U(x+u) \, U^\dagger(x) \, \partial_\mu (U(x) \, U^\dagger(x+u)) \\
&= A^{V_u}(x),
\end{aligned}$$

with $V_u(x) \equiv U(x+u) U^\dagger(x)$. One can treat rotations in the same way, with $V_R(x) \equiv U(R^{-1}x) U^\dagger(x)$.

(c) We write

$$\forall u, \; U(\tau + \beta + u_0, \mathbf{x} + \mathbf{u}) U^\dagger(\tau + \beta, \mathbf{x}) = U(\tau + u_0, \mathbf{x} + \mathbf{u}) U^\dagger(\tau, \mathbf{x}),$$

$$\forall R, \quad U(\tau + \beta, R^{-1}\mathbf{x}) U^\dagger(\tau + \beta, \mathbf{x}) = U(\tau, R^{-1}\mathbf{x}) U^\dagger(\tau, \mathbf{x}),$$

or, equivalently,

$$\forall u, \; U^\dagger(\tau + \beta, \mathbf{x}) U(\tau, \mathbf{x}) = U^\dagger(\tau + \beta + u_0, \mathbf{x} + \mathbf{u}) U(\tau + u_0, \mathbf{x} + \mathbf{u}),$$

$$\forall R, \; U^\dagger(\tau + \beta, \mathbf{x}) U(\tau, \mathbf{x}) = U^\dagger(\tau + \beta, R^{-1}\mathbf{x}) U(\tau, R^{-1}\mathbf{x}).$$

It follows that

$$U^\dagger(\tau + \beta, \mathbf{x}) U(\tau, \mathbf{x}) = V,$$

for a certain $V \in SU(N)$.

(d) We write

$$\begin{aligned} W(\tau + \beta, \mathbf{x}) &= U(\tau + \beta, \mathbf{x}) M e^{-ir_j H_j} e^{-i\frac{\tau}{\beta}r_j H_j} \\ &= U(\tau, \mathbf{x}) V^\dagger M e^{-ir_j H_j} e^{-i\frac{\tau}{\beta}r_j H_j} \\ &= U(\tau, \mathbf{x}) M M^\dagger V^\dagger M e^{-ir_j H_j} e^{-i\frac{\tau}{\beta}r_j H_j} \\ &= U(\tau, \mathbf{x}) M e^{-i\frac{\tau}{\beta}r_j H_j} = W(\tau, \mathbf{x}). \end{aligned}$$

Moreover

$$W e^{i\frac{\tau}{\beta}r_j H_j} \partial_\mu (W e^{i\frac{\tau}{\beta}r_j H_j})^\dagger = U M \partial_\mu (U M)^\dagger = U \partial_\mu U^\dagger = -\bar{A}_\mu.$$

In other words

$$\beta \bar{A}_\mu^{W^\dagger} = -e^{i\frac{\tau}{\beta}r_j H_j} \beta \, \partial_\mu e^{-i\frac{\tau}{\beta}r_j H_j} = ir_j H_j \delta_{\mu 0}.$$

We have thus shown that the most general periodic pure gauge configurations compatible with the properties (Hom) and (Iso) belong to the same $\mathcal{G}_0$-orbits as constant temporal backgrounds.

Problems of Chap. 6

6.1 Field-Strength Tensor and Background Covariant Derivatives (⋆)
We write

$$
\begin{aligned}
F^a_{\mu\nu} &= \partial_\mu A^a_\nu - \partial_\nu A^a_\mu + f^{abc} A^b_\mu A^c_\nu \\
&= \partial_\mu (\bar{A}^a_\nu + a^a_\nu) - \partial_\nu (\bar{A}^a_\mu + a^a_\mu) + f^{abc} (\bar{A}^b_\mu + a^b_\mu)(\bar{A}^c_\nu + a^c_\nu) \\
&= \partial_\mu \bar{A}^a_\nu + \partial_\mu a^a_\nu - \partial_\nu \bar{A}^a_\mu - \partial_\nu a^a_\mu \\
&\quad + f^{abc} \bar{A}^b_\mu \bar{A}^c_\nu + f^{abc} \bar{A}^b_\mu a^c_\nu + f^{abc} a^b_\mu \bar{A}^c_\nu + f^{abc} a^b_\mu a^c_\nu \\
&= \partial_\mu \bar{A}^a_\nu - \partial_\nu \bar{A}^a_\mu - f^{abc} \bar{A}^b_\mu \bar{A}^c_\nu \\
&\quad + \partial_\mu a^a_\nu + f^{abc} \bar{A}^b_\mu a^c_\nu - \partial_\nu a^a_\mu - f^{abc} \bar{A}^b_\nu a^c_\mu + f^{abc} a^b_\mu a^c_\nu \\
&= \bar{F}^a_{\mu\nu} + \bar{D}_\mu a^a_\nu - \bar{D}_\nu a^a_\mu + f^{abc} a^b_\mu a^c_\nu .
\end{aligned}
$$

6.2 Covariant Derivative in a Cartan-Weyl Basis (⋆)
We have $X = i X^\kappa t^\kappa$ as well as $\bar{A}_\mu = i \delta_{\mu 0} T r^j t^j$. Then

$$
\begin{aligned}
\bar{D}_\mu X &= \partial_\mu X - [\bar{A}_\mu, X] \\
&= i \partial_\mu X^\kappa t^\kappa - i^2 \delta_{\mu 0} T r^j X^\kappa [t^j, t^\kappa] \\
&= i \partial_\mu X^\kappa t^\kappa - i^2 \delta_{\mu 0} T r^j \kappa^j X^\kappa t^\kappa , \\
&= i (\partial_\mu - i \delta_{\mu 0} T r^j \kappa^j) X^\kappa t^\kappa .
\end{aligned}
$$

6.3 Background Independence at Zero Temperature (⋆⋆)
Writing the potential as a function of $\hat{r} = Tr$ and applying (6.21) n times, we find

$$
V(\hat{r} + nT \bar{\alpha}^{(k)}; T) = V(\hat{r}; T) .
$$

Now, taking the $T \to 0$ limit in the form $T = u/n$ with $n \to \infty$ and $u \in \mathbb{R}$, we arrive at

$$
V(\hat{r} + u \bar{\alpha}^{(k)}; T \to 0) = V(\hat{r}; T \to 0) .
$$

Finally, since the vectors $\bar{\alpha}^{(k)}$ form a basis of the restricted background space, we have

$$
V(\hat{r} + \delta \hat{r}; T \to 0) = V(\hat{r}; T \to 0) , \quad \forall \delta \hat{r} ,
$$

as announced.

6.4 SU(3) Polyakov Loops (⋆)

Apply Eqs. (6.35) and (6.36) with $N = 3$ and ρ running over $(1/2, 1/(2\sqrt{3}))$, $(-1/2, 1/(2\sqrt{3}))$, and $(0, -1/\sqrt{3})$. We have

$$
\ell(r) = \frac{1}{3}\left[e^{i\frac{r_3}{2}+i\frac{r_8}{2\sqrt{3}}} + e^{-i\frac{r_3}{2}+i\frac{r_8}{2\sqrt{3}}} + e^{-i\frac{r_8}{\sqrt{3}}} \right]
$$

$$
= \frac{e^{-i\frac{r_8}{\sqrt{3}}} + 2e^{i\frac{r_8}{2\sqrt{3}}}\cos(r_3/2)}{3},
$$

and similarly for $\bar{\ell}(r)$ with $r_8 \to -r_8$.

6.5 Summation Formula (⋆⋆)

(a) We write

$$
\sum_{k=0}^{N-1} n_{q-i\frac{2\pi}{N}k} = \sum_{k=0}^{N-1}\sum_{p=1}^{\infty} e^{-\beta q p}e^{i\frac{2\pi}{N}kp} = \sum_{p=1}^{\infty}e^{-\beta q p}\sum_{k=0}^{N-1}e^{i\frac{2\pi}{N}kp},
$$

where we have inverted the order of the summations. Now, the inner sum over k vanishes unless p is a multiple of N, in which case it equals N. Therefore, the sum becomes $N\sum_{p=1}^{\infty} e^{-N\beta q p} = N n_{Nq}$, as announced.

(b) We write

$$
\sum_{k=0}^{N-1} T\sum_{n\in\mathbb{Z}} f\left(i\omega_n + iT\frac{2\pi}{N}k\right) = N\frac{T}{N}\sum_{k=0}^{N-1}\sum_{n\in\mathbb{Z}} f\left(i2\pi\frac{T}{N}(Nn+k)\right).
$$

Excluding the factor of N, we can interpret the double sum in the RHS as a single Matsubara sum at a reduced temperature T/N. Therefore, the result of the sum in the LHS is $NS(T/N)$.

Problems of Chap. 7

7.1 Color Conservation (⋆)

We have

$$
\mathrm{ad}_{t_{0(j)}}\,\mathrm{ad}_{t_\alpha} t_\beta = [t_{0(j)}, [t_\alpha, t_\beta]]
$$

$$
= -[t_\alpha, [t_\beta, t_{0(j)}]] - [t_\beta, [t_{0(j)}, t_\alpha]]
$$

$$
= \beta[t_\alpha, t_\beta] - \alpha[t_\beta, t_\alpha]
$$

$$
= (\alpha + \beta)\mathrm{ad}_{t_\alpha} t_\beta.
$$

7.2 Symmetry of $\Delta(Q_\kappa, K_\lambda, L_\tau)$ (⋆)

We write

$$
\begin{aligned}
\Delta(Q_\kappa, K_\lambda, L_\tau) &= Q_\kappa^2 K_\lambda^2 - (Q_\kappa \cdot K_\lambda)^2 \\
&= K_\lambda^2 Q_\kappa \cdot P^\perp(K_\lambda) \cdot Q_\kappa \\
&= K_\lambda^2 (-L_\tau - K_\lambda) \cdot P^\perp(K_\lambda) \cdot (-K_\lambda - L_\tau) \\
&= K_\lambda^2 L_\tau \cdot P^\perp(K_\lambda) \cdot L_\tau \\
&= L_\tau^2 K_\lambda^2 - (L_\tau \cdot K_\lambda)^2 = \Delta(L_\tau, K_\lambda, Q_\kappa).
\end{aligned}
$$

The same applies to the permutation $K_\lambda \leftrightarrow L_\tau$ and, therefore, to any possible permutation of $(Q_\kappa, K_\lambda, L_\tau)$.

7.3 Contractions (⋆⋆⋆)

(a) We write

$$
\begin{aligned}
&-\left[Q_\kappa \cdot P^\perp(L_\tau) \cdot K_\lambda \right] \mathrm{tr}\left[P^\perp(K_\lambda) P^\perp(Q_\kappa) \right] \\
&= -\left[Q_\kappa \cdot P^\perp(L_\tau) \cdot (-L_\tau - Q_\kappa) \right] \mathrm{tr}\left[\delta_{\mu\nu} - \frac{Q_\mu^\kappa Q_\nu^\kappa}{Q_\kappa^2} - \frac{L_\mu^\tau L_\nu^\tau}{L_\tau^2} + \frac{Q_\mu^\kappa Q_\rho^\kappa L_\rho^\tau L_\nu^\tau}{Q_\kappa^2 L_\tau^2} \right] \\
&= Q_\kappa \cdot P^\perp(L_\tau) \cdot Q_\kappa \left[d - 2 + \frac{(Q_\kappa \cdot L_\tau)^2}{Q_\kappa^2 L_\tau^2} \right] \\
&= \frac{\Delta(Q_\kappa, K_\lambda, L_\tau)}{L_\tau^2} \left[d - 2 + \frac{(Q_\kappa \cdot L_\tau)^2}{Q_\kappa^2 L_\tau^2} \right].
\end{aligned}
$$

Similarly, we write

$$
\begin{aligned}
&-2 L_\tau \cdot P^\perp(Q_\kappa) \cdot P^\perp(L_\tau) \cdot P^\perp(K_\lambda) \cdot L_\tau \\
&= -2 \left[L_\tau - \frac{L_\tau \cdot Q_\kappa}{Q_\kappa^2} Q_\kappa \right] \cdot P^\perp(L_\tau) \cdot \left[L_\tau - K_\lambda \frac{K_\kappa \cdot L_\tau}{K_\kappa^2} \right] \\
&= -2 \frac{L_\tau \cdot Q_\kappa}{Q_\kappa^2} Q_\kappa \cdot P^\perp(L_\tau) \cdot K_\lambda \frac{K_\kappa \cdot L_\tau}{K_\kappa^2} = 2 \Delta(Q_\kappa, K_\lambda, L_\tau) \frac{(K_\kappa \cdot L_\tau)(L_\tau \cdot Q_\kappa)}{Q_\kappa^2 K_\kappa^2 L_\tau^2}.
\end{aligned}
$$

(b) Since $\Delta(Q_\kappa, K_\lambda, L_\tau)$ is symmetric under permutation of its arguments, we have

$$
\begin{aligned}
&\mathcal{B}_{\mathrm{sym}}(Q_\kappa, K_\lambda, L_\tau) \\
&= \frac{d-2}{3} \left[\frac{1}{Q_\kappa^2} + \frac{1}{K_\lambda^2} + \frac{1}{L_\tau^2} \right] \Delta
\end{aligned}
$$

$$+\frac{1}{3}\frac{(Q_\kappa \cdot K_\lambda)^2 + (K_\lambda \cdot L_\tau)^2 + (L_\tau \cdot Q_\kappa)^2}{Q_\kappa^2 K_\lambda^2 L_\tau^2}\Delta$$

$$+\frac{1}{3}\frac{2(Q_\kappa \cdot K_\lambda)(K_\kappa \cdot L_\tau) + 2(K_\kappa \cdot L_\tau)(L_\tau \cdot Q_\kappa) + 2(L_\tau \cdot Q_\kappa)(Q_\kappa \cdot K_\lambda)}{Q_\kappa^2 K_\kappa^2 L_\tau^2}\Delta.$$

The numerators of the last two terms combine into

$$(Q_\kappa \cdot K_\lambda)^2 + (K_\lambda \cdot L_\tau)^2 + (L_\tau \cdot Q_\kappa)^2$$

$$+2(Q_\kappa \cdot K_\lambda)(K_\kappa \cdot L_\tau) + 2(K_\kappa \cdot L_\tau)(L_\tau \cdot Q_\kappa) + 2(L_\tau \cdot Q_\kappa)(Q_\kappa \cdot K_\lambda)$$

$$= (Q_\kappa \cdot K_\lambda + K_\lambda \cdot L_\tau + L_\tau \cdot Q_\kappa)^2.$$

On the other hand, by using $Q_\kappa + K_\lambda + L_\tau = 0$, we can write

$$Q_\kappa \cdot K_\lambda + K_\lambda \cdot L_\tau + L_\tau \cdot Q_\kappa = \frac{1}{2}\left[(Q_\kappa + K_\lambda + L_\tau)^2 - Q_\kappa^2 - K_\lambda^2 - L_\tau^2\right]$$

$$= -\frac{1}{2}\left[Q_\kappa^2 + K_\lambda^2 + L_\tau^2\right].$$

So, we eventually arrive at

$$\mathcal{B}_{\text{sym}}(Q_\kappa, K_\lambda, L_\tau) = \frac{d-2}{3}\left[\frac{1}{Q_\kappa^2} + \frac{1}{K_\lambda^2} + \frac{1}{L_\tau^2}\right]\Delta$$

$$+\frac{1}{12}\frac{(Q_\kappa^2 + K_\lambda^2 + L_\tau^2)^2}{Q_\kappa^2 K_\kappa^2 L_\tau^2}\Delta.$$

This rewrites

$$\mathcal{B}_{\text{sym}}(Q_\kappa, K_\lambda, L_\tau) = \frac{1}{3}\left\{\left(d-\frac{3}{2}\right)\left[\frac{1}{Q_\kappa^2} + \frac{1}{K_\lambda^2} + \frac{1}{L_\tau^2}\right] + \frac{Q_\kappa^4 + K_\lambda^4 + L_\tau^4}{4Q_\kappa^2 K_\kappa^2 L_\tau^2}\right\}\Delta.$$

7.4 Reduction Formula (★★★)
Since $\Delta(Q_\kappa, K_\lambda, L_\tau) = Q_\kappa^2 K_\lambda^2 - (Q_\kappa \cdot K_\lambda)^2$, we write

$$Q_\kappa^2 K_\lambda^2 G_\alpha(Q_\kappa)G_\beta(K_\lambda)G_\gamma(L_\tau)$$

$$= (Q_\kappa^2 + \alpha - \alpha)(K_\lambda^2 + \beta - \beta)G_\alpha(Q_\kappa)G_\beta(K_\lambda)G_\gamma(L_\tau)$$

$$= G_\gamma(L_\tau) - \alpha G_\alpha(Q_\kappa)G_\gamma(L_\tau) - \beta G_\beta(K_\lambda)G_\gamma(L_\tau)$$

$$+\alpha\beta G_\alpha(Q_\kappa)G_\beta(K_\lambda)G_\gamma(L_\tau).$$

Moreover

$$Q_\kappa \cdot K_\lambda \, G_\alpha(Q_\kappa) G_\beta(K_\lambda) G_\gamma(L_\tau)$$

$$= \frac{1}{2}(L_\tau^2 - Q_\kappa^2 - K_\lambda^2) G_\alpha(Q_\kappa) G_\beta(K_\lambda) G_\gamma(L_\tau)$$

$$= \frac{1}{2}(L_\tau^2 + \gamma - Q_\kappa^2 - \alpha - K_\lambda^2 - \beta + \alpha + \beta - \gamma) G_\alpha(Q_\kappa) G_\beta(K_\lambda) G_\gamma(L_\tau)$$

$$= \frac{1}{2}\Big(G_\alpha(Q_\kappa) G_\beta(K_\lambda) - G_\beta(K_\lambda) G_\gamma(L_\tau) - G_\alpha(Q_\kappa) G_\gamma(L_\tau)\Big)$$

$$+ \frac{\alpha + \beta - \gamma}{2} \, G_\alpha(Q_\kappa) G_\beta(K_\lambda) G_\gamma(L_\tau).$$

From this, it follows that

$$(Q_\kappa \cdot K_\lambda)^2 \, G_\alpha(Q_\kappa) G_\beta(K_\lambda) G_\gamma(L_\tau)$$

$$= \frac{Q_\kappa \cdot K_\lambda}{2} \Big(G_\alpha(Q_\kappa) G_\beta(K_\lambda) - G_\beta(K_\lambda) G_\gamma(L_\tau) - G_\alpha(Q_\kappa) G_\gamma(L_\tau)\Big)$$

$$+ \frac{\alpha + \beta - \gamma}{4} \Big(G_\alpha(Q_\kappa) G_\beta(K_\lambda) - G_\beta(K_\lambda) G_\gamma(L_\tau) - G_\alpha(Q_\kappa) G_\gamma(L_\tau)\Big)$$

$$+ \frac{(\alpha + \beta - \gamma)^2}{4} \, G_\alpha(Q_\kappa) G_\beta(K_\lambda) G_\gamma(L_\tau).$$

We can put the first line in a more symmetrical form by making use of $Q_\kappa + K_\lambda + L_\tau = 0$ to write

$$-\frac{Q_\kappa \cdot K_\lambda}{2} G_\beta(K_\lambda) G_\gamma(L_\tau) = \frac{K_\lambda^2}{2} G_\beta(K_\lambda) G_\gamma(L_\tau) + \frac{K_\lambda \cdot L_\tau}{2} G_\beta(K_\lambda) G_\gamma(L_\tau)$$

$$= \frac{1}{2} G_\gamma(L_\tau) - \frac{\beta}{2} G_\beta(K_\lambda) G_\gamma(L_\tau)$$

$$+ \frac{K_\lambda \cdot L_\tau}{2} G_\beta(K_\lambda) G_\gamma(L_\tau)$$

and similarly

$$-\frac{Q_\kappa \cdot K_\lambda}{2} G_\alpha(Q_\kappa) G_\gamma(L_\tau) = \frac{1}{2} G_\gamma(L_\tau) - \frac{\alpha}{2} G_\alpha(Q_\kappa) G_\gamma(L_\tau)$$

$$+ \frac{Q_\kappa \cdot L_\tau}{2} G_\alpha(Q_\kappa) G_\gamma(L_\tau).$$

We then arrive at

$$
(Q_\kappa \cdot K_\lambda)^2 \, G_\alpha(Q_\kappa) G_\beta(K_\lambda) G_\gamma(L_\tau)
$$

$$
= G_\gamma(L_\tau) + \frac{Q_\kappa \cdot K_\lambda}{2} G_\alpha(Q_\kappa) G_\beta(K_\lambda) + \frac{K_\lambda \cdot L_\tau}{2} G_\beta(K_\lambda) G_\gamma(L_\tau)
$$

$$
+ \frac{Q_\kappa \cdot L_\tau}{2} G_\alpha(Q_\kappa) G_\gamma(L_\tau)
$$

$$
+ \frac{\alpha + \beta - \gamma}{4} G_\alpha(Q_\kappa) G_\beta(K_\lambda)
$$

$$
+ \frac{\gamma - \alpha - 3\beta}{4} G_\beta(K_\lambda) G_\gamma(L_\tau)
$$

$$
+ \frac{\gamma - 3\alpha - \beta}{4} G_\alpha(Q_\kappa) G_\gamma(L_\tau)
$$

$$
+ \frac{(\alpha + \beta - \gamma)^2}{4} G_\alpha(Q_\kappa) G_\beta(K_\lambda) G_\gamma(L_\tau) .
$$

Combining this with the expression above, we obtain

$$
\Delta(Q_\kappa, K_\lambda, L_\tau) G_\alpha(Q_\kappa) G_\beta(K_\lambda) G_\gamma(L_\tau)
$$

$$
= -\frac{Q_\kappa \cdot K_\lambda}{2} G_\alpha(Q_\kappa) G_\beta(K_\lambda) - \frac{K_\lambda \cdot L_\tau}{2} G_\beta(K_\lambda) G_\gamma(L_\tau)
$$

$$
- \frac{Q_\kappa \cdot L_\tau}{2} G_\alpha(Q_\kappa) G_\gamma(L_\tau)
$$

$$
+ \frac{\gamma - \alpha - \beta}{4} G_\alpha(Q_\kappa) G_\beta(K_\lambda) + \frac{\alpha - \beta - \gamma}{4} G_\beta(K_\lambda) G_\gamma(L_\tau)
$$

$$
+ \frac{\beta - \gamma - \alpha}{4} G_\alpha(Q_\kappa) G_\gamma(L_\tau)
$$

$$
- \frac{\alpha^2 + \beta^2 + \gamma^2 - 2\alpha\beta - 2\beta\gamma - 2\gamma\alpha}{4} G_\alpha(Q_\kappa) G_\beta(K_\lambda) G_\gamma(L_\tau) .
$$

which leads to the announced formula.

7.5 Gauging Away the Background in the Polyakov Loop Operator (⋆)

We have

$$
\bar{A}^U = e^{-g\bar{A}\tau} \bar{A} e^{g\bar{A}\tau} - \frac{1}{g} e^{-g\bar{A}\tau} \frac{d}{d\tau} e^{g\bar{A}\tau}
$$

$$
= e^{-g\bar{A}\tau} \bar{A} e^{g\bar{A}\tau} - e^{-g\bar{A}\tau} \bar{A} e^{g\bar{A}\tau} = 0 .
$$

Since $(\bar{A} + a)^U = \bar{A}^U + a^U = a^U \equiv UaU^\dagger$, it follows that

$$L_{\bar{A}+a}(\mathbf{x}) = U^\dagger(\beta) L_{(\bar{A}+a)^U}(\mathbf{x}) U(0) = U^\dagger(\beta) L_{a^U}(\mathbf{x}).$$

Problems of Chap. 8

8.1 Polyakov Loop in the First Fundamental Representation of SU(4) (★★)

We denote the defining weights as ρ_1, ρ_2, ρ_3 and ρ_4, with $\rho_1 + \rho_2 + \rho_3 + \rho_4 = 0$. Then, recalling that $\alpha_j \equiv \rho_j - \rho_4$ (for $j \neq 4$) provide an independent set of roots, we write:

$$
\begin{aligned}
\ell_4(r) &= \frac{1}{4}\left[e^{i\rho_1 \cdot r} + e^{i\rho_2 \cdot r} + e^{i\rho_3 \cdot r} + e^{i\rho_4 \cdot r} \right] \\
&= \frac{e^{\frac{i}{4}4\rho_4 \cdot r}}{4}\left[1 + e^{i(\rho_1 - \rho_4) \cdot r} + e^{i(\rho_2 - \rho_4) \cdot r} + e^{i(\rho_3 - \rho_4) \cdot r} \right] \\
&= \frac{e^{\frac{i}{4}(3\rho_4 - \rho_1 - \rho_2 - \rho_3) \cdot r}}{4}\left[1 + e^{i(\rho_1 - \rho_4) \cdot r} + e^{i(\rho_2 - \rho_4) \cdot r} + e^{i(\rho_3 - \rho_4) \cdot r} \right] \\
&= \frac{e^{-\frac{i}{4}(\alpha_1 + \alpha_2 + \alpha_3) \cdot r}}{4}\left[1 + e^{i\alpha_1 \cdot r} + e^{i\alpha_2 \cdot r} + e^{i\alpha_3 \cdot r} \right] \\
&= \frac{e^{-i\frac{\pi}{2}(x_1 + x_2 + x_3) \cdot r}}{4}\left[1 + e^{2\pi i x_1} + e^{2\pi i x_2} + e^{2\pi i x_3} \right].
\end{aligned}
$$

8.2 Polyakov Loop in the Second Fundamental Representation of SU(4) (★★)

Using similar remarks as in the previous problem, we write:

$$
\begin{aligned}
\ell_6(r) &= \frac{1}{6}\Big[e^{i(\rho_1 + \rho_2) \cdot r} + e^{i(\rho_1 + \rho_3) \cdot r} + e^{i(\rho_1 + \rho_4) \cdot r} \\
&\qquad + e^{i(\rho_2 + \rho_3) \cdot r} + e^{i(\rho_2 + \rho_4) \cdot r} + e^{i(\rho_3 + \rho_4) \cdot r} \Big] \\
&= \frac{1}{6}\Big[e^{i(\rho_1 + \rho_2) \cdot r} + e^{i(\rho_1 + \rho_3) \cdot r} + e^{-i(\rho_2 + \rho_3) \cdot r} \\
&\qquad + e^{i(\rho_2 + \rho_3) \cdot r} + e^{-i(\rho_1 + \rho_3) \cdot r} + e^{-i(\rho_1 + \rho_2) \cdot r} \Big] \\
&= \frac{1}{3}\mathrm{Re}\Big[e^{-i(\rho_1 + \rho_2) \cdot r} + e^{-i(\rho_1 + \rho_3) \cdot r} + e^{-i(\rho_2 + \rho_3) \cdot r} \Big] \\
&= \frac{1}{3}\mathrm{Re}\, e^{2i\rho_4 \cdot r}\Big[e^{-i(\rho_1 + \rho_2 + 2\rho_4) \cdot r} + e^{-i(\rho_1 + \rho_3 + 2\rho_4) \cdot r} + e^{-i(\rho_2 + \rho_3 + 2\rho_4) \cdot r} \Big] \\
&= \frac{1}{3}\mathrm{Re}\, e^{\frac{i}{2}(\rho_4 - \rho_1 - \rho_2 - \rho_3) \cdot r}\Big[e^{i(\rho_1 - \rho_4) \cdot r} + e^{i(\rho_2 - \rho_4) \cdot r} + e^{i(\rho_3 - \rho_4) \cdot r} \Big] \\
&= \frac{1}{3}\mathrm{Re}\, e^{-i\pi(x_1 + x_2 + x_3)}\Big[e^{2\pi i x_1} + e^{2\pi i x_2} + e^{2\pi i x_3} \Big].
\end{aligned}
$$

Problems of Chap. 9

9.1 Euclidean Dirac Matrices in the Weyl Basis ($\star$)

If $\gamma_\mu = \gamma_0$ or γ_2, we have

$$\gamma_2\gamma_0\gamma_\mu\gamma_0\gamma_2 = -\gamma_2\gamma_0\gamma_0\gamma_2\gamma_\mu = -\gamma_\mu = -\gamma_\mu^{\mathrm{t}},$$

whereas if $\gamma_\mu = \gamma_1$ or γ_3, we have

$$\gamma_2\gamma_0\gamma_\mu\gamma_0\gamma_2 = \gamma_2\gamma_0\gamma_0\gamma_2\gamma_\mu = \gamma_\mu = -\gamma_\mu^{\mathrm{t}},$$

so the result is $-\gamma_\mu^{\mathrm{t}}$ in all cases, as announced.

Similarly, if $\gamma_\mu = \gamma_0$ or γ_2, we have

$$\gamma_3\gamma_1\gamma_\mu\gamma_1\gamma_3 = \gamma_3\gamma_1\gamma_1\gamma_3\gamma_\mu = \gamma_\mu = \gamma_\mu^{*},$$

whereas if $\gamma_\mu = \gamma_1$ or γ_3, we have

$$\gamma_3\gamma_1\gamma_\mu\gamma_1\gamma_3 = -\gamma_3\gamma_1\gamma_1\gamma_3\gamma_\mu = -\gamma_\mu = \gamma_\mu^{*},$$

so the result is γ_μ^{*} in all cases, as announced.

9.2 Charge Conjugation ($\star\star$)

(a) We write

$$\bar{\psi}^C \equiv \overline{\psi^C} = \overline{\gamma_0\gamma_2\bar{\psi}^{\mathrm{t}}} = \overline{\bar{\psi}^{\mathrm{t}}}\,\bar{\gamma_2}\,\bar{\gamma_0} = \psi^{\mathrm{t}}\bar{\gamma_2}\,\bar{\gamma_0}.$$

Now, in the considered Weyl representation, we have $\bar{\gamma_0} = \gamma_0$ and $\bar{\gamma_2} = -\gamma_2$, and we arrive at $\bar{\psi}^C = -\psi^{\mathrm{t}}\gamma_2\gamma_0$.

(b) Since the YM part of the action is invariant under charge conjugation, we just need to look at the matter part. Focusing on the contribution from one flavor, we have

$$\bar{\psi}_f^C(x)\,(\slashed{\partial} - g\slashed{A}^C + M_f - \mu\gamma_0)\,\psi_f^C(x)$$
$$= -\psi_f^{\mathrm{t}}(x)\,\gamma_2\gamma_0\,(\slashed{\partial} + g\slashed{A}^{\mathrm{t}} + M_f - \mu\gamma_0)\,\gamma_0\gamma_2\,\bar{\psi}_f^{\mathrm{t}}(x)$$
$$= \psi_f^{\mathrm{t}}(x)\,(\slashed{\partial}^{\mathrm{t}} + g\slashed{A}^{\mathrm{t}} - M_f - \mu\gamma_0^{\mathrm{t}})\,\bar{\psi}_f^{\mathrm{t}}(x).$$

Upon getting rid of the transposition, we have to take into account the Grassmannian nature of the quark fields. Moreover, for the derivative operator $\slashed{\partial}$ to act on ψ_f rather than on $\bar{\psi}_f$, we consider an integration by parts which adds an extra sign. We eventually arrive at

$$\bar{\psi}_f^C(x)\,(\partial\!\!\!/ - g A\!\!\!/^C + M_f - \mu\gamma_0)\,\psi_f^C(x)$$

$$= \bar{\psi}_f(x)\,(\partial\!\!\!/ - g A\!\!\!/ + M_f + \mu\gamma_0)\,\psi_f(x)\,.$$

This shows that the only effect of charge conjugation is to flip the sign of μ in the matter contribution to the QCD action. Since the YM part is μ-independent, its invariance under C can also be interpreted in the same terms, so one eventually arrives at $S[A^C, \psi^C, \bar{\psi}^C; \mu] = S[A, \psi, \bar{\psi}; -\mu]$.

9.3 Hermicity (★★)

We proceed as in the previous problem.

(a) We first write

$$\bar{\psi}^\mathcal{K} \equiv \overline{\psi^\mathcal{K}} = \overline{\gamma_1\gamma_3\psi} = \bar{\psi}\bar{\gamma_3}\,\bar{\gamma_1} = \bar{\psi}\bar{\gamma_3}\,\bar{\gamma_1}\,.$$

Now, in the considered Weyl representation, we have $\bar{\gamma_1} = -\gamma_1$ and $\bar{\gamma_3} = -\gamma_3$, and we arrive at $\bar{\psi}^\mathcal{K} = \bar{\psi}\gamma_3\gamma_1$.

(b) Since the YM part of the action is invariant under $\mathcal{K}$, we just need to look at the matter part. Focusing on the contribution from one flavor, we have

$$\bar{\psi}_f^\mathcal{K}(x)\,(\partial\!\!\!/ - g A\!\!\!/^\mathcal{K} + M_f - \mu\gamma_0)\,\psi_f^\mathcal{K}(x)$$

$$= \bar{\psi}_f(x)\,\gamma_3\gamma_1\,(\partial\!\!\!/ - g A_\mu^*\gamma_\mu + M_f - \mu\gamma_0)\,\gamma_1\gamma_3\,\psi_f(x)$$

$$= \bar{\psi}_f(x)\,(\partial\!\!\!/^* - g A_\mu^*\gamma_\mu^* + M_f - \mu\gamma_0^*)\,\psi_f(x)$$

$$= \bar{\psi}_f(x)\,(\partial\!\!\!/ - g A_\mu\gamma_\mu + M_f - \mu^*\gamma_0)^*\,\psi_f(x)\,.$$

This shows that the only effect of $\mathcal{K}$ on the matter part of the QCD action is to complex conjugate the action (we have already seen that we can assume that $\psi^* = \psi$ and $\bar{\psi}^* = \bar{\psi}$) while replacing μ by μ^*. Since the YM part is μ-independent and real, its invariance under $\mathcal{K}$ can also be interpreted in the same terms, so one eventually arrives at $S[A^\mathcal{K}, \psi^\mathcal{K}, \bar{\psi}^\mathcal{K}; \mu] = S[A, \psi, \bar{\psi}; \mu^*]^*$.

9.4 Dirac Operator (★)

From the derivation in Prob. 9.2, we have seen that

$$-\gamma_2\gamma_0 M[A^C; \mu]\gamma_0\gamma_2 = (\partial\!\!\!/^t + g A\!\!\!/^t - M_f - \mu\gamma_0^t)\,.$$

To rewrite this in terms of a full transposition, we need to take into account the presence of the derivative operator ∂. This introduces an extra sign, and we thus arrive at

$$-\gamma_2\gamma_0 M[A^C; \mu]\gamma_0\gamma_2 = (-\partial\!\!\!/ + g A\!\!\!/ - M_f - \mu\gamma_0)^t$$

$$= -(\partial\!\!\!/ - g A\!\!\!/ + M_f + \mu\gamma_0)^t = -M[A; -\mu]^t\,.$$

This shows that $\gamma_2\gamma_0\, M[A^C; \mu]\,\gamma_0\gamma_2 = M[A; -\mu]^t$, as announced. Similarly, from the derivation in Prob. 9.3, we have seen that $\gamma_3\gamma_1\, M[A^\mathcal{K}; \mu]\,\gamma_1\gamma_3 = M[A; \mu^*]^*$.

Combining these two results, one finds

$$\gamma_5 \, M[A; \mu] \, \gamma_5 = \gamma_0 \gamma_1 \gamma_2 \gamma_3 \, M[A; \mu] \, \gamma_0 \gamma_1 \gamma_2 \gamma_3$$

$$= \gamma_3 \gamma_1 \gamma_2 \gamma_0 \, M[(A^K)^C; \mu] \, \gamma_0 \gamma_2 \gamma_1 \gamma_3$$

$$= \gamma_3 \gamma_1 \, M[(A^K); -\mu]^t \, \gamma_1 \gamma_3$$

$$= M[(A; -\mu^*]^\dagger .$$

9.5 Partition Function (⋆⋆)

Upon a change of variables under the functional integral of the form $A \to A^C$, we obtain

$$Z[J; \bar{A}_0, \mu] = \int \mathcal{D}_{\mathrm{gf}}[A^C; \bar{A}_0] \, \Delta[A^C; \mu] \, \exp\left\{ -S_{YM}[A^C] + \int d^d x \, J^j(x)(A_0^C)^j(x) \right\}$$

$$= \int \mathcal{D}_{\mathrm{gf}}[A^C; \bar{A}_0] \, \Delta[A; -\mu] \, \exp\left\{ -S_{YM}[A] - \int d^d x \, J^j(x) A_0^j(x) \right\},$$

where we have used the invariance of the YM action under charge conjugation, the transformation rule of the fermion determinant and the fact that $(A_0^C)^j = -A_0^j$. Then, writing $\bar{A}_0 = (\bar{A}_0^C)^C$ and assuming that the gauge-fixed measure invariant under the simultaneous charge conjugation of the background and dynamical fields, we finally arrive at

$$Z[J; \bar{A}_0, \mu] = \int \mathcal{D}_{\mathrm{gf}}[A; \bar{A}_0^C] \, \Delta[A; -\mu] \, \exp\left\{ -S_{YM}[A] - \int d^d x \, J^j(x) A_0^j(x) \right\}$$

$$= Z[-J; \bar{A}_0^C, -\mu] = Z[-J; -\bar{A}_0, -\mu],$$

where we have used that, because the background is also taken along the diagonal part of the algebra, $(\bar{A}_0^C)^j = -\bar{A}_0^j$.

Similarly, upon a change of variables under the functional integral of the form $A \to A^K$, we obtain

$$Z[J; \bar{A}_0, \mu] = \int \mathcal{D}_{\mathrm{gf}}[A^K; \bar{A}_0] \, \Delta[A^K; \mu] \, \exp\left\{ -S_{YM}[A^K] + \int d^d x \, J^j(x)(A_0^K)^j(x) \right\}$$

$$= \int \mathcal{D}_{\mathrm{gf}}[A^K; \bar{A}_0] \, \Delta[A; \mu^*]^* \, \exp\left\{ -S_{YM}^*[A] - \int d^d x \, J^j(x) A_0^j(x) \right\},$$

where we have used the invariance of the YM action under K, the fact that the YM action is real, the transformation rule of the fermion determinant, and the fact that $(A_0^K)^j = -A_0^j$. Then, writing $\bar{A}_0 = (\bar{A}_0^K)^K$ and assuming that the gauge-fixed measure is complex conjugated, Eq. (9.32), under the simultaneous transformation of the background and dynamical fields, we finally arrive at

$$Z[J; \bar{A}_0, \mu] = \int \mathcal{D}_{\mathrm{gf}}[A; \bar{A}_0^K]^* \, \Delta[A; \mu^*]^* \exp\left\{-S_{YM}[A] - \int d^d x \, J^j(x) A_0^j(x)\right\}$$

$$= Z[-J^*; \bar{A}_0^K, \mu^*]^* = Z[-J^*; -\bar{A}_0^*, \mu^*]^*,$$

where we have used that, because the background is also taken along the diagonal part of the algebra, $(\bar{A}_0^K)^j = -(\bar{A}_0^j)^*$, and we have allowed for the possibility of both complex background and sources.

Finally, by combining the two obtained identities, we arrive at

$$Z[J; \bar{A}_0, \mu] = Z[J^*; \bar{A}_0^*, -\mu^*]^*.$$

Problems of Chap. 10

10.1 Matter Contribution to the SU(3) Potential (⋆⋆)

(a) The defining weights of SU(3) are $(1, 1/\sqrt{3})/2$, $(-1, 1/\sqrt{3})/2$, and $(0, -1/\sqrt{3})$. It follows that the various scalar products $r \cdot \rho$ that appear are $(r_3 + r_8/\sqrt{3})/2$, $(-r_3 + r_8/\sqrt{3})/2$, and $-r_8/\sqrt{3}$. It is then straightforward to arrive at Eq. (10.6).

(b) It is convenient to keep the defining weights implicit. By combining first the logarithms involving the same sign of $\varepsilon_q \pm \mu$, we find

$$\ln\left[1 + e^{-\beta(\varepsilon_{q,f}\mp\mu)\pm ir\cdot\rho_1}\right]\left[1 + e^{-\beta(\varepsilon_{q,f}\mp\mu)\pm ir\cdot\rho_2}\right]\left[1 + e^{-\beta(\varepsilon_{q,f}\mp\mu)\pm ir\cdot\rho_3}\right]$$

$$= \ln\left[1 + e^{-\beta(\varepsilon_{q,f}\mp\mu)}\left(e^{\pm ir\cdot\rho_1} + e^{\pm ir\cdot\rho_2} + e^{\pm ir\cdot\rho_3}\right)\right.$$

$$+ e^{-2\beta(\varepsilon_{q,f}\mp\mu)}\left(e^{\pm ir\cdot(\rho_1+\rho_2)} + e^{\pm ir\cdot(\rho_2+\rho_3)} + e^{\pm ir\cdot(\rho_3+\rho_1)}\right)$$

$$\left. + e^{-3\beta(\varepsilon_{q,f}\mp\mu)}\, e^{\pm ir\cdot(\rho_1+\rho_2+\rho_3)}\right].$$

We have thus found the expected expression with

$$A_\pm = e^{\pm ir\cdot\rho_1} + e^{\pm ir\cdot\rho_2} + e^{\pm ir\cdot\rho_3}.$$

$$B_\pm = e^{\pm ir\cdot(\rho_1+\rho_2)} + e^{\pm ir\cdot(\rho_2+\rho_3)} + e^{\pm ir\cdot(\rho_3+\rho_1)}.$$

$$C_\pm = e^{\pm ir\cdot(\rho_1+\rho_2+\rho_3)}.$$

(c) It immediately follows from $\rho_1 + \rho_2 + \rho_3 = 0$ that $C_\pm = 1$. Moreover

$$B_\pm = e^{\pm ir\cdot(-\rho_3)} + e^{\pm ir\cdot(-\rho_1)} + e^{\pm ir\cdot(-\rho_2)} = A_\mp.$$

It follows that the above combination of logarithms simplifies to

$$\ln\left[1 + e^{-\beta(\varepsilon_{q,f}\mp\mu)\pm ir\cdot\rho_1}\right]\left[1 + e^{-\beta(\varepsilon_{q,f}\mp\mu)\pm ir\cdot\rho_2}\right]\left[1 + e^{-\beta(\varepsilon_{q,f}\mp\mu)\pm ir\cdot\rho_3}\right]$$

$$= \ln \left[1 + A_\pm e^{-\beta (\varepsilon_{q,f} \mp \mu)} + A_\mp e^{-2\beta (\varepsilon_{q,f} \mp \mu)} + e^{-3\beta (\varepsilon_{q,f} \mp \mu)} \right].$$

To complete the calculation, one just needs to remark that $A_+ = 3\ell$ and $A_- = 3\bar{\ell}$.

10.2 Matter Contribution to the SU(N) Potential ($\star$)

Combining the logarithms in Eq. (10.5), one arrives at

$$\ln \prod_{j=1}^{N} \left[1 + e^{-\beta (\varepsilon_{q,f} \mp \mu) \pm i r \cdot \rho_j} \right]$$

$$= \ln \left[1 + \sum_{\nu=1}^{N} e^{-\nu \beta (\varepsilon_{q,f} \mp \mu)} \sum_{j_1 < \cdots < j_\nu} e^{\pm i \left(\rho_{j_1} + \cdots + \rho_{j_\nu} \right)} \right].$$

We have seen in Chap. 8 that the fundamental Polyakov loops play a role in the characterization of center symmetry. It is no doubt that they will enter the generalization that we are after. At leading order, they are given by

$$\ell_{C_N^\nu}(r) = \frac{1}{C_N^\nu} \sum_{j_1 < \cdots < j_\nu} e^{i \left(\rho_{j_1} + \cdots + \rho_{j_\nu} \right) \cdot r},$$

$$\ell_{\bar{C}_N^\nu}(r) = \frac{1}{C_N^\nu} \sum_{j_1 < \cdots < j_\nu} e^{-i \left(\rho_{j_1} + \cdots + \rho_{j_\nu} \right) \cdot r}.$$

We then arrive at

$$\delta V_{1loop}^{SU(N)} (\{\ell_{C_N^\nu}, \ell_{\bar{C}_N^\nu}\}; T, \mu)$$

$$= -\frac{T}{\pi^2} \int_0^\infty dq \, q^2 \left\{ \ln \left[1 + \sum_{\nu=1}^{N} C_N^\nu \ell_{C_N^\nu} e^{-\nu \beta (\varepsilon_{q,f} - \mu)} \right] + \right.$$

$$\left. + \ln \left[1 + \sum_{\nu=1}^{N} C_N^\nu \ell_{\bar{C}_N^\nu} e^{-\nu \beta (\varepsilon_{q,f} + \mu)} \right] \right\}.$$

At vanishing chemical potential, one can restrict to $\ell_{\bar{C}_N^\nu} = \ell_{C_N^\nu}$ because the minimum should obey this constraint from charge conjugation invariance.

10.3 Gluonic Contribution to the SU(3) Potential (★★★)

(a) Combining the logarithms in Eq. (6.18), we obtain

$$\ln \prod_{j=1}^{8} \left[1 - e^{-\beta\varepsilon_q + ir\cdot\kappa_j} \right]$$

$$= \ln (1 - e^{-\beta\varepsilon_q})^2 \prod_{j=1}^{3} \left[1 - e^{-\beta\varepsilon_q + ir\cdot\alpha_j} \right] \prod_{j=1}^{3} \left[1 - e^{-\beta\varepsilon_q - ir\cdot\alpha_j} \right].$$

Next, using $\alpha_1 + \alpha_2 + \alpha_3 = 0$, we write

$$\prod_{j=1}^{3} \left[1 - e^{-\beta\varepsilon_q + ir\cdot\alpha_j} \right] = 1 - e^{-3\beta\varepsilon_q} - e^{-\beta\varepsilon_q} \sum_{j=1}^{3} e^{ir\cdot\alpha_j} + e^{-2\beta\varepsilon_q} \sum_{j=1}^{3} e^{-ir\cdot\alpha_j}.$$

$$\prod_{j=1}^{3} \left[1 - e^{-\beta\varepsilon_q - ir\cdot\alpha_j} \right] = 1 - e^{-3\beta\varepsilon_q} - e^{-\beta\varepsilon_q} \sum_{j=1}^{3} e^{-ir\cdot\alpha_j} + e^{-2\beta\varepsilon_q} \sum_{j=1}^{3} e^{ir\cdot\alpha_j}.$$

Then

$$\prod_{j=1}^{3} \left[1 - e^{-\beta\varepsilon_q + ir\cdot\alpha_j} \right] \prod_{j=1}^{3} \left[1 - e^{-\beta\varepsilon_q - ir\cdot\alpha_j} \right]$$

$$= (1 - e^{-3\beta\varepsilon_q})^2$$

$$+ e^{-\beta\varepsilon_q} (e^{-\beta\varepsilon_q} - 1)(1 - e^{-3\beta\varepsilon_q}) \left(\sum_{j=1}^{3} e^{ir\cdot\alpha_j} + \sum_{j=1}^{3} e^{-ir\cdot\alpha_j} \right)$$

$$+ e^{-2\beta\varepsilon_q} (e^{-2\beta\varepsilon_q} + 1) \sum_{j=1}^{3} e^{ir\cdot\alpha_j} \sum_{k=1}^{3} e^{-ir\cdot\alpha_k}$$

$$- e^{-3\beta\varepsilon_q} \left[\left(\sum_{j=1}^{3} e^{ir\cdot\alpha_j} \right)^2 + \left(\sum_{k=1}^{3} e^{-ir\cdot\alpha_k} \right)^2 \right].$$

(b) We now write

$$\sum_{j=1}^{3} e^{ir\cdot\alpha_j} + \sum_{j=1}^{3} e^{-ir\cdot\alpha_j} = \sum_{j\neq k} e^{ir\cdot(\rho_j-\rho_k)}$$

$$= \sum_{j=1}^{3} e^{ir\cdot\rho_j} \sum_{k=1}^{3} e^{-ir\cdot\rho_k} - 3$$

$$= 9\ell_3(r)\ell_{\bar{3}}(r) - 3 \, .$$

Moreover

$$\sum_{j=1}^{3} e^{ir\cdot\alpha_j} \sum_{k=1}^{3} e^{-ir\cdot\alpha_k} = 3 + e^{ir\cdot(\rho_1+\rho_2-2\rho_3)} + e^{-ir\cdot(\rho_1+\rho_2-2\rho_3)}$$

$$+ e^{ir\cdot(\rho_2+\rho_3-2\rho_1)} + e^{-ir\cdot(\rho_2+\rho_3-2\rho_1)}$$

$$+ e^{ir\cdot(\rho_3+\rho_1-2\rho_2)} + e^{-ir\cdot(\rho_3+\rho_1-2\rho_2)}$$

$$= 3 + e^{3ir\cdot\rho_1} + e^{3ir\cdot\rho_2} + e^{3ir\cdot\rho_3}$$

$$+ e^{-3ir\cdot\rho_1} + e^{-3ir\cdot\rho_2} + e^{-3ir\cdot\rho_3}$$

$$= 3 + 3\ell_3(3r) + 3\bar{\ell}_3(3r) \, .$$

We can finally write

$$\left(\sum_{j=1}^{3} e^{ir\cdot\alpha_j}\right)^2 + \left(\sum_{k=1}^{3} e^{-ir\cdot\alpha_k}\right)^2$$

$$= \left(\sum_{j=1}^{3} e^{ir\cdot\alpha_j} + \sum_{j=1}^{3} e^{-ir\cdot\alpha_j}\right)^2$$

$$- 2\sum_{j=1}^{3} e^{ir\cdot\alpha_j} \sum_{k=1}^{3} e^{-ir\cdot\alpha_k} \, .$$

(c) Let us first write

$$9\ell_3(r)^2 = \sum_{j=1}^{3}\sum_{k=1}^{3} e^{ir\cdot(\rho_j+\rho_k)}$$

$$= \sum_{j=1}^{3} e^{2ir\cdot\rho_j} + \sum_{j\neq k} e^{ir\cdot(\rho_j+\rho_k)}$$

$$= 3\ell_3(2r) + 2\sum_{j=1}^{3} e^{-ir\cdot\rho_j} = 3\ell_3(2r) + 6\ell_{\bar{3}}(r),$$

that is,

$$\ell_3(2r) = 3\ell_3(r)^2 - 2\ell_{\bar{3}}(r).$$

Similarly

$$27\ell_3(r)^3 = \sum_{j=1}^{3}\sum_{k=1}^{3}\sum_{l=1}^{3} e^{ir\cdot(\rho_j+\rho_k+\rho_l)}$$

$$= \sum_{j=1}^{3} e^{3ir\cdot\rho_j} + 3\sum_{j\neq k} e^{ir\cdot(2\rho_j+\rho_k)} + 6e^{ir\cdot(\rho_1+\rho_2+\rho_3)}$$

$$= \sum_{j=1}^{3} e^{3ir\cdot\rho_j} + 3\left(\sum_{j=1}^{3}\sum_{k=1}^{3} e^{ir\cdot(2\rho_j+\rho_k)} - \sum_{j=1}^{3} e^{3ir\cdot\rho_j}\right) + 6$$

$$= 6 - 6\ell_3(3r) + 27\ell_3(2r)\ell_3(r)$$

$$= 6 - 6\ell_3(3r) + 27(3\ell_3(r)^2 - 2\ell_{\bar{3}}(r))\ell_3(r)$$

$$= 6 - 6\ell_3(3r) + 27(3\ell_3(r)^2 - 2\ell_{\bar{3}}(r))\ell_3(r),$$

that is,

$$\ell_3(3r) = 1 + 9\ell_3(r)^3 - 9\ell_3(r)\ell_{\bar{3}}(r).$$

(d) Combining the previous results, we can now write

$$\prod_{j=1}^{3}\left[1 - e^{-\beta\varepsilon_q + ir\cdot\alpha_j}\right]\prod_{j=1}^{3}\left[1 - e^{-\beta\varepsilon_q - ir\cdot\alpha_j}\right]$$

$$= (1 - e^{-3\beta\varepsilon_q})^2 + e^{-\beta\varepsilon_q}(e^{-\beta\varepsilon_q} - 1)(1 - e^{-3\beta\varepsilon_q})\left(9\ell_3(r)\ell_{\bar{3}}(r) - 3\right)$$

$$+ e^{-2\beta\varepsilon_q}(e^{-2\beta\varepsilon_q} + 1)\left(9 + 27\ell_3(r)^3 + 27\ell_{\bar{3}}(r)^3 - 54\ell_3(r)\ell_{\bar{3}}(r)\right)$$

$$-e^{-3\beta\varepsilon_q}\left((9\ell_3(r)\ell_{\bar 3}(r)-3)^2-18-54\ell_3(r)^3-54\ell_{\bar 3}(r)^3+108\ell_3(r)\ell_{\bar 3}(r)\right),$$

that is,

$$\prod_{j=1}^{3}\left[1-e^{-\beta\varepsilon_q+ir\cdot\alpha_j}\right]\prod_{j=1}^{3}\left[1-e^{-\beta\varepsilon_q-ir\cdot\alpha_j}\right]$$

$$=(1-e^{-3\beta\varepsilon_q})^2+e^{-\beta\varepsilon_q}(e^{-\beta\varepsilon_q}-1)\left(9\ell_3(r)\ell_{\bar 3}(r)-3\right)$$

$$+e^{-2\beta\varepsilon_q}(e^{-2\beta\varepsilon_q}+1)\left(9+27\ell_3(r)^3+27\ell_{\bar 3}(r)^3-54\ell_3(r)\ell_{\bar 3}(r)\right)$$

$$-e^{-3\beta\varepsilon_q}\left(81\ell_3(r)^2\ell_{\bar 3}(r)^2-27-54\ell_3(r)^3-54\ell_{\bar 3}(r)^3+54\ell_3(r)\ell_{\bar 3}(r)\right).$$

After a straightforward calculation, we finally arrive at

$$\ln\prod_{j=1}^{8}\left[1-e^{-\beta\varepsilon_q+ir\cdot\kappa_j}\right]$$

$$=\ln\left[1+e^{-8\beta\varepsilon_q}-(9\ell_3\ell_{\bar 3}-1)\left(e^{-\beta\varepsilon_q}+e^{-7\beta\varepsilon_q}\right)\right.$$

$$-(81\ell_3^2\ell_{\bar 3}^2-27\ell_3\ell_{\bar 3}+2)\left(e^{-3\beta\varepsilon_q}+e^{-5\beta\varepsilon_q}\right)$$

$$+(27\ell_3^3+27\ell_{\bar 3}^3-27\ell_3\ell_{\bar 3}+1)\left(e^{-2\beta\varepsilon_q}+e^{-6\beta\varepsilon_q}\right)$$

$$\left.+(162\ell_3^2\ell_{\bar 3}^2-54\ell_3^3-54\ell_{\bar 3}^3+18\ell_3\ell_{\bar 3}-2)e^{-4\beta\varepsilon_q}\right].$$

10.4 Roberge-Weiss Symmetry and Phase of the Polyakov Loop (⋆)

The Roberge-Weiss symmetry is a combination of charge conjugation followed by a center transformation and an abelian transformation. The Polyakov loop transforms thus as

$$\ell\to\bar\ell\to e^{-i2\pi/3}\bar\ell.$$

Now, in the case of an imaginary chemical potential, $\bar\ell=\ell^*$. It follows that the Polyakov loop transforms as $\ell\to e^{-i2\pi/3}\ell^*$ under the Roberge-Weiss symmetry. If the symmetry is not broken, we must thus have $\ell=e^{-i2\pi/3}\ell^*$. Denoting $\ell=\rho e^{i\theta}$, this implies that $e^{i\theta}=e^{i\pi/3}$.

10.5 Universality of the Upper Boundary Line in the Columbia Plot (★★★)

(a) We have

$$0 = V'(\ell) = V'_{\text{glue}}(\ell) - \frac{6}{\pi^2} N_f \beta^6 M^2 K_2(\beta M),$$

$$0 = V''(\ell) = V''_{\text{glue}}(\ell),$$

$$0 = V'''(\ell) = V'''_{\text{glue}}(\ell).$$

The last two equations fix the critical values for T and ℓ. Since the quark contribution is absent from these two equations, we deduce that the critical temperature (as well as the critical value of ℓ) is essentially constant along the boundary line. From the first equation, one then deduces the critical value of M. The right-hand side of this equation is the sum of two terms, the first of which does not depend on N_f. We then deduce that the second term cannot depend on N_f either, that is,

$$N_f R_{N_f}^2 K_2(R_{N_f}) = N'_f R_{N'_f}^2 K_2(R_{N'_f}),$$

where R_{N_f} is the critical value of βM.

(b) In the presence of a chemical potential, the asymptotic large mass form obtained in (a) becomes

$$V(\ell, \bar{\ell}) = V_{\text{glue}}(\ell, \bar{\ell}) - \frac{3}{\pi^2} N_f \beta^6 M^2 K_2(\beta M)(e^{-\beta\mu}\ell + e^{\beta\mu}\bar{\ell}).$$

The boundary line in the Columbia plot is determined from $\partial V/\partial \ell = 0$, $\partial V/\partial \bar{\ell} = 0$, and two other equations involving the second and third derivatives of the potential only. These two equations are then independent of N_f, M, or μ. On the other hand, the two first equations write

$$\frac{\partial V_{\text{glue}}}{\partial \ell} = \frac{3}{\pi^2} N_f \beta^6 M^2 K_2(\beta M)e^{-\beta\mu} \quad \text{and} \quad \frac{\partial V_{\text{glue}}}{\partial \bar{\ell}} = \frac{3}{\pi^2} N_f \beta^6 M^2 K_2(\beta M)e^{\beta\mu}.$$

Taking the ratio of these two equations, one finds an equation that does not depend on N_f or M. Together, with the previous two, it determines the critical values of β, ℓ, and $\bar{\ell}$ for a given μ. Moreover, since $e^{\beta\mu}\partial V_{\text{glue}}/\partial\ell$ does not depend on N_f, we deduce that the universal relation for the ratios R_{N_f} still hods true.

Problems of Chap. 11

11.1 Twisted Gauge Fields (★★)

(a) We would like to evaluate $U_c\varphi U_c^\dagger = e^{-i\tau T\pi\sigma^3/2}\varphi e^{i\tau T\pi\sigma^3/2}$. Consider then

$$\frac{d}{d\tau}e^{-i\tau T\pi\sigma^3/2}\varphi e^{i\tau T\pi\sigma^3/2} = -iT\pi e^{-i\tau T\pi\sigma^3/2}\left[\frac{\sigma_3}{2},\varphi\right]e^{i\tau T\pi\sigma^3/2}.$$

Decomposing $\varphi = i\varphi^\kappa t^\kappa$ along a Cartan Weyl basis with $[\sigma_3/2, t^\kappa] = \kappa t^\kappa$, we find

$$\frac{d}{d\tau}e^{-i\tau T\pi\sigma^3/2}\varphi^\kappa t^\kappa e^{i\tau T\pi\sigma^3/2} = -i\kappa T\pi e^{-i\tau T\pi\sigma^3/2}\varphi^\kappa t^\kappa e^{i\tau T\pi\sigma^3/2}.$$

Since this is valid for any φ, we obtain

$$\frac{d}{d\tau}e^{-i\tau T\pi\sigma^3/2}t^\kappa e^{i\tau T\pi\sigma^3/2} = -i\kappa T\pi e^{-i\tau T\pi\sigma^3/2}t^\kappa e^{i\tau T\pi\sigma^3/2},$$

which integrates to

$$e^{-i\tau T\pi\sigma^3/2}t^\kappa e^{i\tau T\pi\sigma^3/2} = e^{-i\kappa\tau T\pi}t^\kappa.$$

It follows that

$$e^{-i\tau T\pi\sigma^3/2}\varphi^\kappa t^\kappa e^{i\tau T\pi\sigma^3/2} = e^{-i\kappa\tau T\pi}\varphi^\kappa t^\kappa.$$

Thus, the components of the field along a Cartan-Weyl basis are transformed as $\varphi^\kappa \to e^{-i\kappa\tau T\pi}\varphi^\kappa$.

(b) We notice that $e^{-i\kappa(\tau=\beta)T\pi} = e^{-i\kappa\pi}$ which equals 1 for $\kappa = 0$ but -1 for $\kappa = \pm 1$. It follows that U_c transforms fields φ_κ that are periodic irrespectively of κ into fields φ_κ that remain periodic for $\kappa = 0$ and that become anti-periodic for $\kappa = \pm 1$.

(c) In the SU(3) case, we should consider the transformation $e^{-i\tau T(4\pi/3)\lambda^3/2}$. Then a given periodic field φ^κ transforms into $e^{-i\kappa_3\tau T4\pi/3}\varphi^\kappa$. We have $e^{-i\kappa_3(\tau=\beta)T4\pi/3} = e^{-i\kappa_3 4\pi/3}$. The transformed field remains periodic for the neutral modes, periodic modulo the phase $e^{\pm i2\pi/3}$ for $\kappa = \pm(1,0), \pm(-1/2, \sqrt{3}/2)$ and $\pm(-1/2, -\sqrt{3}/2)$.

11.2 Center-Symmetric Landau Gauge and Gribov Copies (★★)

(a) Consider $U \in \mathcal{G}$ such that $\bar{A}_c^U = \bar{A}_c$ and A such that $\bar{D}_\mu[\bar{A}_c](A_\mu - \bar{A}_{c,\mu}) = 0$. We have

$$\bar{D}_\mu[\bar{A}_c](A_\mu^U - \bar{A}_{c,\mu}) = \bar{D}_\mu[\bar{A}_c^U](A_\mu^U - \bar{A}_{c,\mu}^U)$$

$$= U\bar{D}_\mu[\bar{A}_c](A_\mu - \bar{A}_{c,\mu})U^\dagger = 0.$$

Therefore, the set of solutions to $F[A] = 0$ is invariant under U.

(b) Since $F[(A^{U_0})^U] = UF[A^{U_0}]U^\dagger$, we have $F^a[(A^{U_0})^U] = \mathcal{R}^{ab}F^b[A^{U_0}]$ for a certain orthogonal matrix $\mathcal{R} \in O(N^2 - 1)$. Then

$$\frac{\left|\det \left.\frac{\delta F[(A^{U_0})^U]}{\delta U_0}\right|_{U_0^{(i)}(A)}\right|^{-1}}{\left|\det \left.\frac{\delta F[A^{U_0}]}{\delta U_0}\right|_{U_0^{(i)}(A)}\right|^{-1}} = \underbrace{|\det \mathcal{R}|^{-1}}_{=1} = \frac{\left|\det \left.\frac{\delta F[A^{U_0}]}{\delta U_0}\right|_{U_0^{(i)}(A)}\right|^{-1}}{\left|\det \left.\frac{\delta F[A^{U_0}]}{\delta U_0}\right|_{U_0^{(i)}(A)}\right|^{-1}} = 1 \, .$$

11.3 Gluonic Determinant (★★)

(a) Evaluate the right-hand side of the matrix relation.

(b) We write $Q^2 P_{\mu\nu}^{\perp}(Q) + m^2 \delta_{\mu\nu} = (Q^2 + m^2) P_{\mu\nu}^{\perp}(Q) + m^2 P_{\mu\nu}^{\parallel}(Q)$, whose inverse is

$$\frac{P_{\mu\nu}^{\perp}(Q)}{Q^2 + m^2} + \frac{P_{\mu\nu}^{\parallel}(Q)}{m^2} \, .$$

We can now use the previous formula to evaluate the determinant, and we find

$$m^2 (Q^2 + m^2)^{d-1} \left[\frac{\bar{Q} \cdot P^{\perp}(Q) \cdot \bar{Q}}{Q^2 + m^2} + \frac{\bar{Q} \cdot P^{\parallel}(Q) \cdot \bar{Q}}{m^2} \right]$$

$$= m^2 (Q^2 + m^2)^{d-1} \left[\frac{\bar{Q}^2 - (\bar{Q} \cdot Q)^2 / Q^2}{Q^2 + m^2} + \frac{(\bar{Q} \cdot Q)^2}{Q^2 m^2} \right]$$

$$= m^2 (Q^2 + m^2)^{d-1} \left[\frac{\bar{Q}^2}{Q^2 + m^2} + \frac{(\bar{Q} \cdot Q)^2}{Q^2} \left(\frac{1}{m^2} - \frac{1}{Q^2 + m^2} \right) \right]$$

$$= m^2 (Q^2 + m^2)^{d-1} \left[\frac{\bar{Q}^2}{Q^2 + m^2} + \frac{(\bar{Q} \cdot Q)^2}{m^2 (Q^2 + m^2)} \right]$$

$$= (Q^2 + m^2)^{d-2} \left[m^2 \bar{Q}^2 + (\bar{Q} \cdot Q)^2 \right] \, .$$

References

1. M. Gell-Mann, Symmetries of baryons and mesons. Phys. Rev. **125**, 1067 (1962)
2. M. Gell-Mann, A schematic model of Baryons and Mesons. Phys. Lett. **8**, 214 (1964)
3. G. Zweig, An SU(3) model for strong interaction symmetry and its breaking. Version 2, in *Developments in the Quark Theory of Hadrons*, vol. 1, ed. by D. Lichtenberg, S. Rosen (1964), pp. 22–101
4. E.D. Bloom et al., High-energy inelastic e p scattering at 6-degrees and 10-degrees. Phys. Rev. Lett. **23**, 930 (1969)
5. M. Breidenbach et al., Observed behavior of highly inelastic electron-proton scattering. Phys. Rev. Lett. **23**, 935 (1969)
6. H. Fritzsch, M. Gell-Mann, H. Leutwyler, Advantages of the color octet gluon picture. Phys. Lett. **47B**, 365 (1973)
7. J. Greensite, An introduction to the confinement problem. Lect. Notes Phys. **821**, 1 (2011)
8. M. Creutz, Monte Carlo study of quantized SU(2) gauge theory. Phys. Rev. D **21**, 2308 (1980)
9. H.D. Politzer, Reliable perturbative results for strong interactions? Phys. Rev. Lett. **30**, 1346 (1973)
10. D.J. Gross, F. Wilczek, Ultraviolet behavior of nonabelian gauge theories. Phys. Rev. Lett. **30**, 1343 (1973)
11. M.G. Alford, K. Rajagopal, F. Wilczek, Color flavor locking and chiral symmetry breaking in high density QCD. Nucl. Phys. B **537**, 443 (1999)
12. E.W. Kolb, M.S. Turner, The early universe. Front. Phys. **69**, 1 (1990)
13. N.K. Glendenning, *Compact Stars: Nuclear Physics, Particle Physics, and General Relativity* (Springer, New York, USA, 1997), 390 p.
14. F. Karsch, E. Laermann, C. Schmidt, The Chiral critical point in three-flavor QCD. Phys. Lett. B **520**, 41 (2001)
15. R.V. Gavai, S. Gupta, The critical end point of QCD. Phys. Rev. D **71**, 114014 (2005).
16. L. McLerran, K. Redlich, C. Sasaki, Quarkyonic matter and chiral symmetry breaking. Nucl. Phys. A **824**, 86 (2009)
17. T. Kojo, Y. Hidaka, L. McLerran, R.D. Pisarski, Quarkyonic chiral spirals. Nucl. Phys. A **843**, 37 (2010).
18. A.J. Mizher, M.N. Chernodub, E.S. Fraga, Phase diagram of hot QCD in an external magnetic field: possible splitting of deconfinement and chiral transitions. Phys. Rev. D **82**, 105016 (2010)
19. M.N. Chernodub, Superconductivity of QCD vacuum in strong magnetic field. Phys. Rev. D **82**, 085011 (2010)
20. D. Kharzeev, K. Landsteiner, A. Schmitt, H.U. Yee, Strongly interacting matter in magnetic fields. Lect. Notes Phys. **871**, pp.1 (2013)
21. M. Buballa, NJL model analysis of quark matter at large density. Phys. Rept. **407**, 205 (2005)
22. C. Ratti, S. Roessner, M.A. Thaler, W. Weise, Thermodynamics of the PNJL model. Eur. Phys. J. C **49**, 213 (2007)

© The Author(s), under exclusive license to Springer Nature Switzerland AG 2022
U. Reinosa, *Perturbative Aspects of the Deconfinement Transition*, Lecture Notes in Physics 1006, https://doi.org/10.1007/978-3-031-11375-8

23. S. Roessner, T. Hell, C. Ratti, W. Weise, The chiral and deconfinement crossover transitions: PNJL model beyond mean field. Nucl. Phys. A **814**, 118 (2008)
24. B.J. Schaefer, J.M. Pawlowski, J. Wambach, The phase structure of the Polyakov–Quark–Meson model. Phys. Rev. D **76**, 074023 (2007)
25. T.K. Herbst, J.M. Pawlowski, B.J. Schaefer, The phase structure of the Polyakov–Quark–Meson model beyond mean field. Phys. Lett. B **696**, 58 (2011)
26. E. Braaten, R.D. Pisarski, Soft amplitudes in hot gauge theories: A general analysis. Nucl. Phys. B **337**, 569 (1990)
27. J.O. Andersen, E. Braaten, E. Petitgirard, M. Strickland, HTL perturbation theory to two loops. Phys. Rev. D **66**, 085016 (2002)
28. J.O. Andersen, L.E. Leganger, M. Strickland, N. Su, Three-loop HTL QCD thermodynamics. JHEP **1108**, 053 (2011)
29. K.G. Wilson, Confinement of Quarks. Phys. Rev. D **10**, 2445 (1974)
30. I. Montvay, G. Munster, *Quantum Fields on a Lattice* (Cambridge Monographs on Mathematical Physics, 1994)
31. C.W. Bernard et al., The QCD spectrum with three quark flavors. Phys. Rev. D **64**, 054506 (2001)
32. F. Karsch, E. Laermann, A. Peikert, Quark mass and flavor dependence of the QCD phase transition. Nucl. Phys. B **605**, 579 (2001)
33. A. Bazavov et al. [HotQCD Collaboration], Equation of state in $(2 + 1)$-flavor QCD. Phys. Rev. D **90**, 094503 (2014)
34. S. Borsanyi, Z. Fodor, C. Hoelbling, S.D. Katz, S. Krieg, K.K. Szabo, Full result for the QCD equation of state with 2+1 flavors. Phys. Lett. B **730**, 99 (2014)
35. S. Borsanyi, G. Endrodi, Z. Fodor, S.D. Katz, K.K. Szabo, Precision SU(3) lattice thermodynamics for a large temperature range. JHEP **1207**, 056 (2012)
36. P. de Forcrand, Simulating QCD at finite density. PoS LAT **2009**, 010 (2009)
37. O. Philipsen, Lattice QCD at non-zero temperature and baryon density. arXiv:1009.4089 [hep-lat]
38. Y. Tanizaki, Y. Hidaka, T. Hayata, Lefschetz-thimble analysis of the sign problem in one-site fermion model. New J. Phys. **18**(3), 033002 (2016)
39. G. Aarts, L. Bongiovanni, E. Seiler, D. Sexty, I.O. Stamatescu, Controlling complex Langevin dynamics at finite density. Eur. Phys. J. A **49**, 89 (2013)
40. J.S. Schwinger, On the Green's functions of quantized fields. 1. Proc. Nat. Acad. Sci. **37**, 452 (1951)
41. J.S. Schwinger, On the Green's functions of quantized fields. 2. Proc. Nat. Acad. Sci. **37**, 455 (1951)
42. F.J. Dyson, The S matrix in quantum electrodynamics. Phys. Rev. **75**, 1736 (1949)
43. L. von Smekal, R. Alkofer, A. Hauck, The infrared behavior of gluon and ghost propagators in Landau gauge QCD. Phys. Rev. Lett. **79**, 3591 (1997)
44. R. Alkofer, L. von Smekal, The infrared behavior of QCD Green's functions: Confinement, dynamical symmetry breaking, and hadrons as relativistic bound states. Phys.Rept.**353**, 281 (2001)
45. C. Wetterich, Exact evolution equation for the effective potential. Phys. Lett. B **301**, 90 (1993)
46. U. Ellwanger, M. Hirsch, A. Weber, The Heavy quark potential from Wilson's exact renormalization group. Eur. Phys. J. C **1**, 563 (1998)
47. F. Freire, D.F. Litim, J.M. Pawlowski, Gauge invariance and background field formalism in the exact renormalization group. Phys. Lett. B **495**, 256 (2000)
48. K. Fukushima, K. Kashiwa, Polyakov loop and QCD thermodynamics from the gluon and ghost propagators. Phys. Lett. B **723**, 360 (2013)
49. L. Fister, J.M. Pawlowski, Confinement from correlation functions. Phys. Rev. D **88**, 045010 (2013)
50. C. Feuchter, H. Reinhardt, Variational solution of the Yang-Mills Schrodinger equation in Coulomb gauge. Phys. Rev. D **70**, 105021 (2004)

51. H. Reinhardt, J. Heffner, Effective potential of the confinement order parameter in the Hamiltonian approach. Phys. Rev. D **88**, 045024 (2013)
52. D. Binosi, J. Papavassiliou, Pinch technique: theory and applications. Phys. Rept. **479**, 1 (2009)
53. P. Boucaud, J.P. Leroy, A. Le Yaouanc, J. Micheli, O. Pene, J. Rodriguez-Quintero, On the infrared behaviour of the Landau-gauge ghost propagator. JHEP **06**, 099 (2008)
54. P. Boucaud, J.P. Leroy, A.L. Yaouanc, J. Micheli, O. Pene, J. Rodriguez-Quintero, The infrared behaviour of the pure Yang-Mills green functions. Few Body Syst. **53**, 387 (2012)
55. A. Cucchieri, T. Mendes, Phys. Rev. Lett. **100**, 241601 (2008); arXiv:1001.2584 [hep-lat]
56. A. Cucchieri, T. Mendes, Constraints on the infrared behavior of the ghost propagator in Yang-Mills theories. Phys. Rev. D **78**, 094503 (2008)
57. A. Cucchieri, T. Mendes, Landau-gauge propagators in Yang-Mills theories at beta = 0: Massive solution versus conformal scaling. Phys. Rev. D **81**, 016005 (2010)
58. V.G. Bornyakov, V.K. Mitrjushkin, M. Muller-Preussker, Infrared behavior and Gribov ambiguity in SU(2) lattice gauge theory. Phys. Rev. D **79**, 074504 (2009)
59. V.G. Bornyakov, V.K. Mitrjushkin, M. Muller-Preussker, SU(2) lattice gluon propagator: Continuum limit, finite-volume effects and infrared mass scale m(IR). Phys. Rev. D **81**, 054503 (2010)
60. I.L. Bogolubsky, E.M. Ilgenfritz, M. Muller-Preussker, A. Sternbeck, Lattice gluodynamics computation of Landau gauge Green's functions in the deep infrared. Phys. Lett. B **676**, 69 (2009)
61. D. Dudal, O. Oliveira, N. Vandersickel, Indirect lattice evidence for the Refined Gribov-Zwanziger formalism and the gluon condensate $\langle A^2 \rangle$ in the Landau gauge. Phys. Rev. D **81**, 074505 (2010)
62. A. Maas, Describing gauge bosons at zero and finite temperature. Phys. Rept. **524**, 203 (2013)
63. T. Kugo, I. Ojima, Local covariant operator formalism of nonabelian gauge theories and Quark confinement problem. Prog. Theor. Phys. Suppl. **66**, 1 (1979)
64. D. Zwanziger, Nonperturbative Landau gauge and infrared critical exponents in QCD. Phys. Rev. D **65**, 094039 (2002)
65. C.S. Fischer, R. Alkofer, Non-perturbative propagators, running coupling and dynamical Quark mass of Landau gauge QCD. Phys. Rev. D **67**, 094020 (2003)
66. C.S. Fischer, H. Gies, Renormalization flow of Yang-Mills propagators. JHEP **0410**, 048 (2004)
67. C.S. Fischer, J.M. Pawlowski, Uniqueness of infrared asymptotics in Landau gauge Yang-Mills theory. Phys. Rev. D **75**, 025012 (2007)
68. A.C. Aguilar, A.A. Natale, A dynamical gluon mass solution in a coupled system of the Schwinger-Dyson equations. JHEP **0408**, 057 (2004)
69. Ph. Boucaud et al., Is the QCD ghost dressing function finite at zero momentum?. JHEP **06** 001 (2006)
70. A.C. Aguilar, J. Papavassiliou, Power-law running of the effective gluon mass. Eur. Phys. J. A **35**, 189 (2008)
71. A.C. Aguilar, D. Binosi, J. Papavassiliou, Gluon and ghost propagators in the Landau gauge: Deriving lattice results from Schwinger-Dyson equations. Phys. Rev. D **78** 025010 (2008)
72. C.S. Fischer, A. Maas, J.M. Pawlowski, On the infrared behavior of Landau gauge Yang-Mills theory. Ann. Phys. **324**, 2408 (2009)
73. J. Braun, H. Gies, J.M. Pawlowski, Quark confinement from color confinement. Phys. Lett. B **684**, 262 (2010)
74. F. Marhauser, J.M. Pawlowski, Confinement in Polyakov gauge. arXiv:0812.1144 [hep-ph].
75. J. Braun, L.M. Haas, F. Marhauser, J.M. Pawlowski, Phase structure of two-flavor QCD at finite chemical potential. Phys. Rev. Lett. **106**, 022002 (2011)
76. J. Braun, A. Eichhorn, H. Gies, J.M. Pawlowski, On the nature of the phase transition in SU(N), Sp(2) and E(7) Yang-Mills theory. Eur. Phys. J. C **70**, 689 (2010)
77. J.M. Pawlowski, The QCD phase diagram: Results and challenges. AIP Conf. Proc. **1343**, 75 (2011)

78. C.S. Fischer, A. Maas, J.A. Muller, Chiral and deconfinement transition from correlation functions: SU(2) vs. SU(3). Eur. Phys. J. C **68**, 165 (2010)
79. C.S. Fischer, QCD at finite temperature and chemical potential from Dyson-Schwinger equations. Prog. Part. Nucl. Phys. **105**, 1 (2019)
80. H. Reinhardt, G. Burgio, D. Campagnari, E. Ebadati, J. Heffner, M. Quandt, P. Vastag, H. Vogt, Hamiltonian approach to QCD in Coulomb gauge - a survey of recent results. Adv. High Energy Phys. **2018**, 2312498 (2018)
81. W.j. Fu, J. M. Pawlowski, F. Rennecke, QCD phase structure at finite temperature and density. Phys. Rev. D **101**(5), 054032 (2020)
82. M. Tissier, N. Wschebor, Infrared propagators of Yang-Mills theory from perturbation theory. Phys. Rev. D **82**, 101701 (2010)
83. M. Tissier, N. Wschebor, An infrared Safe perturbative approach to Yang-Mills correlators. Phys. Rev. D **84**, 045018 (2011)
84. V.N. Gribov, Quantization of non-Abelian gauge theories. Nucl. Phys. B **139**, 1 (1978)
85. M. Peláez, U. Reinosa, J. Serreau, M. Tissier, N. Wschebor, Small parameters in infrared quantum chromodynamics. Phys. Rev. D **96**(11), 114011 (2017)
86. M. Peláez, U. Reinosa, J. Serreau, M. Tissier, N. Wschebor, Spontaneous chiral symmetry breaking in the massive Landau gauge: realistic running coupling. Phys. Rev. D **103**(9), 094035 (2021)
87. J. Maelger, U. Reinosa, J. Serreau, Localized rainbows in the QCD phase diagram. Phys. Rev. D **101**(1), 014028 (2020)
88. G. Curci, R. Ferrari, On a Class of Lagrangian Models for Massive and Massless Yang-Mills Fields. Nuovo Cim. A **32**, 151 (1976)
89. E.S. Fradkin, I.V. Tyutin, Feynman rules for the massless yang-mills field renormalizability of the theory of the massive yang-mills field. Phys. Lett. B **30** 562 (1969)
90. M. Peláez, M. Tissier, N. Wschebor, Three-point correlation functions in Yang-Mills theory. Phys. Rev. **D88**, 125003 (2013)
91. U. Reinosa, J. Serreau, M. Tissier, N. Wschebor, Yang-Mills correlators at finite temperature: A perturbative perspective. Phys. Rev. D **89**(10), 105016 (2014)
92. U. Reinosa, J. Serreau, M. Tissier, N. Wschebor, Deconfinement transition in SU(N) theories from perturbation theory. Phys. Lett. B **742**, 61 (2015)
93. U. Reinosa, J. Serreau, M. Tissier, N. Wschebor, Deconfinement transition in SU(2) Yang-Mills theory: A two-loop study. Phys. Rev. D **91**, 045035 (2015)
94. U. Reinosa, J. Serreau, M. Tissier, N. Wschebor, Two-loop study of the deconfinement transition in Yang-Mills theories: SU(3) and beyond. Phys. Rev. D **93**, 105002 (2016)
95. U. Reinosa, J. Serreau, M. Tissier, Perturbative study of the QCD phase diagram for heavy quarks at nonzero chemical potential. Phys. Rev. D **92**, 025021 (2015)
96. U. Reinosa, J. Serreau, M. Tissier, A. Tresmontant, Yang-Mills correlators across the deconfinement phase transition. Phys. Rev. D **95**(4), 045014 (2017)
97. U. Reinosa, J. Serreau, M. Tissier, N. Wschebor, How nonperturbative is the infrared regime of Landau gauge Yang-Mills correlators? Phys. Rev. D **96**(1), 014005 (2017)
98. J. Maelger, U. Reinosa, J. Serreau, Perturbative study of the QCD phase diagram for heavy quarks at nonzero chemical potential: Two-loop corrections. Phys. Rev. D **97**(7), 074027 (2018)
99. J. Maelger, U. Reinosa, J. Serreau, Universal aspects of the phase diagram of QCD with heavy quarks. Phys. Rev. D **98**(9), 094020 (2018)
100. D.M. van Egmond, U. Reinosa, J. Serreau, M. Tissier, A novel background field approach to the confinement-deconfinement transition. SciPost Phys. **12**(3), 087 (2022) doi: 10.21468/SciPostPhys.12.3.087 [arXiv:2104.08974 [hep-ph]].
101. M.L. Bellac, *Thermal Field Theory* (Cambridge Monographs on Mathematical Physics)
102. M. Laine, A. Vuorinen, Basics of thermal field theory. Lect. Notes Phys. **925**, pp.1 (2016). [arXiv:1701.01554 [hep-ph]]
103. L.D. Faddeev, V.N. Popov, Feynman diagrams for the Yang-Mills field. Phys. Lett. B **25**, 29 (1967)

104. S. Pokorski, *Gauge Field Theories* (Cambridge University Press)
105. C. Becchi, A. Rouet, R. Stora, Renormalization of gauge theories. Ann. Phys. **98**, 287 (1976)
106. J. Zinn-Justin, Renormalization of gauge theories. Lect. Notes Phys. **37**, 1 (1975)
107. S. Weinberg, *Quantum Field Theory*, vol. 2
108. A. Niemi, Gribov vacuum copies and interpolation in the Coulomb and Landau gauges Of Su(n) Yang-mills theories. Nucl. Phys. B **189**, 115 (1981)
109. H. Neuberger, Nonperturbative BRS invariance. Phys. Lett. B **175**, 69 (1986)
110. H. Neuberger, Nonperturbative BRS invariance and the Gribov problem. Phys. Lett. B **183**, 337 (1987)
111. L. von Smekal, D. Mehta, A. Sternbeck, A.G. Williams, Modified lattice Landau gauge. PoS Lattice **2007**, 382 (2007)
112. L. von Smekal, A. Jorkowski, D. Mehta, A. Sternbeck, Lattice Landau gauge via stereographic projection. PoS Confinement **8**, 048 (2008)
113. D. Zwanziger, Local and renormalizable action from the Gribov horizon. Nucl. Phys. B **323**, 513 (1989)
114. D. Zwanziger, Renormalizability of the critical limit of lattice gauge theory by BRS invariance. Nucl. Phys. B **399**, 477 (1993)
115. M.A.L. Capri et al., Exact nilpotent nonperturbative BRST symmetry for the Gribov-Zwanziger action in the linear covariant gauge. Phys. Rev. D **92**(4), 045039 (2015)
116. M.A.L. Capri et al., Local and BRST-invariant Yang-Mills theory within the Gribov horizon. Phys. Rev. D **94**(2), 025035 (2016)
117. D. Dudal, J.A. Gracey, S.P. Sorella, N. Vandersickel, H. Verschelde, A Refinement of the Gribov-Zwanziger approach in the Landau gauge: Infrared propagators in harmony with the lattice results. Phys. Rev. D **78**, 065047 (2008)
118. D. Dudal, J.A. Gracey, S.P. Sorella, N. Vandersickel, H. Verschelde, The Landau gauge gluon and ghost propagator in the refined Gribov-Zwanziger framework in 3 dimensions. Phys. Rev. D **78**, 125012 (2008)
119. J. Serreau, M. Tissier, Lifting the Gribov ambiguity in Yang-Mills theories. Phys. Lett. B **712**, 97 (2012)
120. L. von Smekal, M. Ghiotti, A.G. Williams, Decontracted double BRST on the lattice. Phys. Rev. D **78**, 085016 (2008)
121. A.P. Young, *Spin Glasses and Random Fields* (vol. 12) (World Scientific, 1997)
122. N. Wschebor, Some non-renormalization theorems in Curci-Ferrari model. Int. J. Mod. Phys. A **23**, 2961 (2008)
123. G. Parisi, N. Sourlas, Random magnetic fields, supersymmetry and negative dimensions. Phys. Rev. Lett. **43**, 744 (1979)
124. M. Tissier, G. Tarjus, Supersymmetry and its spontaneous breaking in the random field ising model. Phys. Rev. Lett. **107**, 041601 (2011)
125. D. Dudal, C.P. Felix, L.F. Palhares, F. Rondeau, D. Vercauteren, The BRST-invariant vacuum state of the Gribov-Zwanziger theory. Eur. Phys. J. C **79**(9), 731 (2019)
126. M. Tissier, Gribov copies, avalanches and dynamic generation of a gluon mass. Phys. Lett. B **784**, 146 (2018)
127. A.K. Cyrol, L. Fister, M. Mitter, J.M. Pawlowski, N. Strodthoff, Landau gauge Yang-Mills correlation functions. Phys. Rev. D **94**(5), 054005 (2016)
128. L. von Smekal, R. Alkofer, A. Hauck, The Infrared behavior of gluon and ghost propagators in Landau gauge QCD. Phys. Rev. Lett. **79**, 3591 (1997)
129. P. Boucaud, J.P. Leroy, A. Le Yaouanc, J. Micheli, O. Pene, J. Rodriguez-Quintero, On the IR behaviour of the Landau-gauge ghost propagator. JHEP **0806**, 099 (2008)
130. L. von Smekal, A. Hauck, R. Alkofer, A solution to coupled Dyson-Schwinger equations for Gluons and Ghosts in Landau gauge. Annals Phys. **267**, 1 (1998); Erratum: [Annals Phys. **269**, 182 (1998)]
131. A.C. Aguilar, D. Binosi, J. Papavassiliou, Gluon and ghost propagators in the Landau gauge: Deriving lattice results from Schwinger-Dyson equations. Phys. Rev. D **78**, 025010 (2008)

132. J. Rodriguez-Quintero, On the massive gluon propagator, the PT-BFM scheme and the low-momentum behaviour of decoupling and scaling DSE solutions. JHEP **1101**, 105 (2011)

133. M. Quandt, H. Reinhardt, J. Heffner, Covariant variational approach to Yang-Mills theory. Phys. Rev. D **89**(6), 065037 (2014)

134. M.Q. Huber, L. von Smekal, Spurious divergences in Dyson-Schwinger equations. JHEP **1406**, 015 (2014)

135. D. Binosi, D. Ibanez, J. Papavassiliou, The all-order equation of the effective gluon mass. Phys. Rev. D **86**, 085033 (2012)

136. A.C. Aguilar, D. Binosi, J. Papavassiliou, The gluon mass generation mechanism: a concise primer. Front. Phys. (Beijing) **11**(2), 111203 (2016)

137. A.C. Aguilar, D. Binosi, J. Papavassiliou, Schwinger mechanism in linear covariant gauges. Phys. Rev. D **95**(3), 034017 (2017)

138. M.Q. Huber, Correlation functions of Landau gauge Yang-Mills theory. Phys. Rev. D **101**(11), 11 (2020).

139. G. Eichmann, J.M. Pawlowski, J.M. Silva, [arXiv:2107.05352 [hep-ph]]

140. A.C. Aguilar, M.N. Ferreira, J. Papavassiliou, Exploring smoking-gun signals of the Schwinger mechanism in QCD. Phys. Rev. D **105**(1), 014030 (2022)

141. J.A. Gracey, M. Peláez, U. Reinosa, M. Tissier, Two loop calculation of Yang-Mills propagators in the Curci-Ferrari model. Phys. Rev. D **100**(3), 034023 (2019)

142. A. Athenodorou, D. Binosi, P. Boucaud, F. De Soto, J. Papavassiliou, J. Rodriguez-Quintero, S. Zafeiropoulos, On the zero crossing of the three-gluon vertex. Phys. Lett. B **761**, 444 (2016)

143. A.G. Duarte, O. Oliveira, P.J. Silva, Lattice gluon and ghost propagators, and the strong coupling in pure SU(3) Yang-Mills theory: finite lattice spacing and volume effects. Phys. Rev. D **94**(1), 014502 (2016)

144. U.M. Heller, F. Karsch, J. Rank, The Gluon propagator at high temperature. Phys. Lett. B **355**, 511 (1995)

145. U. Heller, F. Karsch, J. Rank, The Gluon propagator at high temperature: Screening, improvement and nonzero momenta. Phys. Rev. D **57**, 1438 (1998)

146. A. Cucchieri, F. Karsch, P. Petreczky, Magnetic screening in hot nonAbelian gauge theory. Phys. Lett. B **497**, 80 (2001)

147. A. Cucchieri, F. Karsch, P. Petreczky, Propagators and dimensional reduction of hot SU(2) gauge theory. Phys. Rev. D **64**, 036001 (2001)

148. A. Cucchieri, A. Maas, T. Mendes, Infrared properties of propagators in Landau-gauge pure Yang-Mills theory at finite temperature. Phys. Rev. D **75**, 076003 (2007)

149. A. Cucchieri, T. Mendes, Electric and magnetic Landau-gauge gluon propagators in finite-temperature SU(2) gauge theory. PoS FACESQCD, 007 (2010)

150. A. Cucchieri, T. Mendes, Electric and magnetic screening masses around the deconfinement transition. PoS Lattice **2011**, 206 (2011)

151. R. Aouane, V.G. Bornyakov, E.M. Ilgenfritz, V.K. Mitrjushkin, M. Müller-Preussker, A. Sternbeck, Landau gauge gluon and ghost propagators at finite temperature from quenched lattice QCD. Phys. Rev. D **85**, 034501 (2012)

152. A. Maas, J.M. Pawlowski, L. von Smekal, D. Spielmann, The gluon propagator close to criticality. Phys. Rev. D **85**, 034037 (2012)

153. P.J. Silva, O. Oliveira, P. Bicudo, N. Cardoso, Gluon mass at finite temperature from Landau gauge gluon propagator in lattice QCD. Phys. Rev. **D89**, 074503 (2014)

154. L. Fister, J.M. Pawlowski, Yang-Mills correlation functions at finite temperature. arXiv:1112.5440 [hep-ph]

155. M.Q. Huber, L. von Smekal, On two- and three-point functions of Landau gauge Yang-Mills theory. PoS Lattice **2013**, 364 (2013)

156. M. Quandt, H. Reinhardt, A covariant variational approach to Yang-Mills Theory at finite temperatures. Phys. Rev. D **92**(2), 025051 (2015)

157. U. Reinosa, Gribov copies, restrictions of the functional integral, and Slavnov-Taylor identities in preparation

158. A.M. Polyakov, Thermal properties of gauge fields and Quark liberation. Phys. Lett. **72B**, 477 (1978)
159. B. Svetitsky, Symmetry aspects of finite temperature confinement transitions. Phys. Rept. **132**, 1 (1986)
160. O. Kaczmarek, F. Karsch, P. Petreczky, F. Zantow, Heavy quark anti-quark free energy and the renormalized Polyakov loop. Phys. Lett. B **543**, 41 (2002)
161. B. Lucini, M. Teper, U. Wenger, Properties of the deconfining phase transition in SU(N) gauge theories. JHEP **0502**, 033 (2005)
162. J. Greensite, The potential of the effective Polyakov line action from the underlying lattice gauge theory. Phys. Rev. D **86**, 114507 (2012)
163. D. Smith, A. Dumitru, R. Pisarski, L. von Smekal, Effective potential for SU(2) Polyakov loops and Wilson loop eigenvalues. Phys. Rev. D **88**(5), 054020 (2013)
164. R.D. Pisarski, Notes on the deconfining phase transition. hep-ph/0203271
165. L.F. Abbott, The background field method beyond one loop. Nucl. Phys. B **185**, 189-203 (1981)
166. L.F. Abbott, Introduction to the background field method. Acta Phys. Polon. B **13**, 33 (1982)
167. S. Weinberg, *The Quantum Theory of Fields Vol. 2: Modern Applications* (Univ. Pr., Cambridge, UK, 489), p. 1996
168. J.B. Zuber, *Invariances in Physics and Group Theory* (2014) (unpublished)
169. D. Epple, H. Reinhardt, W. Schleifenbaum, Confining solution of the Dyson-Schwinger equations in Coulomb gauge. Phys. Rev. D **75**, 045011 (2007)
170. R. Alkofer, C.S. Fischer, F.J. Llanes-Estrada, Dynamically induced scalar quark confinement. Mod. Phys. Lett. A **23**, 1105 (2008)
171. C.S. Fischer, Deconfinement phase transition and the quark condensate. Phys. Rev. Lett. **103**, 052003 (2009)
172. C.S. Fischer, J.A. Mueller, Chiral and deconfinement transition from Dyson-Schwinger equations. Phys. Rev. D **80**, 074029 (2009)
173. H. Reinhardt, J. Heffner, The effective potential of the confinement order parameter in the Hamilton approach. Phys. Lett. B **718**, 672 (2012)
174. M. Quandt, H. Reinhardt, Covariant variational approach to Yang-Mills theory: effective potential of the Polyakov loop. Phys. Rev. D **94**(6), 065015 (2016)
175. B. Lucini, M. Panero, SU(N) gauge theories at large N. Phys. Rept. **526**, 93 (2013)
176. N. Weiss, The effective potential for the order parameter of gauge theories at finite temperature. Phys. Rev. D **24**, 475 (1981)
177. D.J. Gross, R.D. Pisarski, L.G. Yaffe, QCD and instantons at finite temperature. Rev. Mod. Phys. **53**, 43 (1981)
178. M. Quandt, H. Reinhardt, Covariant variational approach to Yang-Mills theory: Thermodynamics. Phys. Rev. D **96**(5), 054029 (2017)
179. B. Beinlich, F. Karsch, A. Peikert, SU(3) latent heat and surface tension from tree level and tadpole improved actions. Phys. Lett. B **390**, 268 (1997)
180. T.K. Herbst, J. Luecker, J.M. Pawlowski, Confinement order parameters and fluctuations. arXiv:1510.03830 [hep-ph]
181. K.J.M. Moriarty, A phase transition in SU(4) four-dimensional Lattice gauge theory. Phys. Lett. **106B**, 130 (1981)
182. F. Green, F. Karsch, The SU(4) deconfining transition at strong coupling: A Monte Carlo study. Phys. Rev. D **29**, 2986 (1984)
183. G.G. Batrouni, B. Svetitsky, The order of the finite temperature phase transition in the SU(4) gauge theory. Phys. Rev. Lett. **52**, 2205 (1984)
184. T.D. Cohen, center symmetry and area laws. Phys. Rev. D **90**, 047703 (2014)
185. A. Dumitru, Y. Guo, Y. Hidaka, C.P.K. Altes, R.D. Pisarski, Effective matrix model for deconfinement in pure gauge theories. Phys. Rev. D **86**, 105017 (2012)
186. S. Gupta, K. Huebner, O. Kaczmarek, Renormalized Polyakov loops in many representations. Phys. Rev. D **77**, 034503 (2008)

187. A. Dumitru, Y. Hatta, J. Lenaghan, K. Orginos, R.D. Pisarski, Deconfining phase transition as a matrix model of renormalized Polyakov loops. Phys. Rev. D **70**, 034511 (2004)

188. K. Fukushima, Y. Hidaka, A model study of the sign problem in the mean-field approximation. Phys. Rev. D **75**, 036002 (2007)

189. S. Roessner, C. Ratti, W. Weise, Polyakov loop, diquarks and the two-flavor phase diagram. Phys. Rev. D **75**, 034007 (2007)

190. B.W. Mintz, R. Stiele, R.O. Ramos, J. Schaffner-Bielich, Phase diagram and surface tension in the three-flavor Polyakov-quark-meson model. Phys. Rev. D **87**(3), 036004 (2013)

191. R. Stiele, J. Schaffner-Bielich, Phase diagram and nucleation in the Polyakov-loop-extended Quark-Meson truncation of QCD with the unquenched Polyakov-loop potential. Phys. Rev. D **93**(9), 094014 (2016)

192. P. Kovács, Zs. Szép, G. Wolf, Existence of the critical endpoint in the vector meson extended linear sigma model. Phys. Rev. D **93**(11), 114014 (2016)

193. A. Folkestad, J.O. Andersen, Thermodynamics and phase diagrams of Polyakov-loop extended chiral models. Phys. Rev. D **99**(5), 054006 (2019)

194. A. Dumitru, R.D. Pisarski, D. Zschiesche, Dense quarks, and the fermion sign problem, in a SU(N) matrix model. Phys. Rev. D **72**, 065008 (2005)

195. H. Nishimura, M.C. Ogilvie, K. Pangeni, Complex saddle points in QCD at finite temperature and density. Phys. Rev. D **90**(4), 045039 (2014)

196. H. Nishimura, M.C. Ogilvie, K. Pangeni, Complex saddle points and disorder lines in QCD at finite temperature and density. Phys. Rev. D **91**(5), 054004 (2015)

197. C.S. Fischer, L. Fister, J. Luecker, J.M. Pawlowski, Polyakov loop potential at finite density. Phys. Lett. B **732**, 273 (2014)

198. C.S. Fischer, J. Luecker, J.M. Pawlowski, Phase structure of QCD for heavy quarks. Phys. Rev. D **91**, 014024 (2015)

199. M. Fromm, J. Langelage, S. Lottini, O. Philipsen, The QCD deconfinement transition for heavy quarks and all baryon chemical potentials. JHEP **1201**, 042 (2012)

200. K. Kashiwa, R.D. Pisarski, V.V. Skokov, Critical endpoint for deconfinement in matrix and other effective models. Phys. Rev. D **85**, 114029 (2012)

201. D. Kroff, U. Reinosa, Gribov-Zwanziger type model action invariant under background gauge transformations. Phys. Rev. D **98**(3), 034029 (2018)

202. A. Roberge, N. Weiss, Gauge theories with imaginary chemical potential and the phases of QCD. Nucl. Phys. B **275**, 734 (1986)

203. P. de Forcrand, O. Philipsen, Constraining the QCD phase diagram by tricritical lines at imaginary chemical potential. Phys. Rev. Lett. **105**, 152001 (2010)

204. P.M. Lo, B. Friman, O. Kaczmarek, K. Redlich, C. Sasaki, Polyakov loop fluctuations in SU(3) lattice gauge theory and an effective gluon potential. Phys. Rev. D **88**, 074502 (2013)

205. N. Haque, A. Bandyopadhyay, J.O. Andersen, M.G. Mustafa, M. Strickland, N. Su, Three-loop HTLpt thermodynamics at finite temperature and chemical potential. JHEP **05**, 027 (2014)

206. J.L. Kneur, M.B. Pinto, T.E. Restrepo, Phys. Rev. D **104**(3), L031502 (2021)

207. D.M. van Egmond, U. Reinosa, Signatures of the Yang-Mills deconfinement transition from the gluon two-point correlator. Phys. Rev. D **106**(7), 074005 (2022)

208. M. Peláez, U. Reinosa, J. Serreau, M. Tissier, N. Wschebor, A window on infrared QCD with small expansion parameters. Rep. Prog. Phys. **84**(12), 124202 (2021)

209. M. Peláez, U. Reinosa, J. Serreau, N. Wschebor, in preparation.

Index

© The Author(s), under exclusive license to Springer Nature Switzerland AG 2022
U. Reinosa, *Perturbative Aspects of the Deconfinement Transition*, Lecture Notes
in Physics 1006, https://doi.org/10.1007/978-3-031-11375-8

Printed in the United States
by Baker & Taylor Publisher Services